# 21st International Conference on Building Materials, Products and Technologies (ICBMPT 2018)



IOP Conference Series: Materials Science and Engineering
Volume 379

Blansko-Ceskovice, Czech Republic
29 – 31 May 2018

ISBN: 978-1-5108-6749-9
ISSN: 1757-8981

**Printed from e-media with permission by:**

Curran Associates, Inc.
57 Morehouse Lane
Red Hook, NY 12571

**Some format issues inherent in the e-media version may also appear in this print version.**



Printed by Curran Associates, Inc. (2018)

For permission requests, please contact the Institute of Physics
at the address below.

Institute of Physics
Dirac House, Temple Back
Bristol BS1 6BE UK

Phone: 44 1 17 929 7481
Fax: 44 1 17 920 0979

techtracking@iop.org

**Additional copies of this publication are available from:**

Curran Associates, Inc.
57 Morehouse Lane
Red Hook, NY 12571 USA
Phone: 845-758-0400
Fax: 845-758-2633
Email: curran@proceedings.com
Web: www.proceedings.com

# TABLE OF CONTENTS

## INORGANIC BINDERS



## SECONDARY RAW MATERIALS

## COMPOSITE MATERIALS



## TECHNOLOGY AND TEST METHODS

IOP Conf. Series: Materials Science and Engineering **379** (2018) 011001 doi:10.1088/1757-899X/379/1/011001

# Preface

The 21st International Conference ICBMPT 2018 – International Conference on Building Materials, Products and Technologies organized by the Research Institute for Building Materials was held in Češkovice, Czech Republic, on 29—31 May 2018. The main topics of ICBMPT 2018 were: Inorganic Binders, Secondary raw materials, Composite materials and Technology and test methods. The aim of the conference was to bring researchers from universities and industrial enterprises together to share their experience and to create an international forum for the exchange of both scientific and practical knowledge in the field of building materials.

41 papers were accepted after strict peer-reviewing by expert referees, we would like to thank all the participants for their valuable contribution to the conference and these proceedings. We would like to express our gratitude to the reviewers for their time, valuable recommendations and constructive evaluation, in order to guarantee high quality. The organizing committee is very proud that more than 90 participants from several European countries were attracted. Due to this fact, the spectrum of the contributions reflected many aspects of the state of the art in this fast-developing field of applications. We hope that the conference made a significant contribution to the knowledge in this scientific field.

IOP Conf. Series: Materials Science and Engineering **379** (2018) 011002 doi:10.1088/1757-899X/379/1/011002

# Peer review statement

All papers published in this volume of *IOP Conference Series: Materials Science and Engineering* have been peer reviewed through processes administered by the proceedings Editors. Reviews were conducted by expert referees to the professional and scientific standards expected of a proceedings journal published by IOP Publishing.

# The influence of starting materials´ solubility on tobermorite structure formation under the hydrothermal conditions

**L Galvánková[1], E Bartoníčková[1], T Opravil[1], J Tkacz[1] and P Ptáček[1]**

[1] Brno University of Technology, Faculty of Chemistry, Institute of Material Chemistry, Purkyňova 118, 612 00, Brno, Czech Republic

Email: xcgalvankova@fch.vut.cz

**Abstract**. Overall, the tobermorite mineral crystallises under the hydrothermal conditions from the $CaO$-$SiO_2$-$H_2O$ system. The system of calcium hydrosilicates species is extremely complex and many factors play an important role. The solubility of starting materials belongs to one of the most important factors. Dissolution rates of $SiO_2$ source are significantly affected by their particle size distribution. Also the starting cation´s concentration in the aqueous solution affects the yield of the hydrothermal reaction. The main aim of this contribution is to study the influence of water-to-solid ratio on tobermorite crystallization. The effect of particle size of various sources of $SiO_2$ was also studied. The cost effective Ca and Si sources was primarily favoured. The C/S ratio of the experiments was set to 0.83. All syntheses were performed at 180 °C in laboratory steel autoclaves. The prepared powdered samples were characterized in order to determine the phase composition by using of X-Ray diffractometry. The thermal behaviour of prepared samples was studied by TG-DTA analyser. Morphology and particle size of synthetized powdered products was studied by scanning electron microscopy.

## 1. Introduction

Tobermorite, a mineral belonging to a group of calcium silicate hydrates (C–S–H), is very rare in nature but plays an important role in several technical applications in building chemistry, mainly in hydration processes of Portland cement and production of autoclaved aerated concrete.

Tobermorite was firstly described by Heddle in 1880 [1]. The natural samples were found in three different localities near towns Tobermory and Dunvegan in Scotland. Generally the mineral could be formed in places with possible hydrothermal activity (i.e. in cavities in alkaline rocks, where was formed by hydrothermal alteration or in places where the hydrothermal fluids react with calcium silicate minerals) [2]. Till now, three main different phases of crystal structure have been described. The type of the structure is associated with the different degree of hydration, 9 Å tobermorite also called riversideite ($Ca_5Si_6O_{16}(OH)_2$), 11 Å tobermorite ($Ca_5Si_6O_{17}\cdot5H_2O$) and 14 Å tobermorite also called plombierite ($Ca_5Si_6O_{16}(OH)_2\cdot7H_2O$). The denomination of phases is related to characteristic $d_{002}$ values in the powder diffraction pattern, 9.3 Å, 11.3 Å and 14.0 Å from the least to the most hydrated form [3]. The most important phase is 11 Å tobermorite which crystallizes in orthorhombic or monoclinic form. The monoclinic form was found by Henmi and Kusachi [4] and was named clinotobermorite. The crystal structure of natural 11 Å tobermorite structure was firstly described by Megaw and Kelsey [5] and then by Hamid [6]. However, the detailed description of the real structure of both orthorhombic and monoclinic form was solved by Merlino et al. [7, 8] using order-disorder theory. The basic structure is build up by central CaO-octahedra layer attached from both sides with infinite silicate double chains composed of condensed dreierketten units along the *b* axis. Each dreierketten unit consists of two paired tetrahedra pointing towards to Ca-layer and one bridging tetrahedron oriented to the interlayer. The composite layers consist of one calcium layer and two silicate layers bounded together by the interlayer containing mainly calcium ions and water molecules. Depending on the dehydration behaviour, the two types of 11 Å tobermorite, normal and anomalous type, can be distinguished. During the dehydration process the normal 11 Å form loses its interlayer water and transforms to a 9 Å tobermorite, while the anomalous one is affected by the dehydration process, but the formed crystalline structure is not 9 Å tobermorite [9]. At the temperatures above 800°C, both types recrystallize to new phase called wollastonite ($CaSiO_3$) [2].

The hydrothermal synthesis of tobermorite in various laboratory conditions has been extensively studied [10-14]. Generally, $Ca(OH)_2$ and $SiO_2$ react to form C–S–H phases under hydrothermal conditions. At the beginning of the reaction, the solution is saturated with lime so the Ca-rich phases like C–S–H (II) (C/S ratio around 1.5) and α–$C_2S$ hydrate are formed. Whereas the temperature in the system start increases, the solubility of CaO decrease and the increase of $SiO_2$ solubility occur simultaneously. The next step of the hydrothermal reaction is the reduction of the C/S ratio to 1.25 and the creation of C–S–H (I) phase. In the final stage of the reaction, the C/S ratio drops to the optimal value for tobermorite formation (i.e. between 0.8–1) and the C–S–H (I) phase recrystallize to 11 Å tobermorite [15]. The whole reaction is schematically shown in equation (1). The course of the reaction is affected by many factors. The reaction temperature and pressure are one of the most important factors, temperatures between 120°C and 200°C and saturated steam pressure has been reported as the optimal. The reaction temperature and the reaction time are strongly depended on the particle size of $SiO_2$ [13, 15]. Generally, with higher temperature and longer reaction time the recrystallization of tobermorite to other calcium silicate hydrate, xonotlite ($Ca_6Si_6O_{17}(OH)_2$), is favoured. Besides mentioned factors, the amount of added water to reaction system plays also an important role in the final yield of the hydrothermal reaction [16].

$$Ca(OH)_2 + SiO_2 \rightarrow C-S-H(II) + \alpha - C_2S\ hydr. \rightarrow C-S-H(I) \rightarrow tobermorite \rightarrow xonotlite \quad (1)$$

The formation of tobermorite under the hydrothermal conditions is beside shown the strong dependence on the reaction temperature, time and especially on the solubility of $SiO_2$ source. Therefore, the main aim of the present work was to study the influence of $SiO_2$ particle size distribution together with the starting concentration in the aqueous solution. The influence of reaction time was also examined.

## 2. Experimental

### *2.1. Materials and synthesis*

For the synthesis of a mineral tobermorite under the hydrothermal conditions, two types of silica sources, silica sand Dorsilit and specially crushed quartz sand, were chosen as starting materials. The average particle size ($d_{50}$) of silica sources is 5.29 μm for silica sand Dorsilit and 20.24 μm for specially crushed quartz sand. As the source of calcium oxide, the grounded $CaCO_3$ calcinated at 1000°C for 1 hour was used. The amount of CaO in calcinated limestone was determined to be higher than 97%. All the starting mixtures were mixed using the same C/S ratio equal to 0.83. The method of synthesis used in this study was inspired by Diamond et al. [17]. Required quantity of CaO was mixed with freshly-boiled distilled water (to supress the undesirable carbonization) and stirred for 3 minutes. Then the calculated amount of $SiO_2$ was added to the mixture and was mixed for 5 more minutes. Used water-to-solid (w/s) ratio was set in a range from 1.5 to 15 ml/g. Hydrothermal reactions were carried out in a Teflon-lined laboratory autoclave placed in a conventional dryer. Prepared slurries were hydrothermally treated at 180°C and autogenous pressure for 1 and 5 days. Synthesized products were dried at 50°C for 12 hours.

### *2.2. Characterization of synthesized products*

The phase composition of synthesized products was determined by X-ray diffraction analysis (XRD). Data were collected by X-ray diffractometer Empyrean (Pananalytical) using Cu Kα radiation. For selected samples, a thermogravimetric and differential thermal analysis was performed using TG-DTA instrument Q 600 (TA Instruments). Analyses was run in Pt crucibles and were heated with the rate of 10°C/min from 20°C to 1000°C under the Ar atmosphere. The morphology of these chosen samples was observed by scanning electron microscopy (SEM) using microscope ZEISS EVO LS 10.

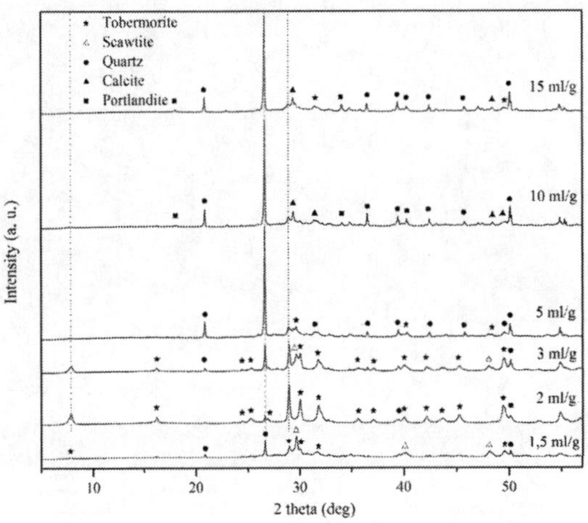


**Figure 1.** XRD patterns of samples synthesised for 1 day using specially crushed quartz sand and different w/s ratios (1.5–15 ml/g).

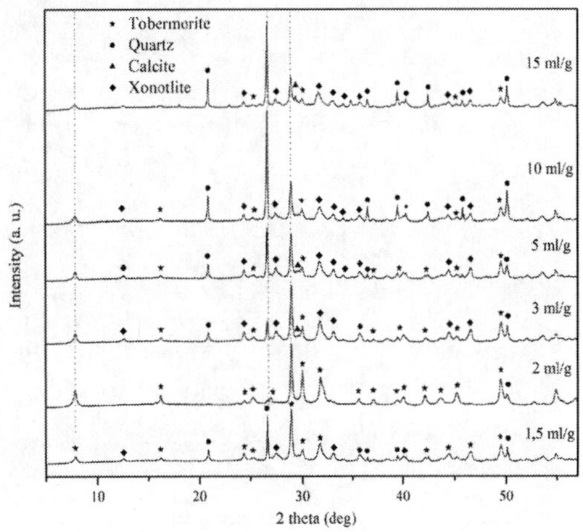


**Figure 2.** XRD patterns of samples synthesised for 5 days using specially crushed quartz sand and different w/s ratios (1.5–15 ml/g).

## 3. Results

### *3.1. Synthesis of tobermorite using specially crushed quartz sand*

The solubility of silica source is affected by the amount of added water to the reaction system. For the examination of the influence of w/s ratio on tobermorite formation, six samples with different w/s ratio were used (1.5, 2, 3, 5, 10 and 15 ml/g). Figure 1 shows the XRD patterns of samples prepared from the mixture of specially crushed sand and calcinated $CaCO_3$ hydrothermally treated for 1 day. It is obvious that after a one day of hydrothermal reaction the different w/s ratio strongly affects the phase composition of prepared samples. With the increasing amount of water in the system the amount of prepared tobermorite strongly decrease. The XRD results show the presence of crystalline tobermorite only in samples prepared with w/s ratio between 1.5–3 ml/g. The highest yield of tobermorite formation after 1 day of hydrothermal treatment was observed with the use of w/s ratio of 2 ml/g. In samples with higher w/s ratio, the starting materials are the only detected. If the reaction time is prolonged from one to five days, the amount of tobermorite in samples is increased. The highest amount of tobermorite is still found in the sample prepared with w/s ratio of 2 ml/g. In some samples, part of crystalline tobermorite was transformed into xonotlite. The unreacted $SiO_2$ is still the main phase even after five days of the treatment. Except the unreacted starting materials, the $CO_2$ bearing compounds like calcite and scawtite ($Ca_7(Si_6O_{18})(CO_3)\cdot 2H_2O$) were presented in some samples. The contamination is caused by the reaction of wet samples with air containing $CO_2$ during the preparing process and mainly during the drying of prepared products [18].

The influence of reaction time on morphology of formed tobermorite particles was observed on samples with the highest amount of tobermorite determined from XRD analysis. Therefore,

IOP Conf. Series: Materials Science and Engineering **379** (2018) 012001 doi:10.1088/1757-899X/379/1/012001

the samples prepared with the used w/s ratio 2 ml/g were analysed by SEM and the images are given in Figure 3 and Figure 4. It could be seen, that the $SiO_2$ particles from the starting material are covered with tobermorite platelet-like crystals. The crystals in sample hydrothermally treated for 1 day are longer and could be better distinguished than the crystals in sample that was autoclaved for 5 days. The higher degree of agglomeration of platelet tobermorite particles could be here also observed. The various length of the hydrothermal reaction showed the differences in the morphology of prepared tobermorite crystals. No other significant phases could be detected from the SEM images.

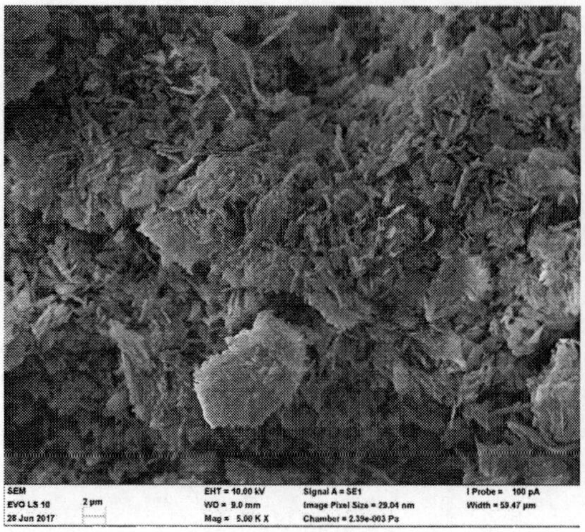

**Figure 3.** SEM images of samples synthesised for 1 day using specially crushed quartz sand and w/s ratio 2 ml/g.

**Figure 4.** SEM images of samples synthesised for 5 days using specially crushed quartz sand and w/s ratio 2 ml/g.

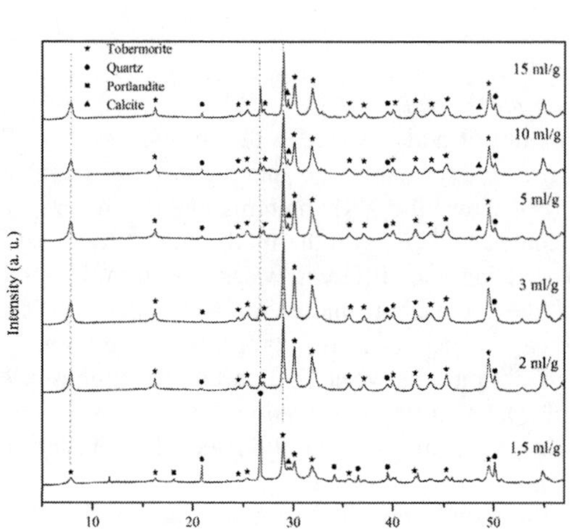


**Figure 5.** XRD patterns of samples synthesised for 1 day using silica sand Dorsilit and different w/s ratios (1.5–15 ml/g).

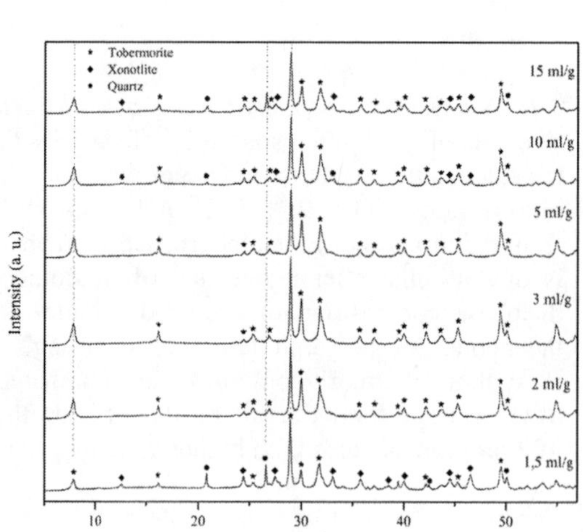


**Figure 6.** XRD patterns of samples synthesised for 5 days using silica sand Dorsilit and different w/s ratios (1.5–15 ml/g).

IOP Conf. Series: Materials Science and Engineering **379** (2018) 012001 doi:10.1088/1757-899X/379/1/012001

### *3.2. Synthesis of tobermorite using silica sand Dorsilit*

Results of the XRD analysis on samples prepared using silica sand Dorsilit are shown in Figure 5 and Figure 6. After one day of hydrothermal reaction, tobermorite is present in all samples. The influence of w/s ratio is consistent with the trend observed in the samples prepared using specially crushed silica sand. It was also determined that with the increasing w/s ratio the amount of prepared tobermorite phase decreased. None of the prepared samples is phase pure and the presence of the starting materials is determined. Prolonging the reaction time from one to five days allowed the reaction to proceed, so the amount of tobermorite phase is higher in all samples independently of the w/s ratio. The only detected crystalline phase, in samples with w/s ratio from 2 to 5 ml/g, was tobermorite. Conversion to xonotlite appears only in samples with higher w/s ratio, but the amount of crystallized xonotlite is almost negligible. No $CO_2$ bearing phases were detected after five days of hydrothermal reaction.

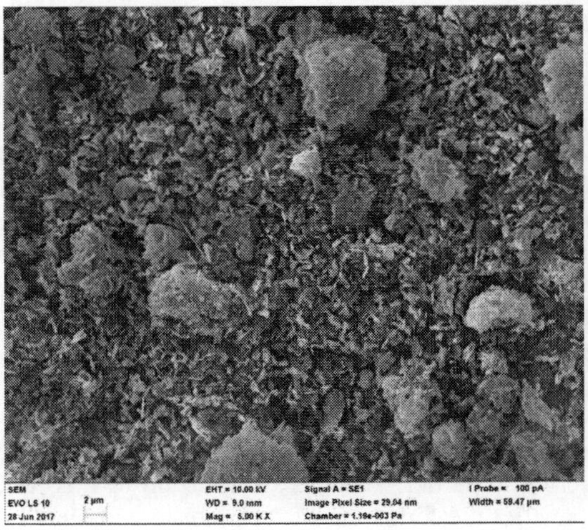

**Figure 7.** SEM images of samples synthesised for 1 day using silica sand Dorsilit and w/s ratio 3 ml/g.

**Figure 8.** SEM images of samples synthesised for 5 days using silica sand Dorsilit and w/s ratio 3 ml/g.

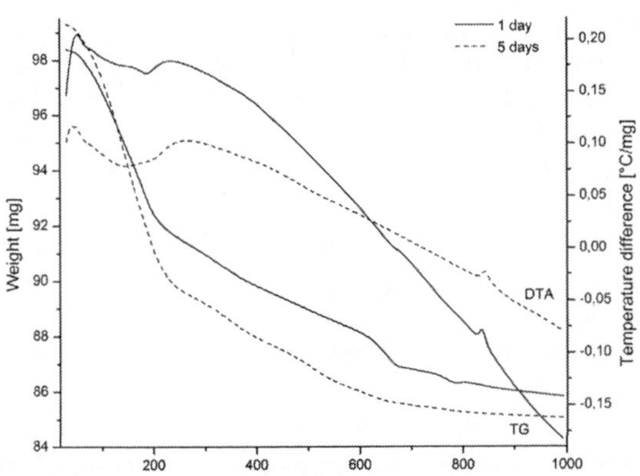


**Figure 9.** TG-DTA curves of samples synthesised for 1 and 5 days using specially crushed quartz sand and w/s ratio 2 ml/g.

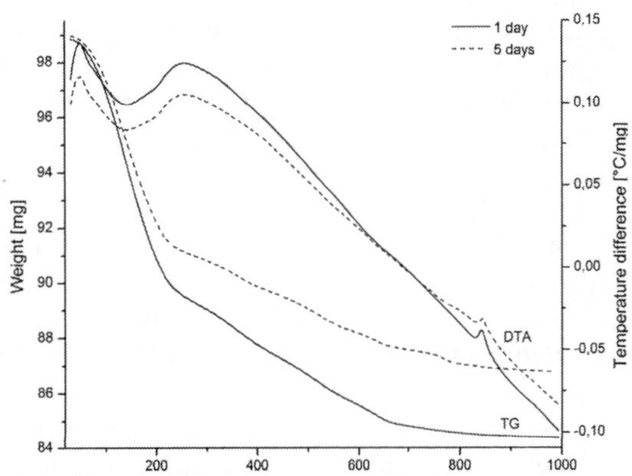


**Figure 10.** TG-DTA curves of samples synthesised for 1 and 5 days using silica sand Dorsilit and w/s ratio 3 ml/g.

The morphology of prepared tobermorite using silica sand Dorsilit as a starting source of $SiO_2$ were studied on the samples with the w/s ratio 3 ml/g. It was found that the prepared tobermorite phases crystalized in the platelet form particles independently of the reaction time (Figure 7 and 8). Observed morphology is almost the same as in case of the synthesis with specially milled quartz sand after five days of hydrothermal reaction.

### *3.3. TG-DTA analysis*

For confirmation of the results from XRD analysis, TG-DTA measurements on samples with the highest yield of tobermorite (detected by XRD) were performed. The influence of particle size on thermal processes was also studied. The main weight loss on a TG curve occurs between 50°C and 250°C. The weight decrease is accompanied by the broad endothermic hump on a DTA curve presumably resulting from the loss of the loosely bound molecular water from tobermorite structure. The weight decrease is comparable in all four samples. In samples prepared using specially crushed silica sand, it is obvious that the amount of tobermorite increases with prolonged reaction time (Figure 9). This confirms the XRD analysis, whilst the results for the samples prepared using silica sand Dorsilit is opposite (Figure 10). The amount of tobermorite phase is lower in sample hydrothermally treated for 5 days. This difference could be caused by the additional carbonation of tobermorite during the sample storage. Between 250°C and 800°C, any remaining molecular water from tobermorite is lost. At about 850°C a sharp exothermic peak is detected indicating the recrystallization of tobermorite into wollastonite. Besides the changes connected with the presence of tobermorite, the presence of calcite is detected by the weight loss between 600°C and 650°C, where the decomposition of calcite takes place. Mentioned weight loss is most obvious in the sample prepared using specially crushed silica sand synthesised for one day.

## 4. Discussion

The formation of tobermorite in hydrothermal conditions is strongly affected by the average particle size of the used source of $SiO_2$. Two sources differing in particle size were used to prove this fact. The experiments reveal that if the particle size is smaller the formation of tobermorite is easier and the yield of the reaction is higher. This result corresponds with previously reported studies [16, 19–21]. The rate of dissolution of quartz particles corresponds to particle size and it is time dependent, the finer particles have quicker dissolution, whilst the coarser one has the slower. So, the overall reaction is accelerated due to an increase in reactivity depending on particle size. According to the literature [21], when the $SiO_2$ source with the particle size higher than 10 μm is used as a starting material, synthesized tobermorite decomposes to xonotlite. The result in this work is basically in good agreement with those found in the literature. Kikuma et al. [16] had proposed an explanation of this phenomenon. During the hydrothermal reaction non-crystalline C–S–H is formed as an intermediate phase before the crystalline calcium silicate hydrates occur. When the less soluble $SiO_2$ source is used, the C/S ratio in the non-crystalline C–S–H is higher and the reaction equation (1) proceeds towards xonotlite instead of tobermorite.

Another not negligible factor for all hydrothermal synthesis, including the formation of tobermorite, is the water-to-solid ratio. The obtained results indicate that the amount of prepared tobermorite significantly decreases with the increasing value of w/s ratio. The calcium and silicate ions concentration in the reaction solution probably plays an important role in tobermorite formation. Thess results ties well with previous study published by Kikuma et al. [16]. They have done an examination on the samples prepared from fine quartz with particle size 2 μm and quicklime. The used w/s ratio was 1.7, 3, 6 and 9 ml/g and the measurements were done by in-situ time-resolved XRD. Based on the obtained results they suggested that the silicate ion concentration is differing according to the w/s ratio. They showed that the lower silicate ion concentration in the solution phase is important for the tobermorite formation under the hydrothermal conditions.

The influence of the particle size of $SiO_2$ source and reaction time on the morphology of prepared tobermorite was also observed. Tobermorite crystallises into the platelet-like crystals. The size of formed crystals decreases with the increasing reaction time and they have a tendency to form aggregates after five days of hydrothermal reaction. The reaction of the coarser $SiO_2$ particles is slower so the aggregates are created after a longer reaction time. The formation of platelets crystals of tobermorite throw the heterogeneous nucleation was also described by Bell and Adair [22].

## 5. Conclusions

Hydrothermal formation of mineral tobermorite in system $CaO–SiO_2–H_2O$ was studied. Two important factors were mainly examined – the average particle size of $SiO_2$ source and the water-to-solid ratio. Based on experiments conducted at 180°C and an autogenous pressure with two different particle size of $SiO_2$ source, it can be shown that the $SiO_2$ source containing smaller particles accelerating the reaction. These could be attributed to a better solubility of finer quartz particles in a hydrothermal system. The formation of tobermorite is also significantly influenced by the changes in w/s ratio. The increase in the w/s ratio probably influences the calcium and silicate ions concentration in a solution, which plays an important role in a process of tobermorite formation. The ideal w/s ratio was almost independent of the used $SiO_2$ sources and lies in a range of 2 to 3 ml/g. Using the optimal w/s ratio and the prolonged reaction time from one to five days resulted in a preparation of samples containing phase pure tobermorite. In samples prepared with higher w/s ratio and longer reaction times the unfavourable formation of xonotlite occurs. This trend was more like observed in samples prepared from coarser $SiO_2$ source, which is probably caused by the higher C/S ratio in a system thanks to its lower solubility.

**Acknowledgements**
This study was supported by the project of Materials Research Centre on Faculty of Chemistry, Brno University of Technology - sustainability and development REG LO1211 with the financial support of National program of sustainability I (The Ministry of Education, Youth, and Sports).

**References**

[1] Heddle M F 1880 *Mineral. Mag.* **4** pp 119-123
[2] Bonaccorsi E and Merlino S 2005 *Rev. Mineral. Geochem.* **57** pp 241-90
[3] Biagioni C, Merlino S and Bonaccorsi E 2015 *Mineral. Mag.* **79**(2) pp 485-95
[4] Henmi C and Kusachi I 1995 *Mineral. Mag.* **56**(384) pp 353-58
[5] Megaw H and Kelsey C 1956 *Nature* **177** pp 390-91
[6] Hamid S *Z. Kristallogr. Cryst. Mater.* **154** pp 189-98
[7] Merlino S, Bonnacorsi E and Armbruster T 2001 *Eur. J. Mineral.* **13** pp 577-90
[8] Merlino S, Bonnacorsi E and Armbruster T 2000 *Eur. J. Mineral.* **12** pp 411-29
[9] Mitsuda T and Taylor H F W 1978 *Mineral. Mag.* **42** pp 229-35
[10] Hong S Y and Glasser F P 2004 *Cem. Concr. Res.* **34** pp 1529-34
[11] Hartmann A, Schulenberg D and Buhl J Ch 2015 *J. Mater. Sci. Eng.* **3** pp 39-55
[12] Shaw S, Clark S M and Henderson C M B 2000 *Chem. Comp.* **167** pp 129-40
[13] Snell D S 1975 *J. AM. Ceram. Soc.* **58**(7-8) pp 292-95
[14] Trankle S, Jahn D, Neumann T, Nicoleau L, Husing N and Volkmer D 2013 *J. Mater. Chem. A* **1** pp 10318-26
[15] Taylor H F W 1964 *The Chemistry of Cements* (London: Academic Press)
[16] Kikuma J, Tsunashima M, Ishikawa T, Matsuno S, Ogawa A, Matsui K and Sato M 2011 *J. Solid State Chem.* **184** pp 2066-74
[17] Diamons S, White J L and Dolch W L 1966 *Amer. Mineral.* **51**(3-4) pp 388-401
[18] Siauciunas R, Rupsyté E, Kitrys S and Galeckas V 2004 *Colloids Surf. A* **244** pp 197-204
[19] Sato H and Grutzeck M 1992 *Mat. Res. Soc. Symp. Proc.* **245** pp 235-240
[20] Bernstein S and Fehr K T 2012 *Prog. Cryst. Growth Charact. Mater.* **58** pp 84-91
[21] El-Hemaly S A S, Mitsuda T and Taylor H F W *Cem. Concr. Res.* **7** pp 429-438
[22] Bell N S and Adair J H 1996 *J. AM. Ceram. Soc.* **79** pp 2175-78

# Application of brick grind dust in systems based on Portland cement.

**J Palovčík[1], T Opravil[1], R Novotný[1] and J Másilko[1]**

[1] Brno university of technology, Faculty of Chemistry, Materials Research Centre, Purkyňova 464/118, Brno Cz-612 00, Czech Republic

Email: xcpalovcik@fch.vut.cz

**Abstract**. This paper deals with possibilities of application brick-grind-dust which is a secondary product from manufacturing of grinded bricks. This product is usually partially returned to brick manufacturing or buried on a land field site. The basic idea of using this material in systems based on cement comes from their mineral composition, origin and particle size. These materials are potentially pozzolan active, which is interesting for use in hydraulic binders. Also red colour of brick dust can be interesting for visual application. First physical and chemical properties of brick-grind-dust was studied, then binders based on the material and Portland cement were prepared. Mechanical properties of prepared binders were studied over time. Hydration of the selected binders was analysed with isothermal calorimetry.

## 1. Introduction

In recent years, the brick industry has been focusing on the production of ground bricks with precise dimension and high level of flatness. These bricks are used for rapid and accurate bricklaying to thin layer of special mortar or polyurethane foam. Secondary product of manufacturing these bricks is a fine dust with some fragments from bricks. This material is partially reused in the production of bricks, but the remaining production is landfilled.

Production of brick is based on drying and burning brick clay, soils and other components in tunnel furnace about 800°C exceptionally up to 1000°C. Brick materials typically contain relatively low and varying amounts of clay minerals (i.e. kaolinite, illite, montmorillonite, chlorite and their mixed structures). The predominant fraction is composed of soil containing silica, mica and rock fragments including carbonates. There is also an iron oxide pigment [1].

Dehydroxylation of clay minerals at these temperatures made these materials potentially pozzolan active. The American Standard ASTM C 125 defines pozzolans as silicate and alumina materials which in themselves bear little or no binding properties. However, in finely ground form, in the presence of water, they react with calcium hydroxide at ambient temperature to produce compounds having binding properties. Pozzolans are chemically referred to as silicates or alumino-silicate inorganic substances, mostly without or with little lime [2].

Poźzolan activity gives information about amount of calcium hydroxide which can react with some pozzolan. Like pozzolan activity is so defined reaction velocity between these substances. Both parameters are depended on the type of pozzolan, surface area, chemical and phase composition (Amorphous content), CaO/pozzolan ratio and so with curing parameters [3].

Hydration products formed by pozzolan activity are very close to product of hydration calcium-silicate or calcium-aluminate phases which is called C-S-H gel and C-A-H gel. Pozzolan reaction can be written in most simply way like schema: $\text{puzzolan} + CH + H \rightarrow C\text{-}S\text{-}H^*$.

There is C-S-H* gel marked with the star like a secondary C-S-H gel in spite of primary C-S-H gel from hydration of calcium-silicates from Portland cement. Pozzolan reaction takes much longer than hydration of clinker minerals. It is reason for less porosity and more compact structure for pozzolan cement [4].

## 2. Materials and methods

Brick-grind-dust was produced by manufacturing bricks called Family in plant Hevlín (HELUZ). Mineralogical composition was analyzed by powder XRD (Empyrean, Panalytical) with Rietveld analysis. For pozzolan reaction is most important amorphous phase content, which is 28%. Others phases contents are said in Table 1.

**Table 1.** Phase composition of brick rind dust.

| Phase | Amorphous | Quartz | Augite | Albite | Muscovite | Orthoclase | Akermanite | Hematite |
|---|---|---|---|---|---|---|---|---|
| (%) | 28.4 | 30.3 | 5.6 | 17.1 | 11.3 | 2.0 | 4.3 | 0.9 |

Particle size distribution was analyzed by six sieve 1.4 mm, 0.8 mm, 0.4 mm, 0.2 mm, 0.105 mm, 0.063 mm and results are shown in Table 2. Fraction less than 0.063 mm was analyzed by laser analysis of particle size (HELOS, Sympatec). Particle size distribution of Portland cement CEM I 42.5 R plant Mokrá was analyzed to by HELOS for comparison with particle size distribution of brick grind dust. Cement has a much smaller grain size scatter, but $D_{90}$ of both samples are in this very close. Modus and $D_{50}$ of dust are approximately half than cement. Smaller particles in case of brick dust can apply the filler effect.

**Table 2.** Sieve analysis of brick-grind-dust.

| Fraction (mm) | <0.063 | 0.063–0.105 | 0.105–0.2 | 0.2–0.4 | 0.4–0.8 | 0.8–1.4 | >1.4 |
|---|---|---|---|---|---|---|---|
| (%) | 66.8 | 14.2 | 4.8 | 3.8 | 5.1 | 3.6 | 1.6 |

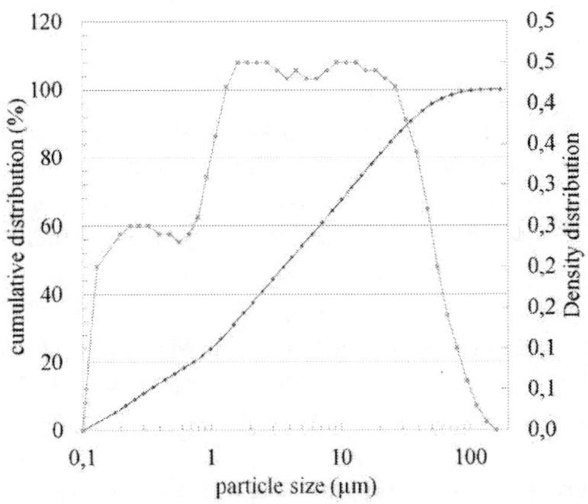


**Figure 1.** Particle size of brick-grind-dust.

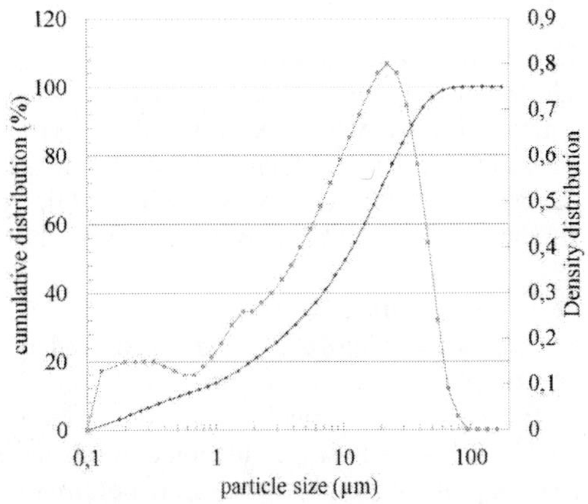


**Figure 2.** Particle size of CEM I 42.5 R.

IOP Conf. Series: Materials Science and Engineering **379** (2018) 012002 doi:10.1088/1757-899X/379/1/012002

**Table 3.** Particle size parameters.

| Sample | $D_{50}$ (μm) | $D_{90}$ (μm) | Modus (μm) |
|---|---|---|---|
| Brick-grind-dust | 4.10 | 33.47 | 11.55 |
| CEM I 42.5 R | 10.35 | 36.56 | 20.51 |

Mechanical properties were tested on three 20 × 20 × 100 mm test specimens from each sample and time on the equipment for testing of building materials (DESTTEST, BETONSYSTEM). The average values of compressive strength are shown in the graph in the Figure 3. Composition of prepared samples is shown in Table 4. Water/cement ratio is marked like W/C and W/S is water/solid components ratio. From the samples for analyzing mechanical properties were, after 28 days, a representative of 5 analyzed to compare the phase composition by XRD.

**Table 4.** Composition of prepared samples.

| Sample | A | B | C | D | E | F | G | H | I | J |
|---|---|---|---|---|---|---|---|---|---|---|
| Brick grind dust[g] | 0 | 100 | 200 | 300 | 400 | 500 | 600 | 700 | 800 | 900 |
| CEM I 42.5 R[g] | 1000 | 900 | 800 | 700 | 600 | 500 | 400 | 300 | 200 | 100 |
| $H_2O$ [g] | 333 | 334 | 347 | 353 | 360 | 370 | 378 | 395 | 413 | 424 |
| W/C | 0.33 | 0.37 | 0.43 | 0.50 | 0.60 | 0.74 | 0.95 | 1.32 | 2.07 | 4.24 |
| W/S | 0.33 | 0.33 | 0.35 | 0.35 | 0.36 | 0.37 | 0.38 | 0.40 | 0.41 | 0.42 |

For the study of hydration processes using isothermal calorimetry (TAM Air, TA Instruments), 4 g samples were prepared. Samples were the same composition as samples for testing mechanical properties or XRD analysis, but amount of water was increased by 20% in all blends to improve homogenization of the system. Increasing the water factor by such an amount would affect the mechanical properties, but the change in the chemistry of hydration is not expected. Reaction should be only accelerated and it does not affect for the comparison of the samples.

Samples were mixed with water for 1 minute outside the calorimeter. Subsequently, together with the reference (distilled water), they were inserted into the calorimeter. The temperature was set at 40°C, which would be sufficient to capture all significant hydration processes within 100 hours. Since the measurement was at 40°C, the start of the recording (about 15 minutes) is subject to considerable heat flow, which is caused by the temperature of the samples from 25 to 40°C. For this reason, the first peak does not represent chemical processes, so it will not be discussed in more detail.

## 3. Results and discussion

The prepared samples were all able to handling after 24 hours, but samples I, J exhibited zero compressive strengths. Generally, the compressive strength (Figure 3) can be read by the trend of a gradual increase in strengths, even after 28 days. It may be a proof of the pozzolan activity of the brickwork. But after 180 days is here little decrease for all samples. Trend of increasing compressive strengths is well visible after 60 days in blends with brick ash contents above 40%. At 40% brick ash (E), the compressive strengths after 28 days were 60–70 MPa scattered, and with each 10% brick ash decreased the strength by approximately 10 MPa. So the system was probably overload with brick dust. Sample J containing 10% of cement showed a compressive strength of more than 13 MPa after 28 days, and sample I with 20% cement showed strength of 26 MPa. These values are for some applications fully sufficient and economically interesting. The problem could be unstable strength after 24 hours, which could complicate the manipulation of products. The solution could be found in technological parameters such as prolonged aging time before deformation, increased curing temperature, or the use of a suitable solidification accelerator.

The bulk weights of the prepared bodies (Figure 4) stored in the humid deposit decreased for the first 10 days, than were then approximately constant. As the brick ash content increased, the final bulk density from 1.95 g·cm$^{-3}$ to the body from pure cement to 1.43 g·cm$^{-3}$ in the 90% brick ash was reduced.

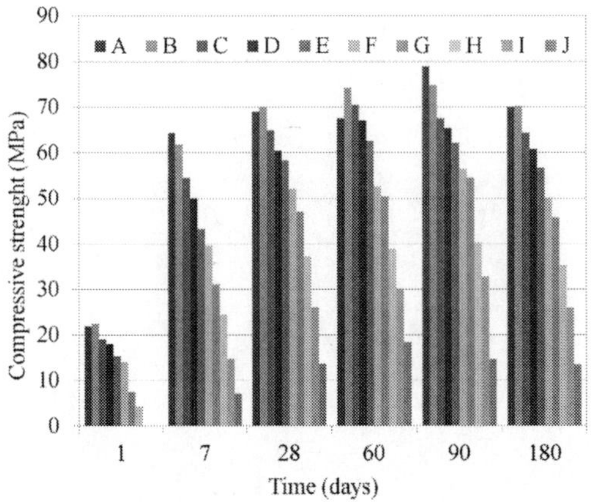


**Figure 3.** Graph of development compressive strength in time.

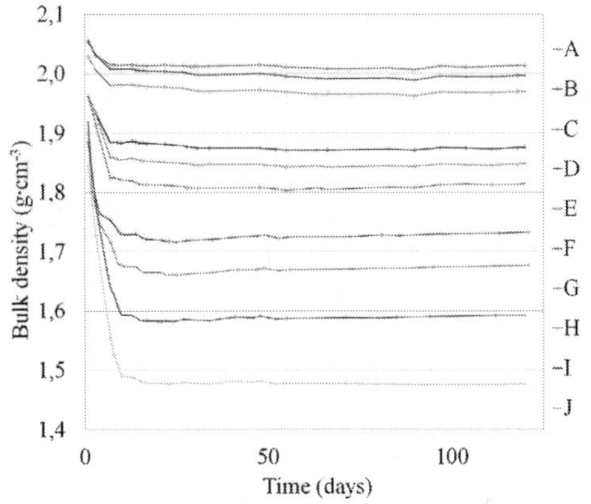


**Figure 4.** Graph of development bulk density in time.

**Figure 5.** Colour change depending on cement/brick dust ratio.

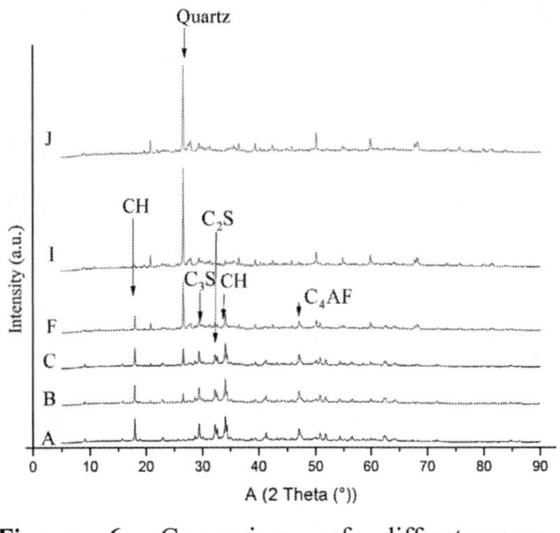


**Figure 6.** Comparison of diffractograms for selected samples.

The color of the samples (Figure 5) was changing with increasing brick ash content from gray to brick-orange. The color of the bodies and the variability of their strengths could be attractive for visual applications or concrete outdoor products.

In the Figure 6 are the individual diffractograms of samples A, B, C, F, I, J showed. The arrows indicate the phases of clinker minerals ($C_3S$, $C_2S$, $C_4AF$), Portlandite (CH) as the product of Portland cement hydration and the potential reactant for pozzolan reaction and quartz as one of the most striking phase occurring in the brick dust. Generally quartz is believed to be unreactive or very few. With a simple comparison of the diffractograms, the peak intensity of the cements and the Portlandite hydration product decrease with increasing brick content. It is observed due to the dilution of the cement. The increase in intensity in same way can be observed for the peak representing quartz. This is the content which increases as the brick ash content increases. Samples contains many phases of both cement and brick. This complicates their phase analysis, however, no new crystaline phases of react product have been proven. The pozzolan reaction is based on the reaction of amorphous phases of silicates or alumino-silicates with $Ca(OH)_2$ to form the C-S-H or C-S-(A)-H gel. Although the intensity of the Portlandite diffraction decreases with the brick ash content, it is rather due to dilution of the cement with the brick ash rather than its consumption for the formation of the C-S-H gel. In order to verify the pozzolan reaction and to understand the hydration of these mixtures, the following experiment, isothermal calorimetry, was proposed.

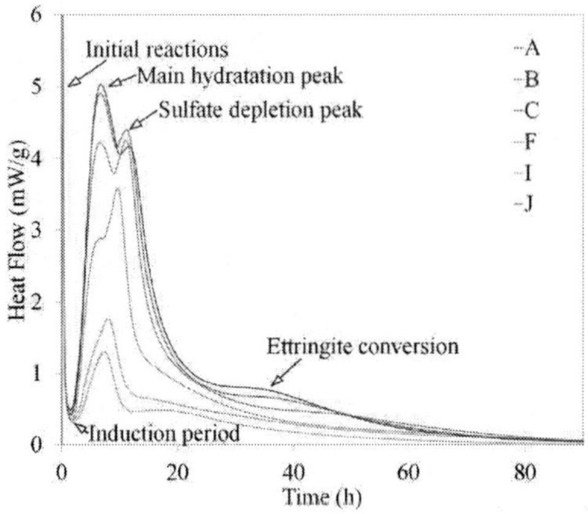


**Figure 7.** Calorimetric curves of samples A, B, C, F, I, J.

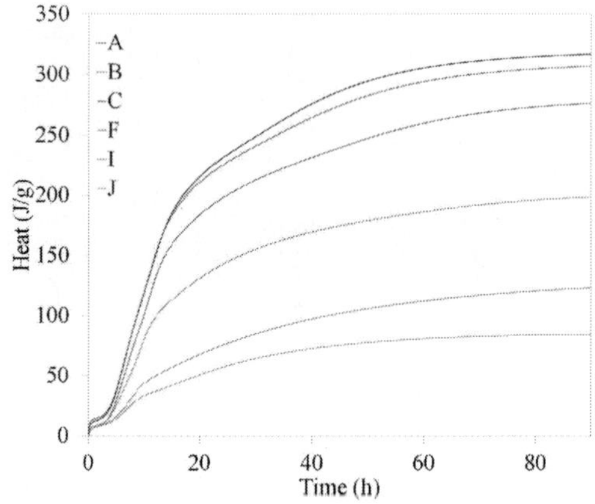


**Figure 8.** Graph of total hydration heat of samples A, B, C, F, I, J.

**Table 5.** Table of approximate times from the calorimetric curve and total hydration heat.

| Sample | Time (h) | | | | Heat (J/g) |
|---|---|---|---|---|---|
| | End of induction period | Main hydration peak | Sulphate depletion peak | Ettringite conversion | |
| A | 2.0 | 6.5 | 11.5 | 35 | 317 |
| B | 2.1 | 6.5 | 10.8 | 36 | 307 |
| C | 2.1 | 6.4 | 10.8 | 46 | 276 |
| F | 2.3 | 6.4 | 9.4 | - | 198 |
| I | 3.0 | - | 7.6 | - | 123 |
| J | 3.5 | - | 7.4 | 20 | 84 |

In the Table 5 are the approximate times of individual hydration steps subtracted from the calorimetric curve (Figure 7) and total hydration heat listed. The induction period of the cement without brick dust lasts for one hour and increases with the addition of brick dust for up to 2.5 hours. It is a sign of prolonging workability, which is related to the slower start-up of the strengths.

The effect of brick ash addition on the main hydration position over time is not significant. The maximum heat flux is reached for samples up to 50% (inclusive) of brick ash in about 6.5 hours. For samples with brick ash content higher than 50%, the main hydration peak overlaps with the sulphate depletion peak, so it is not possible to determine exactly its position.

Brick dust does not affect the main hydration peak with only a dilution effect because the drop in heat flux is not directly proportional to the loss of cement in the samples. Due to particle size of brick ash, the effect of hydration is influenced by the "filler effect". The explanation of the effect filler is based on two mechanisms. In the first instance, a cement containing filler like inert powder facilitates hydration, since, while maintaining the W/C ratio, the inert material does not engage in hydration. It does not form hydration products, and there is more room for hydration of the clinker minerals. The second mechanism is to introduce a new surface with an inert material on which hydration products can form. When it is an inert material of quartz, an amorphous layer may be formed on the surface by grinding, which may additionally be a pozzolan active [5].

The most significant change in the calorimetric curve due to the addition of brick ash is observed on the third hydration peak representing sulfate depletion [6]. This is also possible in the area where diffusion-controlled chemical reactions revert to the heat flux directive to positive values. This peak appears on a calorimetric curve with increasing brick ash content still earlier (11.5 hours for sample A and 7.4 hours for sample J). Gradually, it moves from the diffusion-controlled area to the C-S-H gel surface-driven region [7]. There is a rapid increase in the intensity of this peak. The intensity is strongly influenced by concentration of sulphate and aluminate ions over time. The main source of sulphate ions is gypsum, which is added to the cement. The ratio between clinker minerals and gypsum in the samples is therefore constant. It is expected that peak shifts cause alumina ions to so-called "active alumina" from brick ash and aluminate ions are introduced into the solution by the influence of the pozzolan reaction. For optimum hydration, this active alumina should be compensated for by addition of gypsum [8]. Optimal gypsum content in cement binders is such that sulphate depletion occurs several hours after the main hydration peak. If the peak is present before the main hydration peak, the strength decreases due to the formation of the defective C-S-H gel. If it is found for more than a few hours, the cement dilution effect by gypsum [9]. In this case, the condition of the optimal gypsum content would be samples A, B, C, sample F is in the transition region, and Samples I and J suffer from a considerable deficiency of gypsum.

The total heat (Figure 8) is almost constant after 90 hours at 40°C, so it is believed that hydration is controlled by diffusion and is very slow and total heat can be compared. Assuming that the brick is only non-reactive filler, the 10% decrease in brick ash should be the one-tenth of the released heat by about 31 J/g. However, the difference is only 10 J/g. Likewise, for sample F with a brick ash content of 50%, the hydration heat should be approximately 158 J/g and in fact is 198 J/g. Increasing the hydration heat is therefore a proof of the ongoing pozzolan reaction. Its thermal expression is not significantly affected by the thermal flow over time. So the reaction is going on for a long time. The benefit of brick dust for Portland cement bonding systems is probably a combination of pozzolan reaction and filler effect.

## 4. Conclusion

Influence of adding brick-grind-dust was showed by the growth of long-term compressive strengths, in case samples with addition of brick dust up to 40%. So decreasing of compressive strengths, in case of samples with more content of brick dust, was not directly proportional to the loss of cement, which was attributed to pozzolan activity,

X-ray diffraction analysis of the prepared samples did not show presence of a new crystalline phase of hydration products of cement and brick, which encouraged the likelihood of the formation of amorphous hydration products of this pozzolan.

The basic comparison of the total hydration heat, confirmed that a pozzolan reaction occurred because the decrease of the hydration heat is not directly proportional to the reduction of the cement content. Another mechanism influencing the positive properties of these alternative binders, resulting from the isothermal calorimetry and from the particle size of brick ash, is filler effect. This theory should be confirmed by other experiments.

**Acknowledgement**

This work was supported by the project “FV10022 in program MPO - TRIO: EFEKTIVNÍ VYUŽITÍ CIHELNÉHO RECYKLÁTU (Ministry of Industry and Trade)”.

**References**

[1] Pytlík P 1995 *Cihlářství* (Brno: CERM)
[2] ASTM C125-13 2007 Standard Terminology Relating to Concrete and Concrete Aggregates

IOP Conf. Series: Materials Science and Engineering **379** (2018) 012002 doi:10.1088/1757-899X/379/1/012002

(American Society for Testing and Materials)
[3] Moropoulou A, Bakolas A and Aggelakopoulou E 2004 Evaluation of pozzolanic activity of natural and artificial pozzolans by thermal analysis *Thermochim. Acta* **420** pp 135–40
[4] CollepardiM 2009 *Moderní beton* (Praha: Ic ČKAIT)
[5] Lothenbach B, Scrivener K and Hooton R D 2011 The Influence of Gypsum on the Hydration and Properties of Portland Cement Pastes *Cem. Concr. Res.* **41** pp 217–29
[6] Lerch W 1946 *Am. Soc. Test. Mater.* **46** pp 1252–97
[7] Bullard J W, Jennings H, Livingaton A, Jennings M, Nonat A, Sherer G, Svhweitzer J, Scrivener K and Thomas J 2011 Mechanisms of cement hydration *Cem. Concr. Res.* **41** pp 1208–23
[8] Novotný R, Bartoníčková E, Švec J and Mončeková M 2016 Influence of Active Alumina on the Hydration Process of Portland Cement *Procedia Eng.* **151** pp 80–6

IOP Conf. Series: Materials Science and Engineering **379** (2018) 012003 doi:10.1088/1757-899X/379/1/012003

# Reactivity of meta-talc in sodium hydroxide medium

**J Bednárek[1], P Ptáček[1], F Šoukal[1], J Havlica[1], R Novotný[1], J Másilko[1] and P Šiler[1]**

[1] Materials Research Centre, Faculty of Chemistry, Brno University of Technology, Purkyňova 118, 612 00, Brno, Czech Republic

Email: xcbednarek@fch.vut.cz

**Abstract**. This work deals with the reactivity of XRD amorphous delaminated and dehydroxylated talc phase (meta-talc), in alkali medium. Talc ore Luzenac H70 was treated in a stirred ball mill (attritor) for 30; 60 and 90 min with subsequent thermal treatment at 700°C; 750°C and 800°C. Reactivity of mechanochemically and thermally treated talc ore in NaOH solution was studied using isothermal calorimetry. An influence of material crystallinity and amount of water coordinated in material structure after milling on its reactivity in alkali medium was studied.

## 1. Introduction

Meta-talc (analogically with metakaolin) is an X-ray diffraction (XRD) amorphous delaminated and dehydroxylated talc phase. This new material was prepared at Materials Research Centre at Faculty of Chemistry Brno University of Technology and patented in 2014 [1]. Contrary to metakaolin, meta-talc cannot be prepared by thermal dehydroxylation because of immediate recrystallization of enstatite ($MgO \cdot SiO_2$) accompanied with a separation of silica-rich phase at 870–1050°C [1], [2]. Ptáček et al. [1] published the method of preparation of meta-talc by the way of exposition of talc ore to a mechanochemical treatment, i.e. prolonged milling in a high energy milling device (vibration mill, planetary mill, stirred ball mill). Tangential strain induced by the mill caused the delamination and the loss of crystallinity of treated material. Subsequent thermal treatment of delaminated talc phase at temperatures up to 800°C can provide the XRD-amorphous dehydroxylated talc phase, which was termed as meta-talc [1].

With an advantage of real time information, isothermal calorimetry has become a useful technique in studying cement hydration kinetics [3], [4], [5]. Isothermal calorimetry measures the thermal power (heat production rate) produced by the hydration reaction of cementitious and other binding materials. This follows the overall reaction of the binder. The timing and shape of the heat flow curve obtained by the calorimetry is an indicator of relative performance of binding systems. In the isothermal calorimeter, the curing temperature is maintained very close to constant in relatively small sample, which allows a high repeatability and reproducibility [6].

Investigations of alkali activation of metakaolin and slag have displayed that calorimetric measurements are suitable for obtaining the information needed for description of alkali activation processes [7], [8]. Sun and Vollpracht [3] observed broad exothermic peak following initial dissolution peak at calorimetric curves of alkali activated metakaolin and slag. They deduced that after an initial period of dissolution process of alkali activated metakaolin, the concentrations of dissolved ionic species reached a threshold for the outset of following reactions. The dormant period, which corresponded to induction period in cement hydration, was supposed to be a time needed for concentration growing up

[3]. Both polymerization and condensation of geopolymer gels are exothermic processes and made the main contribution to this second peak [3], [9]. In the case of slag the second peak indicated the extensive precipitation of reaction products, such as C-A-S-H [3], [10]. Contrary to that, calorimetric output of NaOH activated fly ash showed only single extensive peak. The slow cumulative heat increase could be assigned to the continuous dissolution and slight transformation of dissolved species into gels [3].

This paper is focused on observation of alkali activation of meta-talc using isothermal calorimetry. Effect of mechanochemical activation period and thermal treatment temperature on the reactivity of meta-talc samples was assessed in order to optimize the preparation of this new material for future experiments.

## 2. Experimental

For the purposes of this work talc ore Luzenac H70 from Imerys was used. Chemical composition and loss of ignition of starting material are reported in Table 1.

**Table 1.** Chemical composition and loss of ignition of talc ore Luzenac H70.

| substance | content (wt. %) |
|---|---|
| $SiO_2$ | 45.29 |
| MgO | 31.72 |
| $Al_2O_3$ | 11.86 |
| $Fe_2O_3$ | 1.27 |
| CaO | 0.52 |
| LOI | 9.06 |

Phase composition, obtained by X-ray diffractometry with use of Rietveld refinement, is shown in Table 2. Unfortunately, due to a preferred orientation of crystals in examined material it is not possible to determine the correct phase composition. Nevertheless it is possible to state that the chosen talc ore consisted mainly of clinochlore – chlorite type phyllosilicate with high Mg content. The second significant phase is talc. The rest phases are present in content of less than 10%.

**Table 2.** Phase composition of talc ore Luzenac H70.

| phase | content (wt. %) |
|---|---|
| clinochlore | 59.4 |
| talc | 29.8 |
| quartz | 9.0 |
| muscovite | 1.5 |
| dolomite | 0.3 |

### *2.1. Preparation of samples*

A batch of 100 g of talc ore was milled in a stirred ball mill Netsch Attritor for 30; 60 and 90 min. Milling balls with diameter of 7 mm were chosen as milling media. The weight of milling balls was 7 300 g. The stirrer rotation speed was 500 rpm. Mechanochemically treated samples were calcinated at 700°C; 750°C and 800°C. The sample was heated with a rate of 5 $°C \cdot min^{-1}$ up to the required temperature. After being held for 1 hour at temperature of calcination, the crucible with sample was cooled with a cooling rate of 5 °C·min$^{-1}$ to 200°C and finally cooled to room temperature in a desiccator. Phase composition of all samples was characterized by X-ray diffractometer Empyrean (PANalytical).

### *2.2. Isothermal calorimetry*
Samples of mechanochemically activated talc ore without thermal treatment and samples after calcination at above mentioned temperatures were treated in alkali medium. A weight of 2 g of sample was mixed with a volume of 1 cm$^3$ of 19 mol·dm$^{-3}$ NaOH solution. Time dependence of heat flow and evolved heat was measured for 60 hours in isothermal calorimeter TAM Air (TA Instruments).

## 3. Results and Discussion

### *3.1. XRD characterization of prepared samples*
As depicted in Figure 1, mechanochemical activation of talc ore led to significant reduction of crystallinity of studied material. Structures of main phyllosilicate phases, clinochlore and talc, were disrupted due to strain whilst quartz phase was not affected during milling due to tougher covalent structure. Calcination at 700°C caused a shift of peak intensity from hydrated clinochlore to a dehydrated one (decrease of intensity of clinochlore peaks at 12.46°, 18.74° and 25.08°). After thermal treatment at 750°C the clinochlore phase was mostly decomposed; peak of dehydrated clinochlore at 6.23° disappeared. At 800°C partial crystallization of high temperature phases, such as enstatite, forsterite or diopside, took place. We can see that sample milled for 30 min had lower content of high temperature crystalline phases after thermal treatment in comparison with longer milled samples. It was caused by larger particle size of this sample which resulted in less proper thermal treatment.

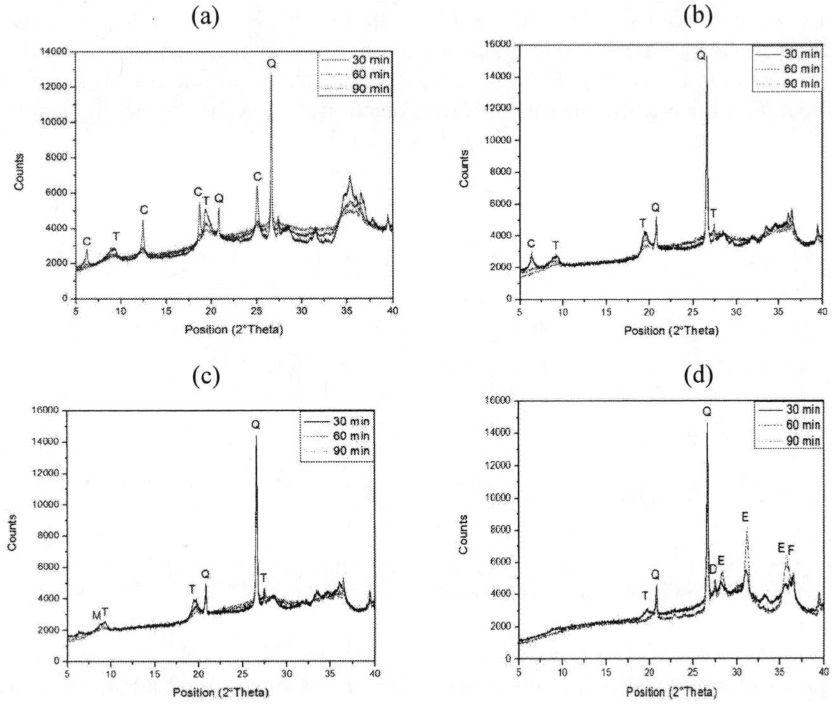


**Figure 1**. Influence of mechanochemical activation and thermal treatment a) none; b) at 700°C; c) at 750°C; d) at 800°C on phase composition of talc ore (C – clinochlore, T – talc, Q – quartz, M –muscovite, E – enstatite, F – forsterite, D – diopside).

### *3.2. Isothermal calorimetry study of meta-talc reactivity in NaOH solution*

Based on the Figure 2, one can consider that reaction between mechanochemically activated talc and NaOH solution takes places quite intensively and about 10 hours after mixing the reaction rate lowers significantly. The reactivity of mechanochemically activated product increases with prolonged milling period due to gradual crystallinity and particle size reduction despite the fact that intensity of heat evolution in first minutes contrarily grows with shorter milling times.

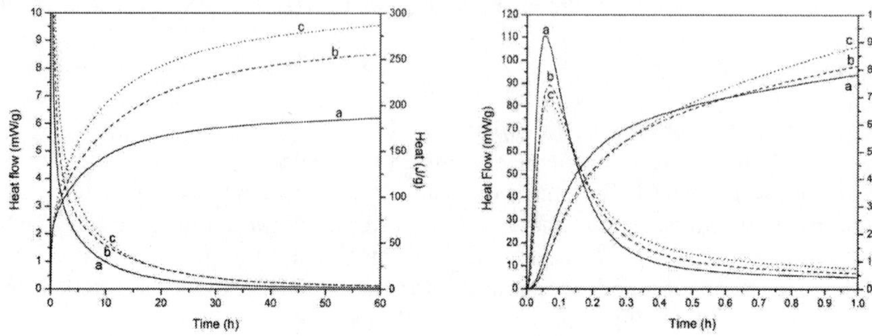


**Figure 2**. Heat evolution of talc ore mechanochemically activated for a) 30 min; b) 60 min; c) 90 min.

Thermal treatment at 700°C led to noticeable changeover in reactivity of delaminated talc ore (Figure 3). Initial dissolution period and generally heat evolution in first hours were less intensive and whole reaction course was considerably more gradual. But on the other hand amount of evolved heat after 60 hours of reaction was comparable with the case of uncalcinated talc ore. We can consider that the transformation of meta-talc treated at 700°C into reaction products was relatively slow. These results are similar to the previous one of alkali activated fly ash [3].

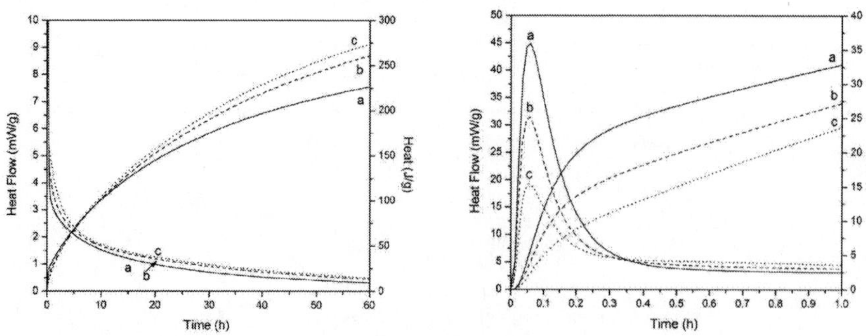


**Figure 3**. Heat evolution of talc ore mechanochemically activated for a) 30 min; b) 60 min; c) 90 min after thermal treatment at 700°C.

Results depicted in Figure 4 indicate that calcination at 750°C brings about further slowing down of alkali activation of studied samples. We can also observe a quite broad additional peak following immediately after initial period. In this case the course of heat evolution curves is more similar to the curves of alkali activation of metakaolin [3]. Overall amount of evolved heat after 60 hours is comparable with previous two cases except of talc ore milled for 30 min. Reactivity of this sample was closer to reactivity of longer milled samples in opposite to the case of samples calcinated at 700°C. It was caused by more intensive removing of water molecules coordinated in material structure after mechanochemical treatment.

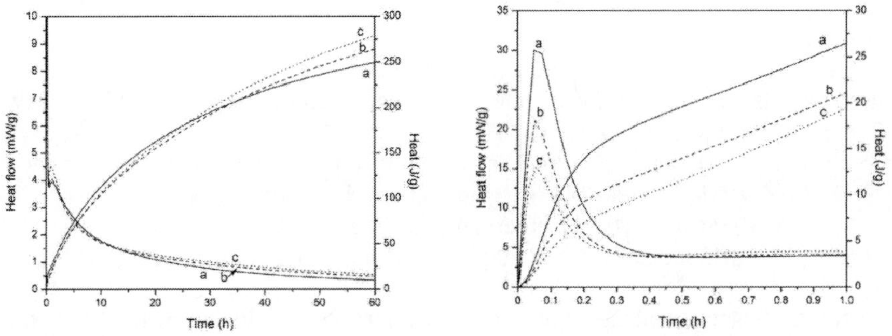


**Figure 4**. Heat evolution of talc ore mechanochemically activated for a) 30 min; b) 60 min; c) 90 min after thermal treatment at 750°C.

As depicted in Figure 5, thermal treatment at 800°C resulted in strong deterioration of material reactivity due to crystallization of stable high temperature phases. In contrary to other three cases sample milled for 30 min was decidedly most reactive due to less proper calcination which corresponds with X-ray diffraction outputs shown in Figure 1.

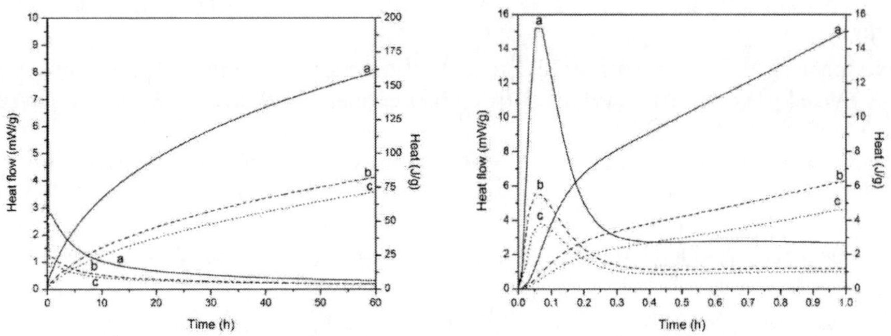


**Figure 5**. Heat evolution of talc ore mechanochemically activated for a) 30 min; b) 60 min; c) 90 min after thermal treatment at 800°C.

## 4. Conclusion

Talc ore Luzenac H70, consisting predominantly of clinochlore and talc, was mechanochemically activated in a stirred ball mill for 30; 60 and 90 minutes and subsequently calcinated at 700°C; 750°C a 800°C.

It was found out that calcination at 700°C and 750°C prolongs reaction period where gradual course of reaction is advantageous for preparation of alkali activated materials. Most significant difference between these two temperatures of thermal treatment were observed in the case of a curve of material activated for 30 minutes where thermal treatment at 750°C lowered the reactivity deficit against samples milled for longer periods. However, calcination at 800°C led to formation of stable crystalline phases (enstatite, forsterite and diopside) and related substantial deterioration of reactivity of all three samples. Rate of reactivity deterioration was dependent on milling period and related particle size of studied material.

**Acknowledgement**
Materials Research Centre at FCH BUT- Sustainability and Development, REG LO1211, with financial support from National Programme for Sustainability I (Ministry of Education, Youth and Sports).

**References**
[1] Ptáček P, Opravil T, Šoukal F, Havlica J, Másilko J and Wasserbauer J 2014 Preparation of dehydroxylated and delaminated talc: Meta-talc *Ceramics International* **39** (Elsevier Ltd and Techna Group S.r.l.) pp 9055–9061
[2] MacKenzie K and Meinhold R 1994 The thermal reactions of talc studied by $^{29}Si$ and $^{25}Mg$ MAS NMR *Thermochimica Acta* **244** (Elsevier Science B.V.) pp 195–203
[3] Sun Z and Vollpracht A 2017 Isothermal calorimetry and in-situ XRD study of the NaOH activated fly ash, metakaolin and slag *Cement and concrete research* **103** (Elsevier Ltd) pp 110–122
[4] Dittrich S, Neubauer J and Goetz-Neunhoeffer F 2013 The influence of fly ash on the hydration of OPC within the first 44 h – A quantitative in situ XRD and heat flow calorimetry study *Cement and Concrete Research* **56** (Elsevier Ltd) pp 129–138
[5] Šiler P, Kolářová I, Novotný R, Másilko J, Pořízka J, Bednárek J, Švec J and Opravil T 2017 Application of isothermal and isoperibolic calorimetry to assess the effect of zinc on cement hydration *Journal of Thermal Analysis and Calorimetry* (*Online* https://doi.org/10.1007/s10973-017-6815-1)
[6] Frølich L, Wadsö L and Sandberg P 2016 Using isothermal calorimetry to predict one day mortar strengths *Cement and Concrete Research* **88** (Elsevier Ltd) pp 108–113
[7] Alonso S and Palomo A 2001 Calorimetric study of alkaline activation of calcium hydroxide – metakaolin solid mixtures *Cement and Concrete Research* **31** (Elsevier Science Ltd) pp 25–30
[8] Ravikumar D and Neithalath N 2012 Reaction kinetics in sodium silicate powder and liquid activated slag binders evaluated using isothermal calorimetry *Thermochimica Acta* **546** (Elsevier B.V.) pp 32–43
[9] Yao X, Zhang Z, Zhu H and Chen Y 2009 Geopolymerization process of alkali-metakaolinite characterized by isothermal calorimetry *Thermochimica acta* **493** (Elsevier B.V.) pp 49–54
[10] Chithiraputhiran S and Neithalath N 2013 Isothermal reaction kinetics and temperature dependence of alkali activation of slag, fly ash and their blends *Constructuon and building materials* **45** pp 233–242

# Hygric and mechanical parameters of ternary binder based plasters lightweighted by expanded perlite

M Čáchová[1], D Koňáková[1], E Vejmelková[1], M Vyšvařil[2] and P Bayer[2]

[1] Department of Materials Engineering and Chemistry, Faculty of Civil Engineering, Czech Technical University in Prague, Thákurova 7, 166 29 Prague 6, Czech Republic
[2] Institute of Chemistry, Faculty of Civil Engineering, Brno University of Technology, Žižkova 17, 602 00 Brno, Czech Republic

Email: monika.cachova@fsv.cvut.cz

**Abstract.** This article is focused on modern plasters with improved thermal insulating ability. The studied plasters were composed of lime, cement, ceramic powder, expanded perlite and silica sand. Regarding the matrix, three kinds of binders were used. Lime was partially replaced by ceramic powder. These waste materials show good pozzolanic reactivity and thus bring economic and ecological benefits. Portland cement was added as a third component to the mixture for the purpose of improving mechanical performance of the studied plasters. In this article, mechanical properties of designed plasters are presented. It was proved that matrix strengthening brought positive effect and eliminated the deterioration of mechanical behaviour; the final values of compressive strength were over 3 MPa. Due to desired high porosity of the studied plasters, their durability can be strongly affected by presence of the water. Therefore, in this article, the hygric parameters were studied too. Namely, the water absorption coefficient, apparent moisture diffusivity, the water vapour diffusion resistance factor and sorption and desorption isotherms belong among studied hygric characteristics. In this article, pore size distribution curves, which affect the discussed parameters, are presented.

## 1. Introduction

Application of lighweigted plasters in civil engineering branch proposes many advantages. The most important is indisputably its contribution to better thermal performance of structure due to their improved thermal insulating ability. The most common way how to lighten plaster is in utilisation of lightweight aggregate. However, an application of those aggregates with higher porosity, but also usually lower mechanical strength, led to deterioration of mechanical performance of final plasters. For example Silva et al. [1] presented that utilisation of 10 wt. % of perlite in lime-cement mortar causes the significant decrease of compressive strength by almost 90% to the value of 0.37 MPa. Due to the 10 wt. % of vermiculite application, the compressive strength was decreased to 1.01 MPa, which means the descent by more than 70%, comparing with mortar with just siliceous aggregate. In the study performed by Palomer et al. [2], the negative effect of both lightweight aggregates was confirmed. The compressive strength of lime-cement mortars decreased due to the 50 vol. % of vermiculite application to final value of 3 MPa, by 51% lower in comparison with reference plaster. 50 vol. % of perlite had also deteriorating effect, however the decrease was just by 20% to the compressive strength of 4.89 MPa. Expanded polystyrene studied in Torres et al. [3] can be named as an example of another lightweighted aggregate. Its application in amount of 1.77 wt. % (of the whole composition) in cement mortar led to decrease by more than 82%. The lightweighted mortar showed compressive strength of

1.71 MPa. Granulated cork was studied by Bras et al. [4]. This material caused also the descent of compressive strength of plasters. When hydraulic lime and 50% of cork were used, the decrease was almost 77% and the compressive strength of this plaster was 0.70 MPa.

This deterioration can be somehow compensated by strengthening of matrix. From ancient time the natural pozzolana (such as example a pumice-stone, tuff, or clays, diatomaceous earth [5]) was used in combination with pure lime for the purpose of obtaining better mechanical performance of a final material. Nowadays, artificial materials are also frequently used, and the most often employed pozzolanic material is the waste material which proposes also the benefits from the ecological point of view. The application of industrial by-products such as ground granulated blast furnace slag or fly ash [6] have already been widely studied and its application in concrete production is common. Waste materials from kaolin production [7] were also studied in current research. Another representative of a clay material is a ceramic powder, which was the matter of the study in our previous work [8]. It also proved its sufficient pozzolanic properties, the kinetics of pozzolanic reaction of this waste material has already been described [9] but its wide application was not so much popularized yet. However, the ceramic powder proposes lower improvement of mechanical performance of studied plasters. Therefore, the combination of three kinds of binders were used in this study. Ternary binder based system have already been studied only in the case of a plaster production [10]. It seems to be a good opportunity applicable also in the case of a lightweight plaster.

Plasters create the exterior face of the majority of low-rise buildings, they does not fulfil only aesthetical role, but also serve as a protective layer. In this context, a plaster is always exposed to environmental deteriorative effects. There are many degradation processes, which can be divided according to the mechanisms of degradation into the physical, chemical and biological degradation. However, in the case of real building materials, degradation processes usually interact together and cannot be simply recognized. The main deteriorating mechanism affecting durability of all building materials is indisputably permeability of a material and presence of water. Water can cause primarily the physical degradation, especially in wintertime its presence causes severe damage due to the repeating freezing and thawing cycles [11, 12]. These conditions lead to the chemical degradation by dissolving or reacting with building materials. Another problem of moisture content is chemical composition of water. In real situations, water is never chemically pure, but brings along dissolved sulphates, chlorides, or gaseous pollutants. These particles can also react with components of building materials and evoke the degradation process; such as alkali-silica reaction, crystallization of salts, or carbonation [13, 14]. Therefore, hygric properties of the plaster are of a great importance and they should be appropriately investigated. Especially in the case of those materials with higher porosity such as lightweighted plasters.

In this article, two kinds of new plasters with improved thermal insulating ability are investigated and a lime-cement plaster as a reference plaster. The pore structure, mechanical properties and hygric properties are the matter of the performed research. Namely, water vapour transport parameters, water liquid properties and moisture storage ability were experimentally determined, compressive strength and achieved results are discussed herein.

## 2. Studied materials

Ternary binder based plasters with improved thermal insulating ability were studied in this article. Designed plasters differ in used aggregates (Table 1). The first material LP III is composed just of expanded perlite, while in the second plaster LP IV is the half of the expanded perlite replaced by the silica sand. The expanded perlite is commercially labelled as EP 180 and it is produced in temperature range 850–1,150°C. This material has low bulk density, specifically about 95 $kg \cdot m^{-3}$, and therefore it has also low thermal conductivity about 0.049 $W \cdot m^{-1} \cdot K^{-1}$ and its compressive strength shows value of 320 kPa. The silica sand was taken from the sand pit of Střeleč (Sklopísek Střeleč a.s). One grading 0–4 mm was employed. The granulometric curves of the aggregate were measured by the standard grading test [15] and they are shown in Figure 2. The matrix of the plasters is composed of three binders. Primarily lime was used. Pure lime hydrate (with 99% of calcium hydroxide) commercially labelled as CL 90 S was produced by the lime plant Čertovy schody a. s. Specific surface of this binder was 1,374 $m^2 \cdot kg^{-1}$. For the economic and ecological purposes, the ceramic powder was used as the second binder. This waste material was taken from the brick plant Hevlín of Heluz cihlářský průmysl, v.o.s and it was produced during grinding process of thermal insulating ceramic blocks, which gave rise to fine material with specific surface area of 641 $m^2 \cdot kg^{-1}$ and sufficient pozzolanic activity (840 mg $Ca(OH)_2$ [$g^{-1}$]) [16]. Portland cement was employed as the third binder. This material was used in half amount and it was chosen for the sake of the matrix strengthening. Specifically CEM I 42.5 R, with specific surface are of 341 $m^2 \cdot kg^{-1}$, originated in Čížkovice (Lafarge company, a. s.), was used. Granulometry of the binders were measured by use of Analysette 22 Micro Tec plus, and the granulometric curves are delineated in Figure 1. The last component of the studied plaster was tap water. The amount was set experimentally with the respect of reaching standard flow [17] of lightened plasters.

The reference plaster was added for comparison with the special lightweight plasters. The reference plaster is composed of the lime (5 kilogram; CL 90 S), Portland cement (1 kilogram; CEM I 42.5 R) and silica sand (18 kilograms; fraction 0.3–4.0 mm). The water binder ratio was 0.22 and the flow was 160/160 mm.

**Table 1.** Mixture composition of the studied plasters.

| Component | LP III | | LP IV | |
|---|---|---|---|---|
| | [kg] | [%] | [kg] | [%] |
| Lime (CL 90 S) | 50 | 29 | 50 | 29 |
| Ceramic powder | 50 | 29 | 50 | 29 |
| Cement (CEM I 42.5 R) | 25 | 14 | 25 | 14 |
| Perlite (Experlit 180) | 50 | 29 | 25 | 14 |
| Silica sand | - | - | 25 | 14 |
| Water/binder ratio | 1.05 | | 1.61 | |
| Flow | 165/165 | | 160/160 | |

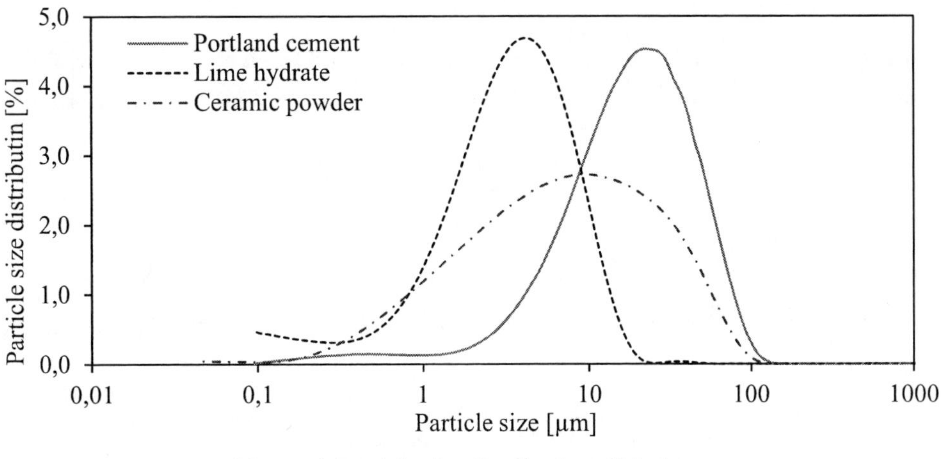


**Figure 1.** Particle size distribution of binders.

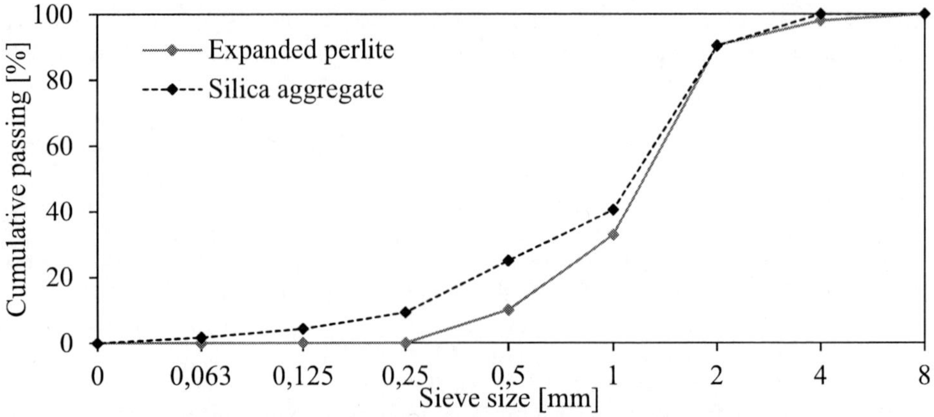


**Figure 2.** Particle size distribution of aggregates of the special plasters.

## 3. Experimental measurements and results

### *3.1. Pore structure.*

The most influencing property for the hygric parameters is indisputably porosity of the studied material. However, more detailed description than just a value of the porosity is required for better comprehension. The pore structure of the studied plasters was determined by use of mercury intrusion porosimetry. For this experiment, the measurement apparatus PASCAL 140 + 440 was used.

Results are presented in Figure 3 and Figure 4. Both plasters had pores in similar range, from 0.1 μm to 10 μm. However, it can be seen that the plaster LP III with expanded perlite had higher porosity and also higher amount of pores in range from 1 μm to 10 μm. This amount of pores is reduced by finer sand application in the case of the plaster LP IV. The reference plaster has shown the lowest pore curves (Figure 3 and Figure 4). This reference graphs confirm the theory of the highly pore system compared to the lightweight plasters.

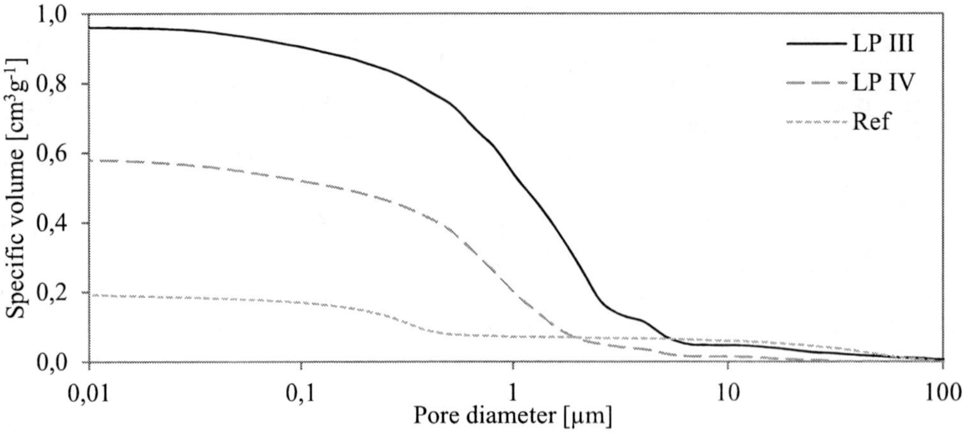


**Figure 3.** Cumulative curve of the studied plasters.

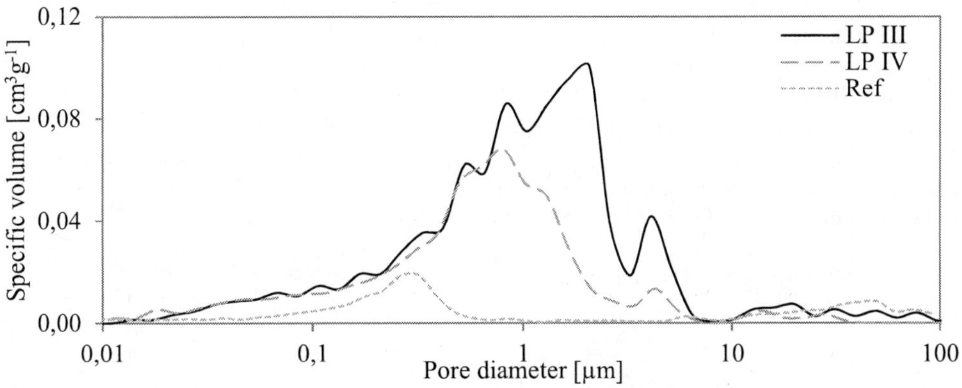


**Figure 4.** Distribution curve of the studied plasters.

### 3.2. *Mechanical properties*

Compressive strengths were measured according to standard methods [18] by using loading device EU40. The experiments were performed at various times to be able to described potentially longer evolution of compressive strength due to the presence of pozzolana in the binder and its prolonged pozzolanic reaction. Achieved results of the compressive strength are presented in Figure 5. Both plasters reached quite high compressive strength, with values over 3 MPa, which were desired results reaching by binder combination. The compressive strength of the plaster LP III with just expanded perlite showed lower value of the compressive strength in comparison with the plaster LP IV with combined aggregate. The difference was in 28 days by about 23%. Regarding the influence of the time, both plasters showed quite high increase; 17% and 14% in the case of the plaster LP III and LP IV respectively. The reference plaster has shown lower mechanical properties than the special plasters with the expandite perlite. In the time evolution the behaviour of the compressive strength (reference plaster) was relatively constant from 0.6 MPa to 0.9 MPa. These results of the compressive strength of the lightweight plaster LP III are more than 3 times higher than plaster Ref.

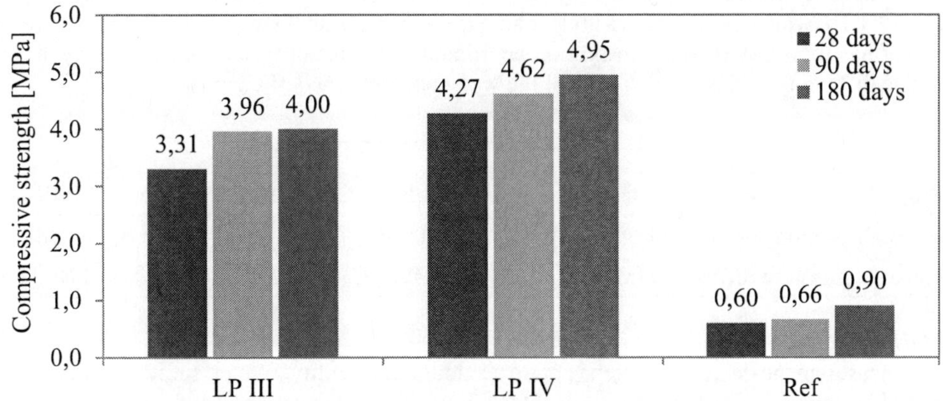


**Figure 5.** Time evolution of the compressive strength.

### 3.3. *Water vapour transport*

For measuring of the water vapour transport, the cup methods described in the standard ČSN 72 7031 were used [19]. Calculated the water vapour resistance factors of the studied plasters are presented in Figure 6.

In both arrangements, the plaster LP III with the expanded perlite reached lower values, which is in accordance with higher porosity of the studied plasters (Figure 3 and Figure 4). The differences were 18% and 31% for dry – cup respectively wet – cup arrangements. Regarding the difference between both arrangements, this was in accordance with results obtained for other porous materials where a significant acceleration of water vapor transport with increasing relative humidity was observed [20]. This phenomenon can be explained by the added capillary condensed water transport in the wet-cup conditions [21]. The reference plaster shows the lowest values of the diffusion resistance factor in the dry cup arrangement and the highest values in the wet cup arrangement.

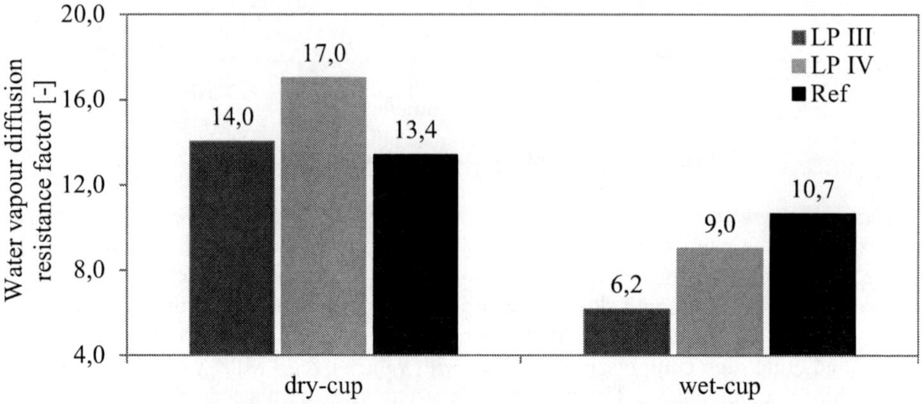


**Figure 6.** The water vapour transport of the studied plasters.

### *3.4. Water liquid transport*

The water absorption coefficient was measured applying sorption experiments and determined from the sorptivity plot [22]. Approximate values of apparent moisture were also calculated [23].

Achieved results are given in Table 2. As in the case of water vapour transport, the plaster LP III with expanded perlite showed higher ability of water liquid transport compared to the plaster LP IV where aggregate combination was used. Water absorption coefficient differed by 29%, while apparent moisture diffusivity varied by 25%. Greater difference is caused by higher amount of bigger pores in the plaster LP III, which has two times greater amount of pores with diameter over 1 μm. The reference plaster Ref has the greatest water absorption coefficient. This result reaches that value of the plaster LP III. The difference at this both plasters of the water absorption coefficient is 15%.

**Table 2.** The water liquid transport parameters.

| Property | Ref | LP III | LP IV |
|---|---|---|---|
| Water absorption coefficient [$kg \cdot m^{-2} \cdot s^{-1/2}$] | 0.393 | 0.334 | 0.238 |
| Apparent moisture diffusivity [$m^2 \cdot s^{-1}$] | 1.40E-07 | 4.22E-07 | 3.17E-07 |

### *3.5. Moisture storage.*

The last measurement describes water vapour accumulation ability. Water sorption isotherms were determined by the special device DVS Advantage 2, which uses dynamic vapor sorption method [23]. In accordance with higher porosity of the plaster LP III, or more precisely with its bigger pores surface area, the plaster with expanded perlite proved higher ability for moisture storage. It is obvious that with increasing relative humidity the difference between both plasters become more remarkable. When the relative humidity was equal to 20%, the difference was 26%. The deviation continuously grew up to 49% in the case of 98% relative humidity. The sorption isotherm of the reference plaster is displayed as the lowest curve. This observation corresponds to the results of the porous system.

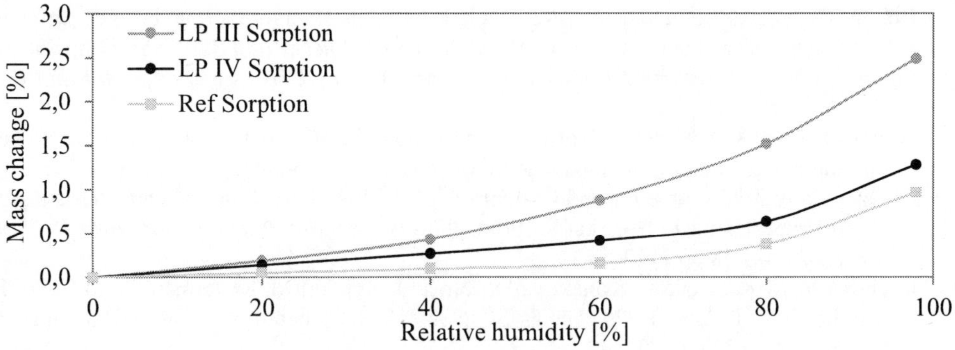


**Figure 7.** The sorption isotherms.

## 4. Summary

In this article, two kinds of the lightweighted plasters were presented. The main emphasis was focused on research of the moisture characteristic, the matrix strengthening and on the lightweight plaster. The lightening was achieved by utilisation of highly porous expanded perlite as main aggregate. The silica sand was employed in the case of the second plaster as the half replacement of the lightweight perlite. Utilisation of the lightweight aggregate usually had led to deterioration of the mechanical behaviour. This effect was confirmed also in the case of the studied plasters. The one with combined aggregate showed greater values of the mechanical properties than the plaster with just expanded perlite. However, the deteriorating effect of the lightweight aggregate was partially compensated by improvement of the matrix of the studied materials. The plasters had ternary based matrix, which composed of the lime, pozzolana and lower amount of the cement. This approach brought positive effect. The application of combined binders had economic and ecological benefits, but especially had positive effect on the mechanical performance of the studied plasters. Despite the high porosity of the designed plasters, the compressive strength reached sufficient values greater than 3 MPa. Utilisation of the expanded perlite led to desired increase of porosity, specifically the amount of pores with diameter from 1 to 10 μm went up. The water vapour permeability of both plasters was high. Satisfactory results were achieved by including higher porosity of the plasters. The difference between both plasters was not so significant. Regarding the water liquid transport, the utilisation of the expanded perlite led to the considerable growth of both water absorption coefficient and the apparent moisture diffusivity comparing to aggregate combination. Moisture storage capability was also influenced by utilisation of the expanded perlite, especially in higher level of the relative humidity. But similar to the water vapour transport, the measured sorption isotherms were not as high as they could be due to high porosity of the studied plasters.

**Acknowledgments**

This research has been supported in the Czech Republic under projects SGS16/199/OHK1/3T/11, P105/12/G059, and No. LO1408 "AdMaS UP - Advanced Materials, Structures and Technologies", supported by Ministry of Education, Youth and Sports under the „National Sustainability Programme I".

**References**

[1] Silva L M, Ribeiro R A, Labricha J A and Ferreira V M 2010 Role of lightweight fillers on the properties of a mixed-binder mortar *Cement & Concrete Composites* **32** pp 19–24

[2] Palomar I, Barluenga G and Puentes J 2015 Lime–cement mortars for coating with improved thermal and acoustic performance *Construction and Building Materials* **75** pp 306–314

[3] Torres M L and García-Ruiz P A 2009 Lightweight pozzolanic materials used in mortars: Evaluation of their influence on density, mechanical strength and water absorption *Cement & Concrete Composites* **31** pp 114–119

[4] Brás A, Gonçalves F and Faustino P 2014 Cork-based mortars for thermal bridges correction in a dwelling: Thermal performance and cost evaluation *Energy and Buildings* **72** pp 296–308
[5] Franco M 2003 Pozzolana and Pozzolanic Cement *Lea's Chemistry of Cement and Concrete (Fourth Edition)* pp 471–635
[6] Grist E R, Paine K A, Heath A, Norman J and Pinder H 2013 Compressive strength development of binary and ternary lime–pozzolan mortars *Materials and Design* **52** pp 514–523
[7] de Azeredo A F N, Struble L J and Carneiro A M P 2014 Microstructural characteristics of lime-pozzolan pastes made from kaolin production wastes *Materials and Structures/Materiaux et Constructions* **48** pp 2123–2132
[8] Čáchová M, Vejmelková E, Koňáková D, Žumár J, Keppert M, Reiterman P and Černý R 2016 Application of Ceramic Powder as Supplementary Cementitious Material in Lime Plasters *Medziagotyra* **22** pp 440–444
[9] Čáchová M, Scheinherrová L, Kobera L, Urbanová M, Brus J and Keppert M 2017 Monitoring of kinetics of Pozzolanic reaction *Key Engineering Materials* **722** pp 126–131
[10] Kočí V, Maděra J, Jerman M, Žumár J, Koňáková D, Čáchová M, Vejmelková E, Reiterman P and Černý R 2016 Application of waste ceramic dust as a ready-to-use replacement of cement in lime-cement plasters: an environmental-friendly and energy-efficient solution *Clean Technologies and Environmental Policy* **18** pp 1725–1733
[11] Kočí J, Maděra J, Rovnaníková P and Černý R 2011 Hygrothermal performance of innovative renovation renders used for different types of historical masonry *WIT Transactions on The Built Environment* **118** pp 683–693
[12] Kočí V, Maděra J and Černý R 2012 Exterior thermal insulation systems for AAC building envelopes: Computational analysis aimed at increasing service life *Energy & Buildings* **47** pp 84–90
[13] Pavlíková M, Pavlík Z, Keppert M and Černý R 2011 Salt transport and storage parameters of renovation plasters and their possible effects on restored buildings' walls *Construction and Building Materials* **25** pp 1205–1212
[14] Park S, Kwon S and Jung S H 2012 Analysis technique for chloride penetration in cracked concrete using equivalent diffusion and permeation *Construction and Building Materials* **29** pp 183–192
[15] ČSN EN 933-1:2012 Tests for geometrical properties of aggregates – Part 1: Determination of particle size distribution – Sieving method (Prague: Czech Standards Institution)
[16] NF P 18-513:2000 Pozzolanic Addition for Concrete – Metakaolin - Definitions, Specifications and Conformity Criteria (Association Française de Normalisation)
[17] ČSN EN 1015-3:2000 Methods of test for mortar for masonry - Part 3: Determination of consistence of fresh mortar (by flow table) (Prague: Czech Standards Institution)
[18] ČSN EN 1015-11:2000 Methods of test for mortar for masonry - Part 11: Determination of flexural and compressive strength of hardened mortar (Prague: Czech Standardization Institute)
[19] Kočí J, Žumár J, Pavlík Z and Černý R 2012 Application of Genetic Algorithm for Determination of Water Vapor Diffusion Parameters of Building Materials *Journal of Building Physics* **35** pp 238–250
[20] Černý R and Rovnaníková P 2002 *Transport processes in concrete* (London: Spon Press)
[21] Vejmelková E, Pavlíková M, Jerman M and Černý R 2009 Free water intake as means of material characterization *Journal of Building Physics* **33** pp 29–44
[22] Kumaran M K 1999 Moisture diffusivity of building materials from water absorption measurements *Journal of Thermal Envelope and Building Science* **22** pp 349–355
[23] Černý R 2013 *Complex System of Methods for Directed Design and Assessment of Functional Properties of Building Materials and Its Applications* (Prague: Czech Technical University in Prague)

IOP Conf. Series: Materials Science and Engineering **379** (2018) 012005 doi:10.1088/1757-899X/379/1/012005

# Influence of fine aggregate on some properties of gypsum mortars

**M Doleželová[1], J Krejsová[1] and A Vimmrová[1]**

[1] Department of Materials Engineering and Chemistry, Faculty of Civil Engineering, Czech Technical University in Prague, Thákurova 7, 166 29 Prague, Czech Republic

Email: magdalena.dolezelova@fsv.cvut.cz

**Abstract**. The importance of gypsum as a building material increases nowadays because of its environmental friendliness, excellent fire properties and good workability. Since the gypsum is mostly used in the form of the paste, the influence of aggregates on properties of gypsum mortars is not investigated very often. The article deals with some basic and mechanical properties of gypsum mortars with different aggregates and compares them with the properties of gypsum paste. Type of the aggregates influences the properties of gypsum mortars significantly. The presence of aggregate does not worsen the adhesive strength of the gypsum material in comparison with gypsum paste and the adhesive strength was even improved for some type of aggregates.

## 1. Introduction

The research of gypsum increases recently, because it can be produced from secondary materials and also because it has very low temperature of calcination (in comparison with the lime or cement production). The gypsum-based pastes were studied mostly and the research of gypsum mortars (e.g. materials with aggregates) is at the beginning nowadays. Gypsum mortars are used to the lesser extent than gypsum pastes, since the presence of aggregates in gypsum is not necessary from the point of view of shrinkage (gypsum does not shrink during setting), however aggregates are used in gypsum mainly for economic reasons. The use of aggregates in gypsum is mostly based on the historical experience or on the knowledge obtained from the study of other binders, especially of cement [1], but the real behaviour of gypsum mortars with common aggregates was not studied sufficiently yet. The authors studied mostly the possibility of waste products utilization [2, 3]. The adhesive strength is determined mostly for the commercial type of mortars [4]. The ready-mixed clay plasters were studied in [5]. Authors found that the procedure, described in EN 1015-12 is not suitable for this type of plasters. There was found only one paper, dealing with the adhesive strength of gypsum mortars [6]. In this study, the ladle furnace slag was used as filler and it was stated, that the adhesive strength decreased with increasing the amount of slag. The goal of the article was to describe the influence of different types of common siliceous aggregates on selected mechanical properties. The four aggregates of different origin, containing different types of particles were chosen for the design of gypsum mortars and the compressive, tensile and adhesive strength of mortars were tested. The results were compared with the properties of pure gypsum paste and they are discussed in this article. The research was performed in order to obtain some basic knowledge about the behaviour of gypsum composites, because it was found, that while the influence of aggregates on the properties and behaviour of cement-based materials was studied thoroughly [7, 8], no such research was performed for gypsum-based materials yet.

## 2. Used materials

Four gypsum mortars and one gypsum paste as a reference were tested. Gypsum (GYPSTREND Ltd.), four types of siliceous aggregates with different origin and particle shape and citric acid were used as dry components of the mixtures. The used gypsum was the common commercial product. The citric acid was used as the setting retarder. The amount of citric acid depends to the amount of gypsum. The specification of sands is in the Table 1. The grain surface roughness of all sands was determined by the confocal laser scanning microscopy (CLSM) as the three dimensional arithmetical mean roughness SRa [μm] [9].

**Table 1.** Materials specification.

| | Sand - S2 | Sand - S3 | Sand - S4 | Sand - S5 |
|---|---|---|---|---|
| Producer | Filtrační písky Ltd. | Sklopísek Střeleč Inc. | KÁMEN Zbraslav Inc. | TAPAS BOREK Ltd. |
| Origin | Quarried | Quarried | Natural crushed | River bed, uncrushed |
| Surface | smooth | rather smooth | rough | smooth |
| *SRa* of grain surface [μm] | 0.235 | 0.361 | 0.771 | 0.215 |
| Remark | Standardized sand (CEN, ČSN EN 196-1) | $SiO_2 > 99$ % | - | - |

### *2.1. Designation of mixtures*

The final compositions of mixtures are in the Table 2. The components of mixtures were dosed by volume. The binder to sand ratio was 1:2. The amount of gypsum was the same for all mixtures. The amount of sand was calculated according to its bulk density to achieve the same volume of sand in each mortar. The volume of standardized sand was chosen as a default value. The detailed description of mixtures design was described in the previous article of authors [10].

**Table 2.** Mixture compositions.

| Mixture designattion | Fine aggregate (%) | Gypsum (%) | Citric acid (%) | Water/binder ratio (-) |
|---|---|---|---|---|
| G | - | 99.92 | 0.08 | 0.45 |
| GS2 | 66.65 | 33.32 | 0.03 | 0.53 |
| GS3 | 65.3 | 34.67 | 0.03 | 0.53 |
| GS4 | 67.15 | 32.82 | 0.03 | 0.53 |
| GS5 | 66.92 | 33.05 | 0.03 | 0.53 |

## 3. Measuring methods

### *3.1. Compressive and tensile strength*

Production of samples for strength measuring was managed according to ČSN EN 13279-2 [11]. One set of test specimens was made from each mixture. Each test set contained three prisms 160 × 40 × 40 mm. Samples were stored in laboratory for 7 days and then they were dried at 50°C to constant weight before the testing. Tensile strength $f_f$ [MPa] was determined by three-point flexural test by mechanical press FP 100 (VEB Industriewerk Ravenstein) and compressive strength $f_c$ [MPa] was tested after the bending test. The strength tests were also performed according to standard ČSN EN 13279-2.

### *3.2. Adhesive strength*

Adhesive strength tests were carried out according to ČSN EN 1015-12 [12]. Aerated autoclaved concrete blocks were used as a substrate. They were wetted and manually coated by mortars/plasters according to the conventional construction techniques. Five circles were cut into the fresh mortar/plaster to form circular test areas of 50 mm in diameter (Figure 1). The day before testing, circular pull-head plates made of stainless steel were glued to the circular test areas with two-component epoxy resin.

Adhesive strength tests were carried out 7 days after plastering using a measuring device COMING plus a.s. Tensile load was applied to the circular test area through the pull-head plate. The measuring assembly is shown in Figure 2. Failure load $F_u$ [N] was recorded and fracture pattern was described. Adhesive strength $f_u$ [$N \cdot mm^{-2}$] was calculated as:

$$f_u = F_u \cdot A^{-1} \quad (1)$$

where A [$mm^2$] is circular test area.

Five values of adhesive strength were calculated for each mixture and evaluated using DDtest (Dean - Dixon test). Each fracture pattern was described as one of the following types or their combination:

a) adhesion fracture at the interface between mortar/plaster and substrate,
b) cohesion fracture in the mortar/plaster itself,
c) cohesion fracture in the substrate.

In the case of fracture patterns b) or c), calculated value of adhesive strength could be considered as lower value than the real value of adhesive strength.

**Figure 1.** Test circles in fresh mortar.

**Figure 2.** Measuring device - COMING plus.

## 4. Results and discussion

The bulk density of all gypsum mortars is around 1780 $kg \cdot m^{-3}$. The bulk density of gypsum paste is lower (1311 $kg \cdot m^{-3}$), because it does not contain any aggregates. The compressive strength and tensile strength can be seen in Figure 3. The samples with roughest particles (i.e. with natural crushed sand GS4) have the highest strength of all materials (22.2 MPa). The strength of gypsum paste is 18 MPa. The strength of all samples with the smooth particles (GS2, GS3, GS5) is lower than the strength of gypsum paste G. The lowest value has the mortar with standardized sand (GS2), which has the smoothest surface of particles from all tested aggregates. It is evident, that the quality of the surface aggregates plays similar role in gypsum as in the cement-based materials [13]. While the aggregate with smooth particles affects the strength of mortars negatively, the aggregates with rough particles cause the increase of the mortars strength.

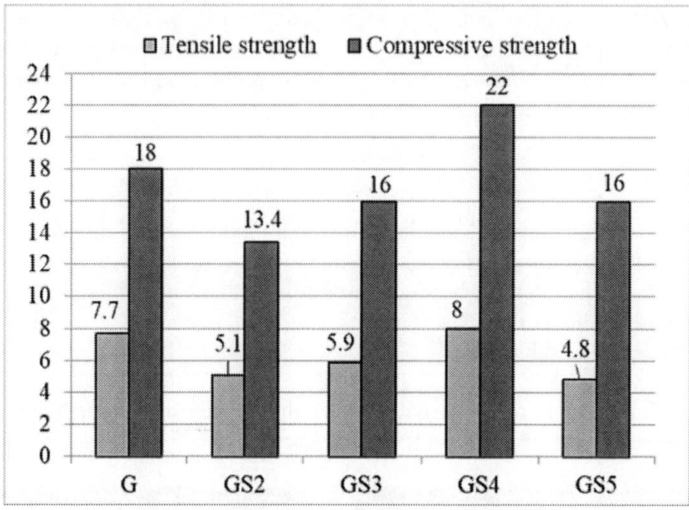


**Figure 3.** Mechanical properties [MPa].

**Table 3.** Adhesive strength at the age of 7 days.

| Mixture desingation | **G** | **GS2** | **GS3** | **GS4** | **GS5** |
|---|---|---|---|---|---|
| Adhessive strength [MPa] | 0.29 ± 0.03 | 0.3 ± 0.2 | 0.5 ± 0.1 | 0.50 ± 0.05 | 0.3 ± 0.1 |

The values of adhesive strength (with the standard deviation) are in the Table 3. The adhesive strength was measured five times for each mortar/paste. No measuring had to be eliminated for the fracture between mortar/paste and the circular pull-head plates. The results of adhesive strength are in Table 3 and the evaluation of fracture pattern is shown in Table 4. The real value adhesive strength could be obtained only when the type **a)** of the fracture occurs (i.e. at the interface between mortar/plaster and substrate). This type of fracture pattern predominated for the gypsum mortars. The samples without the sand (G) were mostly fractured in the paste (type b) itself, so the tensile strength of paste was lower than its adhesive strength.

The area of fracture in the mortar (type b) was in the range from 14% to 36% for mortars. Fracture in the substrate (type **c**) occurred only rarely. The area of fracture **c)** was maximally 12%. It can be supposed that the measured values of samples GS2, GS3are very close to the real adhesive strength, because the fractures were predominantly in the interface between gypsum mortars and substrate. The adhesive strength of mortar with crushed sand (GS4) could be mildly higher in reality, considering the type of fracture. The samples of gypsum paste without sand (G) account the biggest inaccuracy. The paste has lowest measured adhesive strength (0.290 MPa), but it could be considered as a minimal value and the real value of adhesive strength of gypsum paste is probably higher.

The material with the roughest aggregates (GS4) has the highest adhesive strength, similarly as the compressive and tensile strength, while the mortars with smoothest particles (GS2, GS5) have the calculated value of adhesive strength nearly the same as gypsum paste. But because the real adhesive strength of gypsum paste is probably higher than calculated value, the real adhesive strength of gypsum mortars with smooth particles is in reality lower than adhesive strength of gypsum paste. The higher adhesive strength of mortar GS3 can be attributed to the higher amount of fine particles in this type of sand, which was confirmed by the laser diffraction analysis [9]. The conclusion could be drawn, that the presence of smooth aggregates slightly worsens the adhesive strength of gypsum material, but utilization of aggregates with rough particles could improve it.

**Table 4.** The evaluation of fracture pattern.

| Fracture pattern | | G | GS2 | GS3 | GS4 | GS5 |
|---|---|---|---|---|---|---|
| at the interface between plaster and substrate | a) | 18% | 80% | 72% | 56% | 67% |
| in the plaster | b) | 82% | 14% | 16% | 36% | 27% |
| in the substrate | c) | 0% | 6% | 12% | 8% | 6% |

## 5. Conclusions

The influence of the surface quality of the aggregate particles on some mechanical properties of gypsum mortars was investigated. Four gypsum mortars with different types of sand and pure gypsum paste were tested and compared. The results confirm that the quality of aggregate surface affects the mechanical properties of gypsum materials in similar way as cement materials. The aggregate with smooth particles causes the decrease of mechanical properties of mortars and on the contrary, the strength of mortar with rough particles is higher than the strength of gypsum paste. The values of adhesive strength were evaluated considering the type of fracture. The gypsum paste has the lowest value of the adhesive strength, but the result does not express the real value accurately, because the fracture occurred mostly in the paste, while the values of adhesive strength of gypsum mortars with smooth particles corresponds with realty. Also the real value of adhesive strength of mortar with rough particles is higher than the measured value. Therefore it can be said, that the presence of smooth particles worsens the adhesive strength of gypsum material while the rough particles improves it. These results are in accordance with the conclusions, obtained for cement-based materials.

**Acknowledgement:** This research was supported by the Czech Science Foundation, Project No. 16-01438S and by the project SGS16/199/OHK1/3T/11.

## References

[1] Scrivener K, Crumbie A and Laugesen P 2004 *Interface Sci.* **12** pp 411–21
[2] Jiménez Rivero A, De Guzmán Báez A and Navarro J G 2014 *Constr. Build. Mater.* **55** pp 146–52
[3] Adamopoulos S, Foti D, Voulgaridis E and Passialis C 2015 *BioResources* **10** pp 5563–72
[4] Feria P, Santos T and Aubert J.E 2015 *J. Mater. Civ. Eng.* **28** (1)
[5] Delinière R, Aubert J, Rojat F and Gasc-Barbier M 2014 *Build. Environ.* **80** pp 11–7
[6] Rodríguez A, Gutiérrez-González S, Horgnies M and Calderón V 2013 *Mater. Des.* **52** pp 987–94
[7] Ollivier J, Maso J and Bourdette B 1995 *Adv. Cem. Based Mater.* **2** pp 30–38
[8] Tasong W, Lynsdale C and Cripps J 1998 *Cem. Concr. Res.* **28** pp 1453–1465
[9] Krejsová J, Doleželová M and Vimmrová A 2017 *Key Eng. Mater.* **760** pp 245–50
[10] Doleželová M, Krejsová J and Vimmrová A 2016 *Key Eng. Mater.* **722** pp 337–42
[11] ČSN EN 13279-2:2017 Gypsum binders and gypsum plasters. Test methods (Prague: UNMZ)
[12] ČSN EN 1015-12:2000 Methods of test for mortar for masonry - Part 12: Determination of adhesive strength of hardened rendering and plastering mortars on substrates Test methods (Prague: UNMZ)
[13] Mindess S, Francis J and Darwin D 2003 *Concrete* Prentice Hall (Upper Saddle River, NJ)

# Characterization of aerial lime-based mortars with addition of biopolymers

**T Žižlavský[1], M Vyšvařil[1] and P Rovnaníková[1]**

[1] Faculty of Civil Engineering, Brno University of Technology, Veveří 331/95, Brno 602 00, Czech Republic

E mail: zizlavsky.t@fce.vutbr.cz

**Abstract.** The paper studies the influence of addition of viscosity enhancing biopolymers (diutan gum, gellan gum, xanthan gum, carrageenan, and sodium salt of alginic acid) in the doses of 0.1%, 0.5% and 1% on the fresh state, and physical-mechanical properties of hardened mortars. The mortars were prepared with the constant water/binder ratio and the binder/aggregate ratio of 1:3 by weight. The properties have been studied after 7 and 28 days of curing time. All the admixtures, with the exception of gellan gum, improved workability (by lowering the value of spread) in comparison with reference mortar. Most of them also increased the amount of air on the fresh mortar, thus decreasing the density of fresh mortar and also the bulk density of hardened specimens. The addition of biopolymers caused certain decrease of strengths, but there was observable trend of decreasing difference between the reference and tested mortars with the increasing amount of time.

## 1. Introduction

The viscosity enhancing admixtures (VEA's) are widely used in the building materials especially in concrete and dry mix mortars. Due to a volume of concrete consumption in comparison with lime-based mortars, the most of the studies focus on the influence of wide range of admixtures on the concrete or cement-based pastes and mortars thus the influence of the VEA on the lime mortars is only marginally examined. The most common biopolymer admixtures are cellulose ethers but the use of guar gum, chitosan ethers, diutan gum, and xanthan gum is expanding [1].

The paper studies the influence of, in the building practice, less-known biopolymers on the properties of aerial lime-based mortars. Namely the sodium salt of alginic acid, gellan gum, xanthan gum, diutan gum and carrageenan. The biopolymers used can be diversified in two basic groups: seaweed biopolymers [alginic acid sodium salt (ALGNA), carrageenan (CG)], and microbial (gellan gum (GeG), xanthan gum (XG), and diutan gum (DG)). All of the biopolymers are currently mostly used in the food and drug industry as thickening agents, stabilizers, and gelling agents. [1, 2] The alginates in cementitious material caused notably lower strength decrease in comparison with the commercial super absorbent polymers (SAP) and met the strength demand of cement nominative strength in most cases even though the increased water demand [3]. The alginate added to SCC as a VEA notably reduced bleeding and improved the fresh state properties of mixture [4]. Carrageenan, in building materials, was tested as admixture for fly-ash geopolymer, where its addition considerably increased strengths, mainly by creating more condensed structure [5]. Use of carrageenan in cement-based material is reported only as a foam-stabiliser for preparing extremely porous cementitious foam, where the low strength was mainly caused by porosity and not the biopolymer addition [6]. Gellan gum is used mainly in food and drug industry. In building materials GeG hydrogel slurry was mixed with gypsum or cement slurry to create porous materials. The GeG hydrogel did not affect the reaction of cement nor gypsum in any observable way [7]. On the other hand xanthan gum is used in SCC as VEA, where in certain doses it can eliminate bleeding, reduce segregation, and it does not affect strength of concrete [1, 8]. Diutan gum (DG) is chemically similar to GeG, having same backbone and differing only by substituents on the side chain, thus having different properties. DG is mainly used as VEA in concrete, where it is used for same applications as XG. In comparison with XG the effects of DG addition are independent on temperature or concentration of monovalent and bivalent ions (e.g. $Ca^{2+}$) in solution thus being more suitable for cement or lime mixtures due to high concentrations of $Ca^{2+}$ cations in the mixtures [4, 8, 9].

The goal of this work is to compare the effects of above-mentioned biopolymers in various doses on the fresh and hardened state properties of aerial lime-based mortar.

## 2. Materials and methods

### *2.1. Mixing procedure and materials*

The mixture consisted of commercial dry hydrated lime (Carmeuse Czech Republic) of the CL 90 S class according to EN 459-1, the siliceous sand fraction 0-4 mm (Českomoravský štěrk a. s., Hulín, Czech Republic) and the admixture of doses 0.1%, 0.5%, and 1% of binder weight. The admixtures used were: sodium salt of alginic acid (Sigma-Aldrich, co), diutan gum (Kelco-crete DG-F, BASF Construction Solutions GmbH), xanthan gum (Kelzan AP-AS), carrageenan (Genuvisco CG-131), and gellan gum (Kelcogel CG-L-A) three last named supplied by Biesterfeld Silcom s.r.o. but all with exception of alginate being products of CP Kelco. Specimens were prepared by dry mixing of hydrated lime with sand in a 1:1 volumetric ratio, which was converted to 1:3 gravimetric ratio to avoid imperfections, then the admixture was added and all was well-mixed. The dry mixture was added to the specified amount of water (constant water/binder radio of 1.15:1). Prepared mortar was used for test of fresh state properties, and from the rest of the mortar, beams of 40 × 40 × 160 mm were prepared.

### *2.2. Methods*

*2.2.1. Fresh state properties.* On the fresh mortar, several properties were tested. Flow table test was used to determine workability. Water retention value was stated by weighting water-absorbent material placed under specified amount of mortar for 5 minutes. Density of fresh mortar was recorded by weighting vessel of 1 $dm^3$ volume filled with mortar. Than content of air in fresh mortar was obtained by using device designed for this type of measurement. All of these tests and instruments were used according to EN 459-2.

*2.2.2. Hardened state properties.* Bulk density of mortars was stated by measuring and weighting beams before strength test. Flexural strength was determined by three point bending test. Two parts from the test were than tested for compressive strength. These tests were carried out according to EN 1015-11. A piece of sample representing the cross section of the inner part of the beam was milled in mortar mill and used to determine amount of $Ca(OH)_2$ in the samples using differential thermal analysis in combination with thermogravimetric analysis to observe the influence of the admixtures on the carbonation of prepared mortars. The hardened state properties were stated on the specimens 7 and 28 days old stored in laboratory conditions (20±5°C, 50±10% RH) until the day of testing.

## 3. Results and discussion

### *3.1. Fresh state properties*

Fresh state properties are summarized in Figures 1 and 2. The gellan addition had almost none effect on any of tested properties, this being probably due to its high gel setting point at room temperatures [10]. Other admixtures acted as thickening agents with different dosage-dependant effects. The most effective was DG, which even in small doses dramatically reduced slump. CG had more linear course in slump reduction, both reaching slump of 100 mm in the highest dose, where 100 mm is also inner diameter of the testing cone. Surprisingly low effect of XG addition compared to expectations from concrete usage is most likely caused by its sensitivity to concentration of certain cations in the solution, where lime mortar contains many more $Ca^{2+}$ cations in comparison with cement slurry [8, 9].

The values of air content in the fresh mortar corresponded to the slump reduction with no significant difference, thus we can conclude that the volume of air in the prepared mortars is due to the workability and not the air entraining function of any of the admixtures used as it was in the case of different biopolymers in the literature [11, 12]. The density of fresh mortar is closely-linked with the air content, thus the same conclusion for the mortars behaviour ensues.

Water retention value of lime mortar is generally quite high in comparison with cementitious materials. Decrease of water retention of lime mortar with addition of small dose (0.1%) of some of the admixtures (CG, GeG, and XG) occurred. This phenomenon has been observed on lime mortars also with other admixtures e.g. chitosan ethers or starch [12]. The values reached in the highest dose by the DG and alginate are comparable to the values of chitosan ethers in the same dosage [12]. The highest values obtained with the admixtures are comparable to the values of starch in significantly higher dose on lime mortar or with hydroxypropylmethyl cellulose and hydroxypropyl guar added to cement mortar [11, 13].

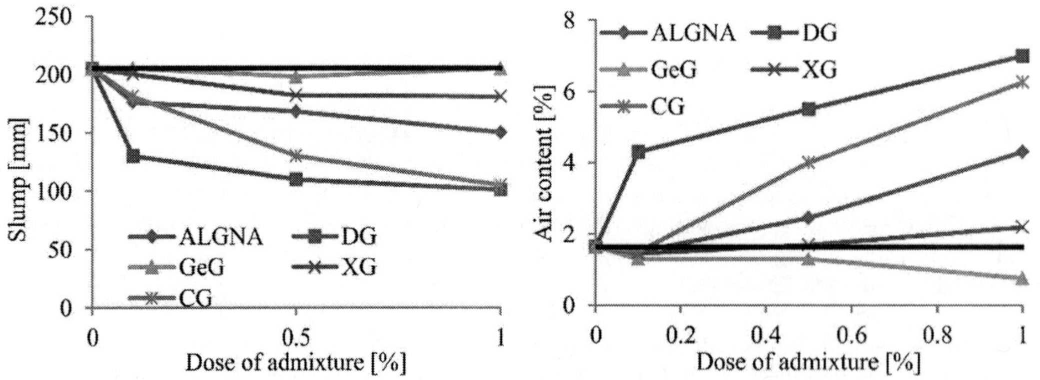


**Figure 1**. Slump and air content of mortars with biopolymer admixtures.

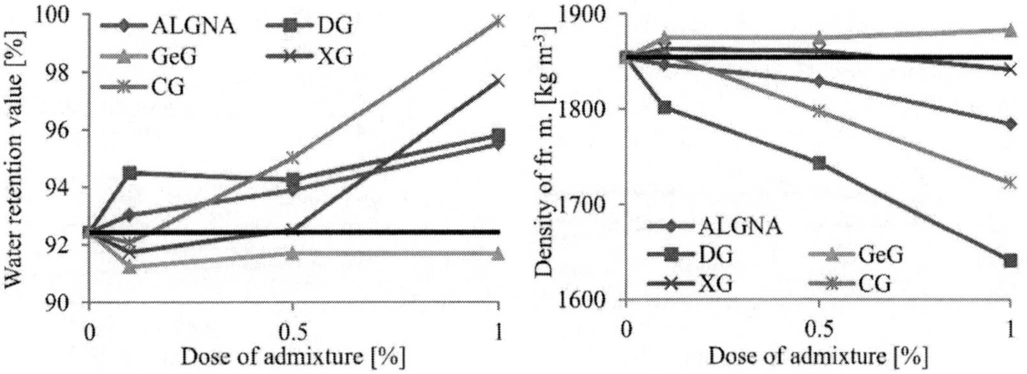


**Figure 2**. Water retention value and density of fresh mortars with admixtures.

### *3.2. Hardened properties of lime mortars*

*3.2.1. Strength properties.* Flexural and compressive strength of specimens had been tested after 7 (Figure 3) and 28 (Figure 4) days of curing time. As seen in Figure 3, the addition of admixture caused notable decrease of early age strength, which is in accordance with results with different admixtures in the literature [11, 14]. The significant improvement, especially in the flexural strength results, is typical for lime mortar [11] and probably caused by different humidity characteristics of the specimens with different admixtures and doses. The dosage dependence in the case of XG and GeG is not as obvious as in the case of other biopolymers.

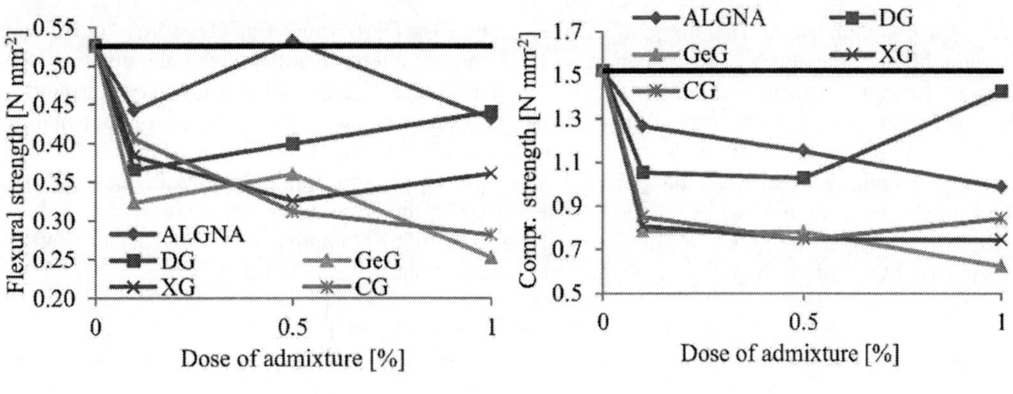


**Figure 3.** Flexural and compressive strength of 7 days old specimens.

IOP Conf. Series: Materials Science and Engineering **379** (2018) 012006 doi:10.1088/1757-899X/379/1/012006

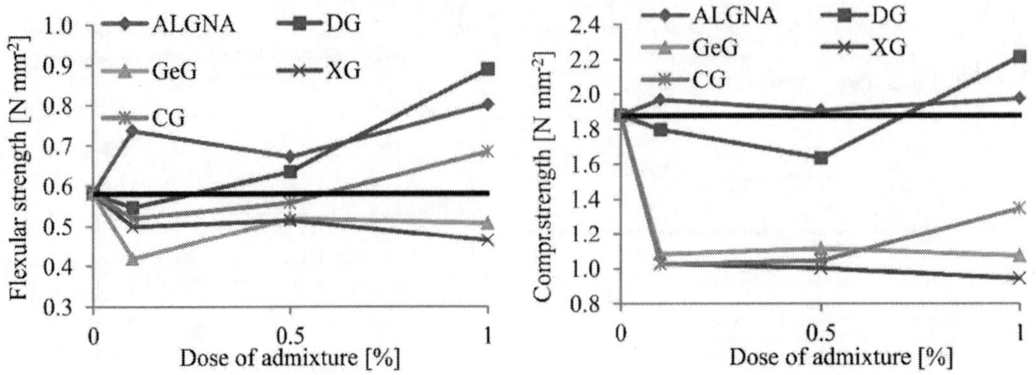


**Figure 4.** Flexural and compressive strength of 28 days old specimens.

Alginates and diutan gum were the best performing admixtures, both showing distinctly lower decrease of strength in the first 7 days, and after 28 days of curing time, the mortars with these admixtures reached similar or even higher values of strength than the refference mortar. The development of strength of mortars with CG is remarkable, at 7 days the strength values were ones of the lowest, but after 28 days CG in highest dose in case of flexural strength surpassed the refference mortar and even for the compressive strength there was perceptible growth of difference with other badly performing admixtures. XG and GeG confirmed their insuitability when used in lime mortar as it was seen even from the fresh mortar properties. Both of them significantly decreased strenghts of mortars with no hint of improvement in later ages.

*3.2.2. Carbonation rate*. The amount of free water, $Ca(OH)_2$, and $CaCO_3$ were studied using diferential thermal analysis in combination with thermogravimetry. Table 1 states the amount of free water in the mortar specimens. As we can see, the humidity of 7 days old mortars vary depending on the type of admixture and its dose. With increasing dose of admixture the humidity of specimen also increased. The samples with alginate were the most humid and XG samples showed the lowest humidity. Humidity of 28 days old samples was varying around similar values in the case of all admixtures. The amount of water retained in the mortar by the admixtures can decrease the short term strength as observed in Figure 3 (lime mortar obtains initial strengths mostly by drying) but also the increase of the strength in 28 days for the carbonation reaction is in solution [15].

**Table 1**. Amount of free water contained in mortar samples after 7 and 28 days of curing time.

| | REF | ALGNA 1 | ALGNA 5 | ALGNA 10 | DG 1 | DG 5 | DG 10 | GeG 1 | GeG 5 | GeG 10 | XG 1 | XG 5 | XG 10 | CG 1 | CG 5 | CG 10 |
|---|---|---|---|---|---|---|---|---|---|---|---|---|---|---|---|---|
| free $H_2O$ [%] 7d | 0.48 | 1.45 | 3.33 | 7.54 | 0.42 | 0.69 | 2.28 | 0.45 | 0.57 | 1.35 | 0.40 | 0.63 | 0.53 | 0.29 | 0.80 | 3.01 |
| free $H_2O$ [%] 28d | 0.45 | 0.35 | 0.28 | 0.26 | 0.30 | 0.39 | 0.35 | 0.38 | 0.39 | 0.41 | 0.43 | 0.45 | 0.39 | 0.27 | 0.31 | 0.27 |

The difference in percentage representation of $Ca(OH)_2$ and $CaCO_3$ is shown in Figure 5. The decrase of amount of $Ca(OH)_2$ as well as increase of $CaCO_3$ are proofs of the ongoing carbonation reaction. From the Figure 5 we can see that only ALGNA in two lower doses slightly slowed carbonation reaction between 7 and 28 days. This can be caused by aprreciable amount of humidity in 7 days old samples because the humidity also partially influences the difussion of $CO_2$ into the mass of specimen. If the strength results shown in Figures 3 and 4 were also taken into account it can be assumed, that the strength of ALGNA specimens is not much affected by the slowed carbonation.

The masive increase of $CaCO_3$ content in the samples with the highest dose of carageenan can also explain the notable difference between 7 and 28 days strength of this mortar. From the results obtained we can assume that the bipolymer addition to the lime mortar did not slow carbonation reaction, and it appeared to be beneficial for it.

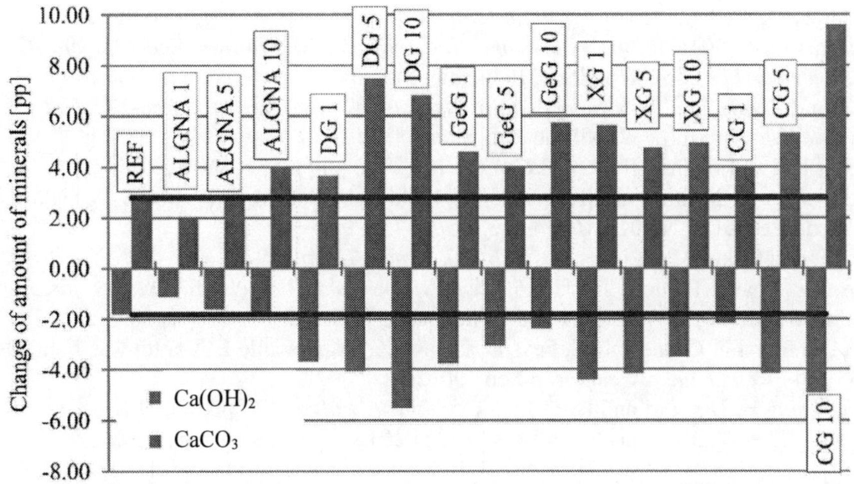


**Figure 5.** Change of amount of $Ca(OH)_2$ and $CaCO_3$ in samples between 7 and 28 days.

## 4. Conclusions

The paper studied the influence of several different biopolymers on the properties of aerial lime-based mortar. The mortars were prepared with constant water/binder and binder/aggregate ratios. The properties of mortars varied with the type and dosage of the admixture.

In the fresh state most of the admixtures acted as thickening agents with different effectiveness. The only exception was gellan gum, for it had no significant impact on the properties of the mortar. The results of air content in the fresh mortar as well as density of fresh mortar were corresponding with the results of the flow table test, thus none of the admixtures showed air-entraining function while used in the aerial lime-based mortar. The water retention value was slightly decreased by the CG and XG in the lowest dose, but with the increasing dosage the values had increased significantly, especially for carrageenan, which performed best in the highest dosage.

All of the admixtures lowered strengths after 7 days; this may be caused by the water-retaining function of the admixtures. After 28 days of curing time the ALGNA, DG, and CG showed results comparable with reference mortar for flexural strength, but the carrageenan mortar still presented significant decrease of compressive strength. Decrease of amount of $Ca(OH)_2$ and increase of amount of $CaCO_3$ in the specimens was higher than in the reference mortar for all specimens with exception of ALGNA in any dosage. This indicates slightly beneficial effect of biopolymer addition on the carbonation reaction.

ALGNA, DG and CG showed interesting results and are to be studied in the future. The effect of addition of XG is discussable and its future research is to be well considered. Gellan gum had no impact on the fresh state properties of lime mortar and significantly decreased strengths of the mortar, thus it is not suitable for the use in the lime mortars in the way presented in this paper.

**Acknowledgements**

This work has been financially supported by the BUT specific research project FAST-J-18-5274.

## References

[1] Plank J 2004 *Appl. Microbiol. Biotechnol.* **66** pp 1–9
Singh NK, Mishra PC, Singh VK and Narang KK 2003 *Cem. Concr. Res.* **33** pp 1319–29
[2] Rinaudo M 2008 *Polym. Int.* **57** pp 397–430
[3] Mignon A, Snoeck D, D'Halluin K, Balcaen L, Vanhaecke F, Dubruel P, Van Vlierberghe S and De Belie N 2016 *Costr. Build. Mater* **110** pp 69–74
[4] Bezerra U T 2016 *Biopolymers and Biotech Admixtures for Eco-Efficient Construction Materials* (Duxford: Woodhead Publishing)
[5] Li Z and Zhang L 2016 *Biopolymers and Biotech Admixtures for Eco-Efficient Construction Materials* (Duxford: Woodhead Publishing)
[6] Akthar F K and Evans J R G 2010 *Cem. Concr. Res.* **40** pp 352–8
[7] Rutkevičius M, Munusami S K, Watson Z, Field A D, Salt M, Stoyanov S D, Petkov J, Mehl G H and Paunov V N 2012 *Mater. Res. Bull.* **47** pp 980–6
[8] Isik I E and Ozkul M H 2014 *Constr. Build. Mater.* **72** pp 239–47
Vazquez A and Pique T M 2016 *Biopolymers and Biotech Admixtures for Eco-Efficient Construction Materials* (Duxford: Woodhead Publishing)
[9] Van Der Vurst F, Grünewald S, Feys D, Lesage K, Vandewalle L, Vantomme J and De Schutter G 2017 *Cem. Concr. Compos.* **82** pp 190–201
Xu L, Gong H, Dong M and Li Y 2015 *Carbohydr. Polym.* **132** pp 620–9
Roca C, Alves V D, Freitas F and Reis M A M 2015 *Front. Microbiol.* **6** 288
[10] Kang K S, Veeder G T, Mirrasoul P J, Kaneko T and Cottrell I W 1982 *Appl. Environ. Microbiol.* **43** pp 1086–91
[11] Izaguirre A, Lanas J and Álvarez J I 2011 *Cem. Concr. Compos.* **33** pp 309–18
Žižlavský T, Vyšvařil M, Bayer P and Rovnaníková P 2018 *Key Eng. Mater.* **760** pp 22–9
Vyšvařil M, Hegrová M and Žižlavský T 2018 *Key Eng. Mater.* **760** pp 257–65
[12] Vyšvařil M and Žižlavský T 2017 *IOP Conf. Ser.: Mater. Sci. Eng.* **251** 012039
[13] Izaguire A, Lanas J and Álvarez J I 2010 *Carbohydr. Polym.* **80** pp 222–8
Lasheras-Zubiate M, Navarro-Blasco I, Fernández J M and Álvarez J I 2012 *Cem. Concr. Compos.* **34** pp 964–73
[14] Izaguirre A, Lanas J and Álvarez J I 2010 *Cem. Concr. Res.* **40** pp 1081–95
[15] Shoh S-M, Ho C-S, Song Y-S and Lin J-P 1999 *Ind. Eng. Chem. Res.* **38** pp 1316–22
Van Balen K and Van Gemert D 1994 *Mater. Struct.* **27** pp 393–8

IOP Conf. Series: Materials Science and Engineering **379** (2018) 012007 doi:10.1088/1757-899X/379/1/012007

# Setting, rheology and packing density of biomass fly ash/Portland cement mixtures

**M Boháč[1], D Kubátová[1], R Nečas[1] and T Staněk[1]**

[1] Research Institute for Building Materials, Hněvkovského 30/65, Brno 617 00, Czech Republic

Email: bohac@vustah.cz

**Abstract.** Paper deals with rheological properties and packing density of fresh mixtures of Portland cement and biomass fly ash. Water demand of paste of normal consistency increases with increase of replacement level of biomass fly ash for Portland cement. Up to 5 wt. % contents of biomass fly ash yield stress and plastic viscosity values decrease, while at contents higher than 5 wt. %, the opposite trend was revealed. Biomass fly ash compared to Portland cement has proportionally lower content of particles with size of tens and first single digit numbers of microns. Deficiency in this region at higher contents of biomass fly ash in cement mixtures results in deflection from ideal packing density curve.

## 1. Introduction

Properties of biomass fly ash (BFA) in cementitious systems has been studied by several authors. Even though the biomass combustion is well-established technology for power and heat production, the environmental management of the ash as a by-product is still a challenge. Quality and quantity of the ash were affected by the biomass used [1], whether it is wood, bark, herbaceous biomass or agriculture waste [2]. In the current trend of power generation, emergence of biomass (forestry and agricultural waste) fueled power plant seems to be a promising source of renewable energy with low operational cost coupled with continuously renewable fuel [3]. A major portion produced (approximately 70%) of the wood waste ash produced is land-filled as a common method of disposal [4, 5].

In the Czech Republic, tens of heat and power plants produce the biomass fly ash in various quantities. Integration of the biomass ashes within the natural cycles has continuously become a necessity on the national level. One of the possible ways is to use BFA as admixture for Portland cement (PC). There are two main scientific approaches how to characterize BFA in cementitious system; BFA as a partial cement replacement material and BFA as a filler. Several authors studied possibilities of use of various BFA in cementitious systems in terms of strengths [6] and hydration [7], but only a few information has been written about workability [8], rheology and packing density. Data on workability of woody BFA quantified by slump tests has shown little reduction or increase in the mix workability, limited setting delay and moderate compressive strength reduction when ground unwashed or washed woody BFA was used at cement replacement levels of 15 wt. % or 30 wt. %.

According to the European Waste Catalogue and hazardous residues list [9], both bottom ash and fly ash coming from the combustion of untreated wood are classified as non-hazardous wastes.

Particle packing is fundamental to the concrete. The packing should be optimized in order to use less binder, that would have been normally required in the concrete. By application of suitable particle packing models to the entire concrete mix, the particle size distribution (PSD) of the entire mix can be adjusted in order to achieve mobility in the fresh state and adequate properties when hardened. Due to the fact that the ingredients of a concrete mix are particles with continuous size distributions, a model should be based on packing of continuous size distributions. The concrete flow and self-compacting ability in the fresh state are usually in conflict with optimal packing [10]. The conflict can be solved by introduction of fine particles i.e. by BFA, silica fume, calcined clays etc. in size of cement and below, into a concrete mix.

## 2. Experimental

### *2.1. Materials and Methods*

Used BFA was filter ash from biomass combustion (Veolia Energie ČR, a.s. Heating plant Krnov). BFA is by-product of grate boiler combustion with 3.72 MPa, operating temperature 445°C and 35 t/hr. Electrical and thermal power of the boiler is 4900 kW and 28 000 kW. Wood chips as basic fuel is used with pellets, sawdust, corn meal, chopped crops as the additional fuel in heating plant. PC is CEM I 52.5 N from manufacturer Českomoravský cement, a.s. (HEIDELBERG Cement Group). PC was replaced by 5, 10, 15 and 20% of BFA. Chemical parameters of BFA are shown in Table 1.

Volume density was determined according to ČSN EN 1097-3. Specific surface area of BFA and PC was measured by $N_2$ absorption according to the Brunauer-Emmett-Teller theory (BET). Nitrogen was used as the adsorbate. The results were compared with the $TiO_2$ standard. Results of BET and volume density are shown in Table 2. Phase composition determined by X-ray diffraction with Rietveld refinement is given in Table 3.

Measurement of particle size distribution was carried out by a laser diffraction method using CILAS 920L laser particle size analyzer. In laser diffraction particle size analysis, a representative cloud or ensemble of particles passes through a broadened beam of laser light which scatters the light onto Fourier lens. This lens focuses the scattered light onto a detector array and, using an inversion algorithm, a particle size distribution is inferred from the collected diffracted light data. Size range of particle size analyzer is 0.3–400 μm, dispersing medium: isopropyl alcohol. Before the measurement, every sample was treated with ultrasound (60 s). PSD of PC and BFA is given in Figure 1. Cumulative particle size distribution was evaluated by EMMA software that calculates and displays the particle size distribution of a mixture of components. The curves of all samples were correlated to modified Andreassen model for ideal packing density with q-value 0.35 and volume density given in Table 2.

Relatively long list of minerals was found in BFA by Rietveld refinement X-ray diffraction method (Table 3). Both high contents of amorphous phase and silica in BFA are promising in terms of pozzolanic activity and use of BFA as supplementary cementitious material (SCM). The high content of $SO_3^{2-}$ on BFA must be balanced by addition of gypsum as the setting retarder. High content of CaO is associated with calcite when all free lime is carbonated during water treatment of BFA. High content of potassium chloride and other chlorides in admixture may cause efflorescence on the product surface. According to ČSN EN 197-1 standard, the chlorides content in cement has to be below 0.1%.

**Table 1.** Chemical parameters of BFA.

| sample | $SiO_2$ | CaO | $Na_2O$ | $K_2O$ | MgO | $Fe_2O_3$ | $Al_2O_3$ | $Cl^-$ | $SO_3^{2-}$ | LOD | LOI |
|---|---|---|---|---|---|---|---|---|---|---|---|
| | | | | | | wt. % | | | | | |
| BFA | 19.39 | 22.01 | 0.43 | 9.66 | 4.15 | 4.15 | 6.16 | 1.93 | 8.35 | 1.64 | 23.41 |

**Table 2.** Volume density and BET of PC and BFA.

| sample | vol. density | BET |
|---|---|---|
| | $kg/m^3$ | $m^2/g$ |
| BFA | 176 | 9.5 |
| PC | 1380 | 6.3 |

**Table 3.** Phase composition of BFA (Rietveld analysis XRD).

| phase (vol. %) | | | | | |
|---|---|---|---|---|---|
| quartz | calcite | hematite | anhydrite | albite | muscovite |
| 4.8 | 10.9 | 3.0 | 4.2 | 1.9 | 2.8 |
| magnetite | arcanite | biotite | sylvite | halite | periclase |
| 0.2 | 2.1 | 2.0 | 2.2 | 0.2 | 2.0 |
| dolomite | rutile | microcline | thenardite | syngenite | amorph. phase |
| 1.1 | 0.3 | 3.5 | 3.7 | 8.4 | 46.7 |

Rheological measurements were carried out on rotational rheometer DHR-1 (TA Instruments). Thixotropy (Pa/s) as a time-dependent rheological property was calculated following stability of the paste test [11]. The thixotropic behaviour was determined by delineating the area ($A_{thix}$) between the curves formed by the points of the static yield stress ($\tau_{o,s}$) and the dynamic yield stress ($\tau_{o,d}$) at different shear rates $\dot{\gamma}$ (Figure 2). Where $T_{max}$ is function of $\tau_{o,s}$ and $T_{min}$ of $\tau_{o,d}$. The extent of shear rate was 25–150 ($s^{-1}$), with steps 25, 50, 75, 100, 125 and 150 ($s^{-1}$) equation (1).

$$A_{thi} = \int_{25}^{150} T_{max}\, d\dot{\gamma} - \int_{25}^{150} T_{min}\, d\dot{\gamma} \tag{1}$$

Yield stress and plastic viscosity were calculated from flow curves using Bingham model equation (2). Where $\tau_0$ is yield stress, $\mu$ is plastic viscosity and $\dot{\gamma}$ is shear rate.

$$\tau = \tau_0 + \mu\dot{\gamma} \tag{2}$$

All rheological tests were done at 25°C. The geometry of coaxial cylinders was used for testing. Normal consistency for each sample was determined and the rheological testing was done at the same consistency for all samples. The course of the setting was evaluated by needle penetration test (Tussenbrock test). During setting, this test basically measures finite value of resistance to penetration of fresh cement paste which is expressed by threshold shear stress $\tau_{thold}$ (kPa) at given times.

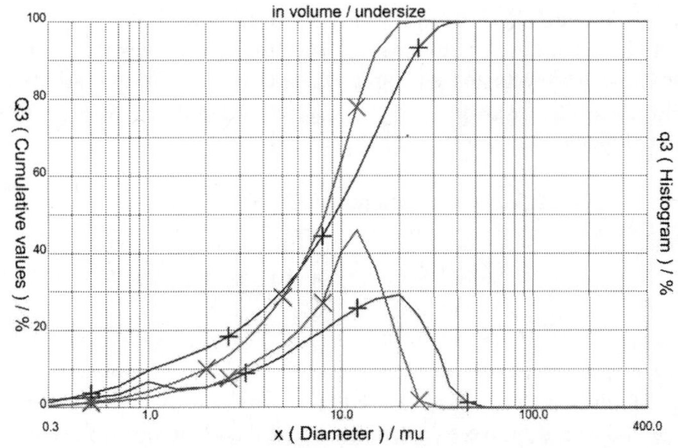


**Figure 1.** PSD of PC 52.5N (+) and BFA (×).

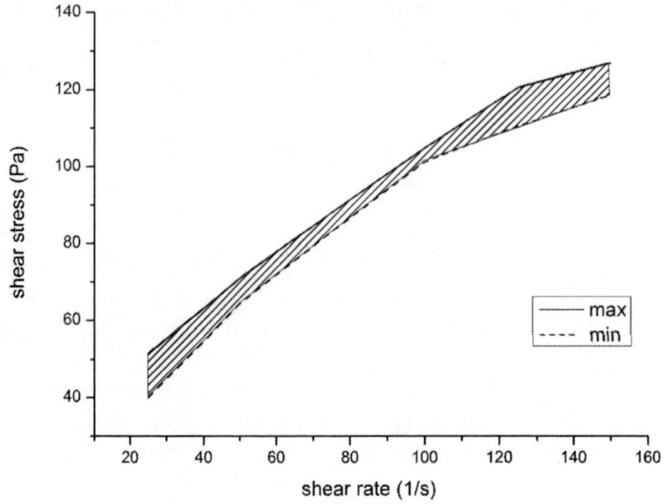


**Figure 2.** Thixotropy (Pa/s) as the delineated area between curves of static and dynamic yield stress.

## *3.* Results and Discussion

### *3.1. Consistency and setting*

Water demand of BFA/PC blended cement pastes increases proportionately with the level of cement replacement by BFA. Higher water demand of pastes with BFA relative to PC is mainly affected by higher specific surface area of BFA compared to OPC particles [3, 12, 13]. Setting and all rheology test were carried on samples with same consistency which was linearly dependent on content of BFA in the mixture (Table 4). Particle size distribution of BFA and PC is given in Figure 1.

The course of setting was monitored by Tussenbrock test. Some analogy to Tussenbrock test can be found with Vicat (ASTM C191) or Gillmore (ASTM C266) needle tests [14] which give results of initial and final setting times but are insufficient in characterizing the course of the setting. Besides the Tussenbrock test, there are alternatives for characterization of the course of setting i.e. state of art modeling of the setting with VCCTL software [15, 16]. Relatively week bonds that are associated with formation of outer C-S-H products and ettringite (AFt) during initial hydration can be broken during setting of cementitious system. The well interconnected structure of newly formed hydrates shows increasing resistance to penetration as the setting proceeds.

**Table 4**. Consistency of PC/BFA mixtures.

| | PC/BFA | | | | |
|---|---|---|---|---|---|
| | 100/0 | 95/05 | 90/10 | 85/15 | 80/20 |
| w/c | 0.4 | 0.45 | 0.49 | 0.54 | 0.58 |

BFA has dilution effect on hydration, which means the amount of formed hydrates is proportionally lower compared to reference pure PC paste. Based on results, it is evident that BFA has retarding effect on setting. Start and end of setting were determined from $\tau_{thold}/t$ curves to find the role of BFA content in studied mixtures. Prolonged workability times and more important slower heat development of PC/BFA mixtures was already discussed in previous studies [3]. The lower heat release during early hydration can reduce thermal expansion risk in massive constructions. Results of setting are given in Figures 3 and 4.

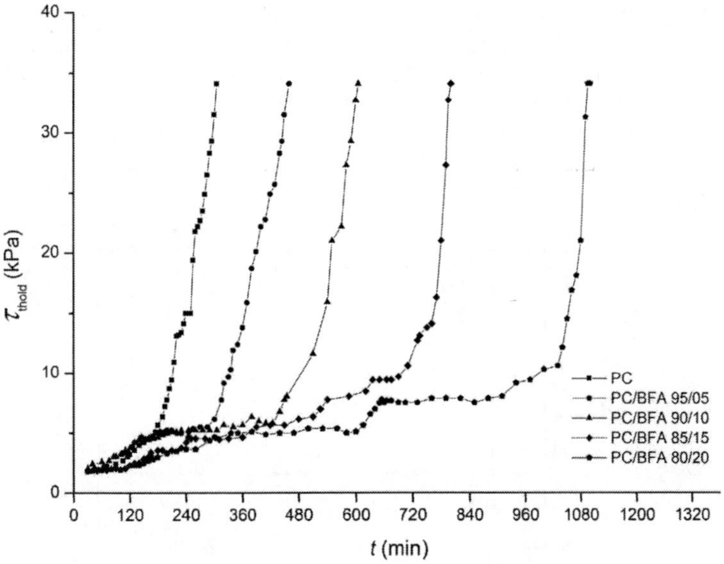


**Figure 3.** Setting of mixtures with BFA.

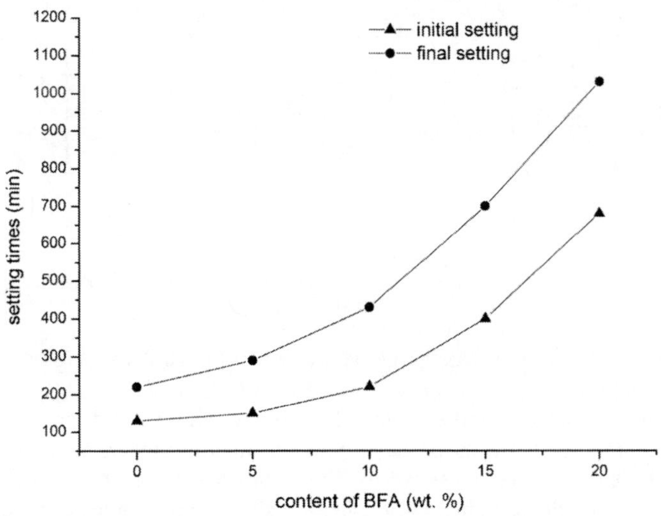


**Figure 4.** Start and end of setting of BFA mixtures.

*3.2. Rheology*

All samples show viscoelastic properties of highly concentrated water solutions that can be characterized as a Bingham fluid. Hysteresis loops of flow curves shear stress vs. shear rate were constructed in the region of 0.1–150 $s^{-1}$. Yield stress (Pa) and plastic viscosity (Pa.s) were calculated from hysteresis flow curves using Bingham model. Since normal consistency of the PC/BFA paste increases with increasing content of BFA, the rheological testing was carried out on samples of same consistency. The inter-particle forces which determine the yield stress are composed of the Van der Waals, electrostatic and steric forces, which are all determined by the particle conformations, surface charges and medium chemistry [17]. Samples with BFA, similarly to reference sample of pure PC, show shear-thinning behaviour in the tested shear rate region. Broad range of bonds are expected from the gradual shape of viscosity/shear rate curves. At higher shear rates (above 25 $s^{-1}$) the hydrodynamic effect gradually prevails. Values of yield stress and plastic viscosity decrease up to 5 wt. % of BFA, while opposite trend was revealed for higher contents of BFA (Figure 5). The decrease of yield stress provides a possible way to workability enhancement for the mixture with small amount of BFA.

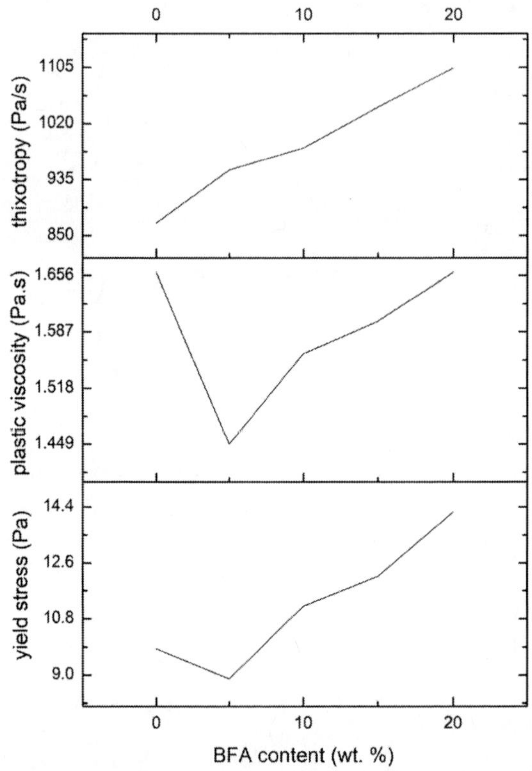


**Figure 5.** Rheological parameters of PC/BFA mixtures.

It was previously concluded, that the water/solid volume ratio is not directly related to yield stress, but instead it is more important how the particles pack. In situations where the particle packing increases due to a broader PSD, less water is needed to fill the voids between the particles, the excess water efficiently separates the particles, reducing flocculation and resulting in lower yield stress [17–19].

Despite the time-dependent rheological parameters can be determined from single flow hysteresis curve, the different approach was chosen for evaluation of thixotropy [11]. This approach allows a better assessment of rheological behavior and stability of the PC/BFA system during the initial hydration and more closely approximates the main use of this material, which is self-compacting concrete.

Thixotropy increase with increased content of BFA shows in this case the poorer stability of fresh mixtures (Figure 5). Thixotropy or structural build-up is closely related to formation of week bonds during formation of gelatinous and crystalline hydration phases. These phases are AFt (ettringite) and outer C-S-H hydrates. Week bonds are broken during shearing motion of inner cylinder. Having in mind the chemical composition of BFA, the high content of alkalis is expected to ruin the sulfate balance of the system.

### *3.3. Packing density*
Compared to PC, BFA has proportionally lower content of particles with size below 10 microns. This deficiency at higher contents of BFA in cement mixtures causes deflection from ideal packing density curve (Figure 6). Number of methods can be used for characterization of packing density. Despite the fact that new methods are purposed recently, there is so far, no generally accepted method of packing density measurement. The packing characteristics of the cementitious materials have great influence on the performance of a concrete mix, especially strengths [20].

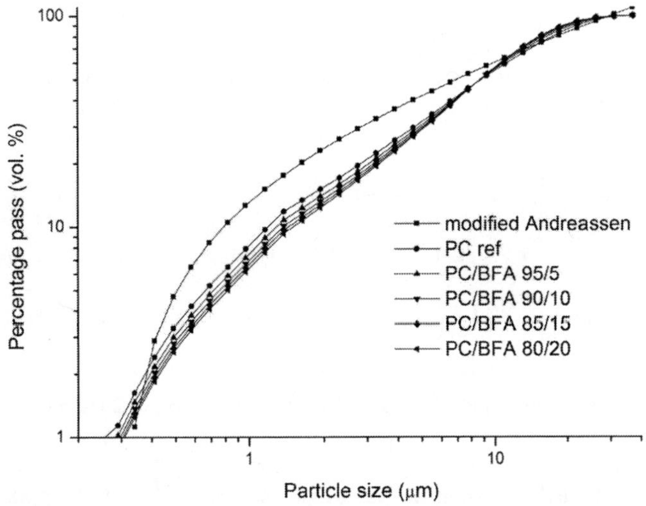


**Figure 6.** Correlation of modified Andreassen model with PC/BFA mixtures.

PSD of PC and BFA at various replacement levels were correlated with modified Andreassen model, as the "ideal" PSD curve for packing density. Modeling of the curve with various admixtures showed that cumulative particle size distribution can be improved by addition of another admixtures, i.e. finely ground blast-furnace slag, silica fume or metakaolin. The statement of the previous study [21] has shown, that sawdust ash, when used as a supplementary cementitious material, decreases the early strength gain of concrete, but when small amount of metakaolin was added to sawdust-ash concrete, the detrimental effects on early-age strength gain was counteracted. Other authors [12] published positive role of woody BFA at a cement replacement level up to 16% by total binder weight on pore structure and its refinement when using in combination with small amounts of silica fume, hence, reducing the chloride diffusivity of mortar mixes produced.

Based on results of rheological testing and modeling of ideal packing density, it can be purposed to use only amounts up to 5 wt.% of woody BFA in PC mixtures. There is a possibility to improve the PC/BFA packing density by addition of admixture with complementary particle size, which means compensate the lack of fine particles in the mixture. With increased replacement level of BFA, the deficiency in particles with size around 1 μm was revealed (Figure 6).

## 4. Conclusion
Evaluation of possibility of reuse of BFA at different replacement levels as a supplementary cementitious material was carried out. For achieving same consistency of mixtures, water demand increases with increasing level of BFA. At the same consistency, setting of PC/BFA pastes is retarded gradually with increasing level of BFA. Used penetration method allows to monitor both setting times and the course of the setting. PC/BFA pastes show shear-thinning behaviour similarly to pure PC reference paste in the tested shear rate region. Broad range of bonds is characterized by the continuous shape of viscosity/shear rate curves. With regards to reference sample, yield stress and plastic viscosity

decrease up to 5 wt. % of BFA and increase continuously when higher contents of BFA are used. Small amounts of BFA improve workability of PC/BFA fresh pastes. The lower stability of fresh mixtures with high content of BFA is characterized by poorer thixotropy and thixotropy rebuilt, which becomes more evident at higher contents of BFA. Based on correlation of rheological testing and modeling of ideal packing density, it can be assumed that only small amounts of woody BFA can improve the performance of PC/BFA mixtures. Further improvement towards ideal packing of particles can be provided by addition of fine admixtures with complementary particle size to PC/BFA mixtures. Deficiency in particles with size around 1 μm was revealed with increasing level of BFA.

**Acknowledgements**
This paper was elaborated with the institutional support for long-term development of research organizations by the Ministry of Industry and Trade of the Czech Republic.

**References**

[1] Loo S V and Koppejan J 2003 *Handbook of Biomass Combustion and Co-firing* (The Netherlands: Twente University Press)
[2] Masia A A T, Buhre B J P, Gupta R P and Wall T F 2007 *Fuel Processing Technology* **88** pp 1071–1081
[3] Ban Ch Ch and Ramli M 2011 *Resources, Conservation and Recycling* **55** pp 669–685
[4] Campbell A G 1990 *TAPPI J* **77** 9 pp 141–143
[5] Etiegni L and Campbell A G 1991 *Bioresource Technol* **37** 2 pp 173–178
[6] Sata V, Tangpagasit J, Jaturapitakkul C and Chindaprasirt P 2012 *Cem Concr Composites* **34** pp 94–100
[7] Tkaczewska E and Małolepszy J 2009 *Constr Build Mater* **23** pp 2694–2700
[8] Berra M, Mangialardi T and Paolini A E 2015 *Constr Build Mater* **76** pp 286–296
[9] Commission of the European Communities. Establishing a list of wastes pursuant to Article 1(a) of Council Directive 75/442/EEC on waste and Council Decision 94/904/EC establishing a list of hazardous waste pursuant to Article 1(4) of Council Directive 91/689/EEC on hazardous waste. Official Journal of the European Communities L 226/3; (2000).
[10] Vogt C 2010 Ultrafine particles in concrete, Influence of ultrafine particles on concrete properties and application to concrete mix design *Doc. Thesis School of Architecture and the Built Environment* (Stockholm: Division of Concrete Structures, Royal Institute of Technology)
[11] Abebe Y A and Lohaus L 2016 Workshop und Kolloquium, Regensburg *Rheologische Messungen an Baustoffen* **25** pp 8–19
[12] Elinwa A U and Ejeh S P 2004 J *Asian Architecture Build Eng* **3** (1) pp 1–7
[13] Elinwa A U and Mahmood Y A 2002 *Cem Concr Composites* **24** pp 219–222
[14] ASTM Annual Book of Standards, Vol. 04.01 Cement; Lime; Gypsum 1999 (PA: American Society for Testing and Materials, West Conshohocken)
[15] Bentz D P and Forney GF 2000 "User's Guide to the NIST Virtual Cement and Concrete Testing Laboratory, Version 1.0," (NISTIR 6583, U.S. Department of Commerce)
[16] Bentz D P 2000 "CEMHYD3D: A Three-Dimensional Cement Hydration and Microstructure Development Modelling Package. Version 2.0," (NISTIR 6485, U.S. Department of Commerce)
[17] Flatt R J and Bowen P 2006 *J Am Ceram Soc* **89** p 1244–1256
[18] Kashani A, San Nicolas R, Qiao G G, van Deventer J S J and Provis J L 2014 *Powder Technology* **266** pp 203–209
[19] Bentz D P, Ferraris C F, Galler M A, Hansen A S and Guynn J M 2012 *Cem Concr Res* ***42*** pp 404–409
[20] Wong H and Kwan A 2008 *Mater Struct* **41** pp 689–701
[21] Elinwa A U, Ejeh S P and Akpabio I O 2005 *Concr Int* **27**(11) pp 49–52

IOP Conf. Series: Materials Science and Engineering **379** (2018) 012008 doi:10.1088/1757-899X/379/1/012008

# Thermal behaviour of inorganic aluminosilicate polymer based on cement kiln dust

**D Kubatová[1], A Rybová[1], A Zezulová[1] and J Švec[2]**

[1] Research Institute for Building Materials, Hněvkovského 30/65, Brno CZ-617 00, Czech Republic

[2] Brno University of Technology, Faculty of Chemistry, Materials Research Centre, Purkyňova 464/118, Brno CZ-612 00, Czech Republic

Email: kubatova@vustah.cz

**Abstract.** Alkali activated materials have shown potential for use in high temperature applications. This paper deals with thermal behaviour of various inorganic polymer formulations, which were prepared by mixing metakaolin, blast furnace slag, fly ash and cement kiln dust as alkaline activator. The cement kiln dust was added in the amount of 10 wt.%. The influence of thermal treatment ranging from 30 to 1000°C on the following characteristics was evaluated: compressive strength, thermal expansion, weight loss, melting point, morphology changes and dehydration. The thermal behaviour was studied by TG-DTA analysis, dilatometry and heating microscope, XRD analysis was used to examine phase changes. All measurements were conducted on samples after 28 days of curing time.

## 1. Introduction

Attempts to reduce the energy cost and environmental effect of Portland cement production have led to the development and application of alternative materials such as alkali-activated binders. The principle of alkaline activation is the reaction of aluminosilicates (blast furnace slag, fly ash, siliceous sediments, metakaolin, etc.) with an alkaline activator, most often in the form of solutions of hydroxides in combination with water glass. The most common activator is sodium hydroxide solution and sodium silicate, preferably with a higher silicate module in water glass [1]. Due to high cost of alkaline activators such as hydroxides and water glass, an effort is made to replace them with waste materials. Cement kiln dust (CKD) is a suitable alternative material due to high content of alkali and negligible price.

CKD is a by-product of Portland cement manufacture. It is generated during the calcining process in the kiln. CKD is a fine cementitious powder, particulate material, readily entrained in the combustion gases moving through the kiln. It is composed primarily of variable mixtures of calcined feed materials, fine cement clinker, fuel combustion by-products, and condensed alkali compounds [2–4]. Due to high alkali content and large amount of free lime, CKD can be used as an activator for pozzolanic materials. CKD was used as an activator for blast furnace slag, fly ash and metakaolin [5]. Addition of CKD reduces the required amount of NaOH. Another research investigated the effects of CKD content on the compressive strength, water absorption, durability etc. [6]. Effects of curing temperature and NaOH addition on hydration and strength development CKD-fly ash (FA) binders were investigated [7]. The results showed significant improvement of comprehensive strength and durability when CKD was used. Bílek et al. [8] studied effects of the CKD and fly ash on workability, mechanical properties, and shrinkage of alkali-activated slag-based composites. Proper dosage of CKD improved the workability but did not affect the shrinkage reduction. The possibility of utilizing CKD as an activator for blast furnace slag (BFS) was investigated by Konsta-Gdoutos and Shah [2]. Kiln dust – slag concrete, with good overall performance regarding setting time, temperature rise, and mechanical properties was successfully produced.

Investigation of thermal properties of alkali-activated materials has explored some of the basic material properties after exposure to high temperature. The key characteristics of thermal stability assessment are thermal expansion, thermal conductivity, spalling, melting point, phase stability, morphology changes etc. [9]. These properties are dependent on various factors like physical and chemical properties of raw materials, alkali sources, curing regime, ratio of sodium silicate to sodium hydroxide and many more.

The objective of this work was to investigate the effect of temperature on the properties of alkali-activated materials containing 10% of CKD and possibility of using as refractory material. The properties and phase changes of pastes after exposure to temperatures up to 1000°C are reported.

## 2. Experimental

### *2.1. Materials*

Cement kiln dust with high calcium content obtained from cement plant Mokrá (Czech Republic) was used as alkali activator for alkali-activated material preparation. Metakaolin (MK), blast furnace slag and fly ash were used as main starting materials. MK Mefisto K05 was supplied by České lupkové závody, a.s. (Czech Republic), finely ground BFS was from Třinec (Czech Republic) and FA was from combined heat and power plant in Skawina (Poland).

Chemical analysis of used materials is given in Table 1. Characteristic particle diameters $D_{10}$, $D_{50}$ and $D_{90}$ are summarized in Table 2.

**Table 1.** Chemical composition of materials used in mixtures (%).

| | $SiO_2$ | $Al_2O_3$ | CaO | $Na_2O$ | $K_2O$ | MgO | $Fe_2O_3$ | $TiO_2$ | LOI |
|---|---|---|---|---|---|---|---|---|---|
| MK | 57.14 | 38.16 | 0.39 | 0.033 | 0.81 | 0.45 | 0.64 | 0.42 | 1.67 |
| BFS | 37.42 | 7.80 | 40.04 | 0.047 | 0.47 | 8.67 | 0.46 | 0.35 | 0.70 |
| FA | 49.36 | 25.71 | 3.52 | 1.77 | 2.90 | 2.39 | 6.84 | - | 5.46 |
| CKD | 8.66 | 1.65 | 32.81 | 0.38 | 23.99 | 0.53 | 0.93 | 0.62 | 20.17 |

**Table 2.** Particle diameters of materials (μm).

| | D10 | D50 | D90 | Mean diameter |
|---|---|---|---|---|
| MK | 1.71 | 6.10 | 10.63 | 6.17 |
| BFS | 1.33 | 9.91 | 30.45 | 13.62 |
| FA | 2.30 | 15.12 | 44.48 | 20.15 |
| CKD | 1.20 | 17.69 | 56.83 | 24.56 |

The XRD analysis of these raw materials was performed. MK contained quartz ($SiO_2$), kaolinite ($Al_2Si_2O_5(OH)_4$), illite ($KAl_2(AlSi_3O_{10})(OH)_2$), rutile ($TiO_2$) and significant amount of amorphous phase. BFS contained calcite ($CaCO_3$), merwinite ($Ca_3MgSi_2O_8$) and melilite ($Ca_4Al_2MgSi_3O_{14}$). In addition, quartz ($SiO_2$) and significant amount of the amorphous phase was present. The majority phase in FA was quartz ($SiO_2$), then mullite ($Al_6Si_2O_{13}$), hematite ($Fe_2O_3$), maghemite ($Fe_2O_3$), anhydrite

($CaSO_4$) and amorphous phase. Sylvite (KCl) was the main component of CKD, lime (CaO), larnite ($Ca_2SiO_4$), arkanite ($K_2SO_4$), quartz ($SiO_2$) and wollastonite ($CaSiO_3$) were also identified.

### *2.2. Methods*

The composition of raw materials and final products was determined by X-ray powder diffraction. The X-ray diffraction analysis was conducted using Bruker D8 Advance apparatus with Cu anode ($\lambda K\alpha$ = 1.54184 Å) and variable divergence slits at Θ-Θ Bragg-Brentano reflective geometry.

Dilatometric measurements were made under air, by means of a contact horizontal dilatometer (Netzsch 402 E/3) on bulk samples (7 × 7 × 50 mm) up to 1000°C. Heating rate was 5 $°C{\cdot}min^{-1}$.

Combined TGA/DTA data were obtained using STA 449 F3 Jupiter (by Netzsch). The samples were tested in Pt crucibles, temperature gradient was 10 $°C{\cdot}min^{-1}$ from 35 to 1100°C. Measurement atmosphere was synthetic air (50 $ml{\cdot}min^{-1}$, $N_2/O_2$ ratio was 80/20).

Material behavior during sintering and melting was investigated by horizontal type heating microscope EM 201 (Leitz). All samples were grinded to powder and then pressed into a cylindrical testing body 3 mm high and 2 mm in diameter. The sample was heated under the heating rate 5 $°C{\cdot}min^{-1}$ up to 1350°C or up to the sample melting point. The picture recorded by CCD camera with 25°C step was used to determine the temperatures of standard points which are defined for heating microscopy and from which the point of sintering (SP), deformation (DP), hemisphere (HP) and flowing (FP) were identified.

Determination of total porosity of prepared material was performed on the apparatus Carlo Erba Strumentazione model 200; measuring range: 3.75–10000 nm. Various levels of pressure were applied onto a sample immersed in mercury.

### *2.3. Sample preparation*

Paste were prepared by mixing solid materials including CDK, MK, BFS or FA with an activating solution in defined ratios. The solid materials were premixed for a few minutes. Subsequently, a 4M NaOH solution was added. The amount of additional water was adjusted to obtain a similar consistency of pastes. Detailed composition of mixtures is given in Table 3. The paste was casted into moulds sized 40 × 40 × 160 mm and 7 × 7 × 50 mm for dilatometric analysis. The samples were then wrapped with plastic foil to avoid moisture evaporation. After 24 h, hardened specimens were demoulded and stored in plastic bags under laboratory conditions for 28 days. The prepared samples were heated in furnace to temperatures 600, 800 and 1000°C for 2 h at a constant heating rate of 5 $°C{\cdot}min^{-1}$.

**Table 3.** Mix proportion of paste (wt. %).

| | C1 | C2 | C3 |
|---|---|---|---|
| CKD | 10 | 10 | 10 |
| BFS | 64.5 | 38 | 43 |
| MK | - | 20 | - |
| FA | - | - | 21 |
| 4M NaOH | 23 | 29 | 26 |
| Water | 2.5 | 3 | - |

## 3. Result and discussion

### *3.1. XRD*

The structure analysis of prepared samples was carried out by XRD. Figure 1 presents XRD traces of prepared sample before and after thermal treatment to 1000°C.

The XRD patterns of all samples had a broad hump, characteristic of an amorphous content. The main crystalline phases of all samples were quartz ($SiO_2$), calcite ($CaCO_3$), larnite ($\beta$-$Ca_2SiO_4$), merwinite ($Ca_3MgSi_2O_8$), åkermanite ($Ca_2MgSi_2O_7$), hydrotalcite ($Mg_6Al_2CO_3(OH)_{16}\cdot 4H_2O$) and aphthitalite ($K_3NaS_2O_8$). The CaO that was present in the CKD completely disappeared after polycondensation and formation of calcium carbonate. The calcite is the main Ca-based product. Hydrotalcite has been identified as one of the main hydration products in alkali-activated blast-furnace slag in earlier works and reflections corresponding to hydrotalcite are present at 7.8° 2θ [10]. The C1 and C3 samples show reflections at 7° 2θ and 49.7° 2θ, which can be attributed to C-S-H (I) [10]. The effect of $Ca^{2+}$ presence and formation of C-S-H gel as a co-product of Ca-added alkali-activated system was discussed in many papers. The reason why C-S-H was not formed in the C2 sample may be due to $Ca^{2+}$ content in the matrix and the difference in the kinetics of dissolution Si and Ca due to different matrix composition [6]. The level of dissolved silicate in the activating solution plays a significant role in determining the effects of calcium on the final reaction product. The formation of calcium silicate hydrates could remove a proportion of the excess silicate from solution [11]. In addition, sylvite (KCl) is present in samples C1 and C3. Mullite ($Al_6Si_2O_{13}$) whose source is fly ash was identified in sample C3.

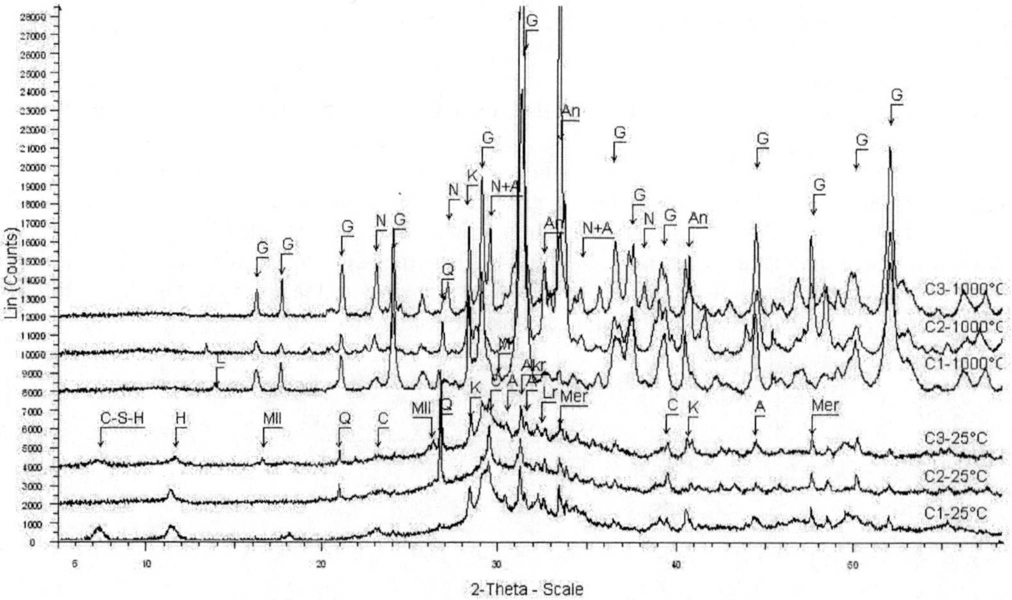


**Figure 1.** XRD traces of alkali activated material by CKD before and after thermal treatment to 1000°C: H = hydrotalcite, Q = quartz, C = calcite, MII = mullite, K = sylvite, A = aphthitalite, Mer = merwinite, G = gehlenite, N = nepheline, An = andradite, M = maghemite, Akr = åkermanite, Lr = larnite.

Thermal treatment of samples at 1000°C has induced some significant changes in samples composition. Samples treated at 400°C were mainly amorphous. Comparing the composition, there was no significant difference between samples treated at room temperature. The peaks of hydrotalcite, aphthitalite, C-S-H (I) were not identified. These phases were decomposed to 400°C. From 600°C to 800°C the amount of the amorphous phase decreased and at 1000°C the amorphous phase was no longer recognizable. At 800°C, the first new phase formation was identified: gehlenite ($Ca_2Al_2SiO_7$), nepheline ($NaAlSiO_4$), maghemite ($Fe_2O_3$), and andradite ($Ca_3Fe_2(Si_3O_{12})$. All samples contained gehlenite. Gehlenite is produced by the decomposition of C-S-H phases at around 800°C in samples containing lime [12]. Peaks of maghemite were found in sample C1. Nepheline [13–15] was identified in sample C3. Nepheline structure is based on a tridymite-like structure [16]. Andradite [13, 17] was found in C2 and C3 samples. Peaks of lazurite [13] ($Na_3Ca(Al_3Si_3O_{12})S$) were identified in sample C1 at 1000°C. The peak intensity of mullite ($Al_6Si_2O_{13}$) in sample C3 decreased due to reaction of mullite with surrounding matrix [13, 18]. All these changes are consistent with the data presented in the literature.

### *3.2. Heat microscope*
The high-temperature characteristics: deformation, sphere, hemisphere and flow temperatures were determined by heat microscope. Measurement shows that the sample C1 shrunk in temperature range from 650°C to 770°C. It represented an area loss about 17%. This could be related to softening or recrystallization of some products. The sintering of the sample started at 1100°C and at 1250°C occurred deformation of the sample followed by melting, formation of hemisphere and final spread over the alumina sample holder. Total shrinkage of sample was over 73%.

Sample C2 started to sinter at 950°C and no other changes were measured up to the final temperature 1300°C. Shrinkage caused by sintering was quantified as 43% area loss of sample.

For sample C3 sintering started at 1030°C and caused 40% area loss of sample. Above 1260°C deformation of the sample started followed by melting, formation of hemisphere and final spread over the alumina sample holder. Total shrinkage of the sample volume is over 77%.

### *3.3. Dilatometry*
Thermal expansion/shrinkage of prepared materials was measured in-situ through dilatometry. Shrinkage or expansion during heating caused both internal and external stresses, which weakened or damaged the structure. The effect of the water/solid ratio, alkali type or source material on the thermal expansion has been observed in previous works [9, 14, 19–21].

Figure 2 presents results of dilatometric analysis of prepared samples. All samples exhibited less thermal stability due to the shrinkage up to 13% that occurred during heating. All samples showed small expansion between 50–175°C as free water is released [22]. Shrinkage between 150–450°C was caused by dehydration of free water and dehydroxylation confirmed by TG as evidenced by the TG/DTG. Slow expansion occurred at 450°C. Another slight shrinkage occurred in sample C1 between 680–750°C likely due to densification and consequently expansion occurs which may be due to recrystallization [14]. Thermal expansion during heating of sodium silicate-fly ash alkali-activated materials has been discussed in the literature [18, 20]. Sharp shrinkage between 450°C and 850°C is probably due to viscous flow in aluminosilicate material causing sintering and subsequent densification of samples [18]. C3 sample has a similar trend of the dilatometric curve, with a difference in the temperature range of 20–150°C resulting in more pronounced expansion due to water excess. Slow expansion started at 450°C. Slight shrinkage due to the densification of the sample is not identified around 700°C as in the case of the sample C1.

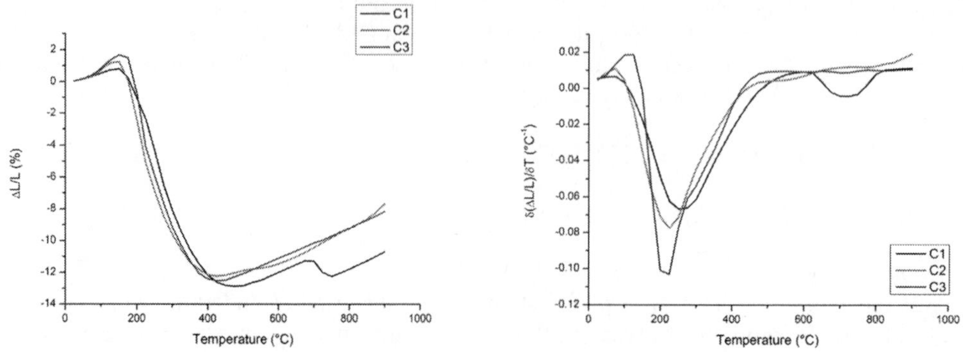


**Figure 2.** Thermal shrinkage of samples C1, C2, C3 (a) Derivative thermal shrinkage (b).

The derivative of dilatometric data is shown in figure 2b, which gives a clear overview of the changes in the rate of axial shrinkage and the onset temperatures. Onset temperature of dehydration of free water is higher for C1 sample than C2 and C3. The rate of shrinkage of this process is lower for C1 than for C2 and C3. The peak around 700°C indicates shrinkage due to densification in sample C1.

### *3.4. Thermogravimetric analysis*

Figure 3a presents results of TG/DTA analysis of prepared materials using CKD activator. The same effects can be observed for all three samples. There was a continuous loss of weight up to 1000°C with the total weight loss around 13%.

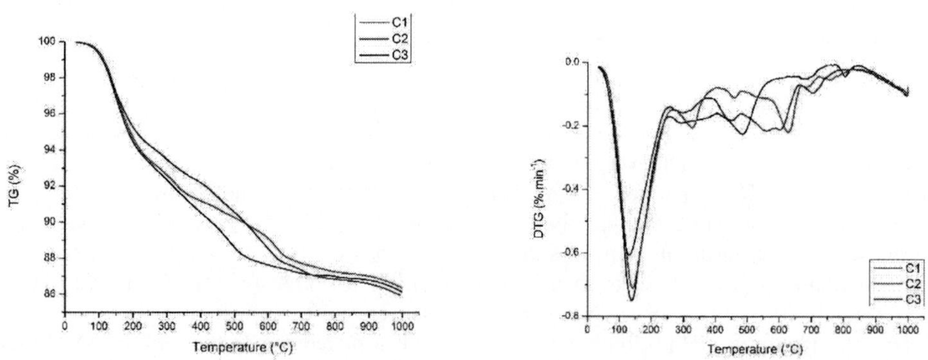


**Figure 3.** Curve of thermogravimetric analyses (TG on the left) and first derivation of the curve (DTG on the right).

The thermogravimetric analysis shows (Figure 3) a major weight loss (around 6%) from room temperature to 220°C due to evaporation of free and physically bounded water [17]. Between 150°C and 250°C, the rate of weight loss for all samples starts to slow down as the physical free water is already evaporated and the interstitial water starts to decompose. The gradual weight loss between approximately 250–400°C is attributed to the dehydroxylation. The weight loss (around 2.5%) between 200°C and 400°C is associated with dehydroxylation and decarboxylation of hydrotalcite [10, 23]. The presence of hydrotalcite was confirmed on the X-ray diffraction pattern. Thermal decomposition of portlandite occurs at about 450°C, the amount of portlandite is around 9.9% in C1, 2.5% in C2 and 7.8% in C3, however, this process is influenced by other simultaneous processes, therefore, the amounts are only approximate. The other small mass loss (around 1.5%) occurs between 550–650°C in samples C2 and C3 and it could be related to the sintering process. Recrystallization appears to cause mass loss between 650°C and 800°C in all samples [17]. The decrease in the amorphous phase in the range of 600°C to 800°C was recorded by XRD. The thermogravimetry data are consistent with the dilatometry analysis.

### *3.5. Compressive strength*

Compressive strength of prepared samples was tested before and after exposure to high temperature in order to assess their strength retention. Figure 4 presents compressive strength of specimens cured at temperature range up to 1000°C. Initial compressive strengths ranged from 40 MPa (C2) to 25 MPa (C3). Samples produced with CKD showed moderate strength. Thermal exposure affected the mechanical strength of tested materials. The changes in mechanical strength occurred due to changes in structure and phase composition. Structural changes include sintering, densification, melting, cracking or pore size changes [14].

In the case of sample C1, the compressive strength decreased under 10 MPa with increasing temperature. Higher susceptibility to strength degradation of sample may be due to a higher content of C-S-H gel which is more vulnerable to dehydroxylation and decarbonation [17]. Sample C2 did not show any decrease in the strength value due to temperature rise, but at 800°C the sample disintegrated as a result of cracking. The cracking is due to migration and evaporation of water in the samples during thermal exposure. Higher strengths are probably due to the content of finer metakaolin particles in the original matrix. The compressive strength of sample C3 gradually increased up to 400°C then it also dropped sharply to 3 MPa. This is mainly attributed to the formation of additional binder gel or sintering un-reacted fly ash particles [9, 17].

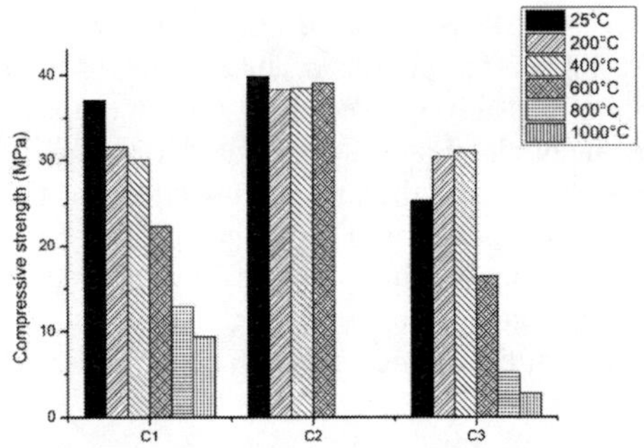


**Figure 4.** Compressive strength before and after heating to 1000 °C of alkali activated materials with CKD.

### *3.6. Porosity*

The results of total porosity of prepared samples are shown in Figure 5. At higher temperatures, multiple factors influence the size and distribution of the pores. Pore structure that changes during thermal exposure has a mixed effect on the mechanical strength [9].

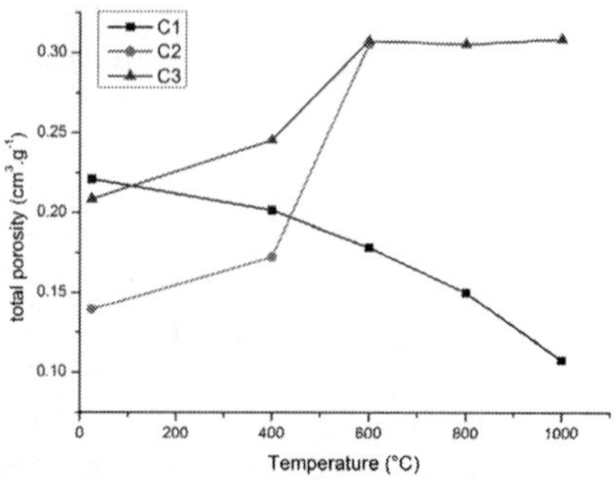


**Figure 5.** Total porosity versus heating temperature for C1, C2, C3 samples.

The total porosity of C2 sample is 50% lower than the other samples, which is probably due to the finer particles of metakaolin in the original matrix. A larger number of open pores in samples C2 and C3 can be assumed from the water adsorption results. Sample C1 has most pores in the range of 10–20 nm, while sample C2 had most pores between 20–40 nm and sample C3 between 40–100 nm.

The total porosity of sample C1 decreases with increasing temperature and the number of pores in the region below 100 nm increases to 600°C. Reduction of total porosity during dehydroxylation and sintering resulted in densification. After heating to 1000°C, small pores in the range of 10–20 nm are not longer detectable and the number of larger pores in the 250–350 nm range increased.

An increase in the number pore size in the 40 nm region occurs with a temperature rise of up to 600°C for sample C2. At higher temperatures, macro cracks and sample destruction occurred. Development of large pores is possibly connected to the decomposition of material due to heating. A number of pores could be the predominant factor related to strength decrease.

For sample C3, the total porosity with the temperature rises and from 600°C the total porosity does not change, only the pores size changes. It was confirmed [9] that fly ash alkali-activated binders have large numbers of interconnected pores which simplify the permeation of water to surface, causing smaller damage to the structure. In general, alkali-activated materials based on metakaolin do not exhibit as great increase in pore volume after heating as fly ash geopolymer.

## 4. Conclusions

Physical characteristic of alkali-activated materials based on cement kiln dust before and after heating to 1000°C have been observed. The thermal stability of all observed samples was rather low due the shrinkage during thermal treatment. The material shrinkage was closely related to volume changes of aluminosilicate caused by dehydration between 25 and 400°C. Based on XRD analysis, it was proved that the prepared materials were amorphous, still containing the crystalline phases initially presented in raw material. New phases formation was identified after heating to 800°C: gehlenite, nepheline and andradite.

Alkali-activated material with MK (C2) showed better mechanical properties compared to material based on FA (C3) and BFS (C1) after 28 days under laboratory conditions. After heating, compressive strength was affected by the varying influence of phase changes, increase of total porosity and damages due to dehydration. The MK sample (C2) exhibited the best compressive strength up to 600°C, but then macro cracks and sample destruction occurred. The compressive strength of sample C3 gradually increased up to 400°C due to formation of additional gel binder.

**Acknowledgments**

This paper was elaborated with the institutional support for long-term development of research organizations by the Ministry of Industry and Trade of the Czech Republic.

## References


[1] Davidovits J 2011 *Geopolymer Chemistry and Applications* (France: Institute Geopolymere) p 623

[2] Konsta-Gdoutos M and Shah SP 2003 *Cement Concrete Res.* **33** 1269

[3] Ata AA, Salem TN and Elkhawas NM 2015 *Constr. Build. Mater.* **93** 950

[4] Al-Jabri KS, Taha RA, Al-Hashmi A and Al-Harthy AS 2006 *Constr. Build. Mater.* **20** 322

[5] Buchwald A and Schuly M 2005 *Cement Concrete Res.* **35** 968

[6] Ahmari S and Zhang L 2013 *Constr. Build. Mater.* **40** 1002

[7] Wanga K, Shahb SP and Mishulovichc A 2004 *Cement Concrete Res.* **34** 299

[8] Bílek V, Pařízek L and Kalina L 2015 *Materials and Technology* **49** 549

[9] Provis JL and van Deventer JSJ 2009 *Geopolymers. Structure, processing, properties and industrial applications* (Oxford: Woodhead Publishing Limited) p 454

[10] Ben Haha M, Le Saout G, Winnefelf F and Lothenbach B 2011 *Cement Concrete Res.* **41** 301

[11] Deventer JSJ, Provis JL, Lukey GC, Duxson P and Luey GC 2007 *J. Hazard. Mater.* **A139** 506

[12] Dombrowski K, Buchwald A and Weil M 2007 *J. Mater. Sci.* **42** 3033

[13] Wattanasiriwech S, Nurgesang FA, Wattanasiriwech D and Timakul P 2017 *Ceram. Int.***43** 16055

[14] Provis JL, Harrex RM, Bernal SA, Duxson P and van Deventer JSJ 2012 *J. Non-Cryst. Solids* **358** 1930

[15] Bakharev T 2006 *Cement Concrete Res.* **36** 1134

[16] Buchwald A, Vicent M, Kriegel R, Kaps Ch, Monzo M and Barba A 2009 *Appl. Clay Sci.* **46** 190

[17] Yang T, Wu Q, Zhu H and Zhang Z 2017 *Constr. Build. Mater.* **155** 475

[18] Rickard WDA, Temuujin J and van Riessen A 2012 *J. Non-Cryst. Solids* **358** 1830

[19] Temuujin J, Minjigamaa A, Lee M, Chen-Tan N and van Riessen A 2011 *Cement Concrete Comp.* **33** 1086

[20] Provis JL, Yong CZ, Duxson P and vand Deventer JSJ 2009 *Colloid. Surface A* **336** 57

[21] Rovnanik P, Bayer P and Rovnanikova P 2013 *Constr. Build. Mater.* **47** 1479

[22] Duxson P, Lukey GC and van Deventer JSJ J. 2006 *Non-Cryst. Solids* **352** 5541

[23] Vagvolgyi V, Palmer SJ, Kristof J, Frost RL and Horvath E 2008 *J. Colloid Interf. Sc.* **318** 302

IOP Conf. Series: Materials Science and Engineering **379** (2018) 012009 doi:10.1088/1757-899X/379/1/012009

# Mechanical activation of frozen lime putties

**D Michoinová[1] and R Nečas[2]**

[1] Czech National Heritage Institute, Valdštejnské nám. 162/3, Prague 118 01
[2] Research Institute for Building Materials, Hněvkovského 30/65, Brno 617 00

Email: michoinova.dagmar@npu.cz, necas@vustah.cz

**Abstract.** The lime putty used for mortar preparation loses its desired properties, especially the plasticity and other rheological parameters, when it freezes. Two kinds of lime putty with 30 wt. % of solid content underwent 5 freeze/thaw cycles and their properties were investigated. Their consistency worsened dramatically after freezing. Mechanical activation was carried out by high-speed mixing of lime putty. This procedure led to a significant improvement in the lime putty plasticity and viscosity. The rheological parameters reached the original values that were determined before freezing. Rheological tests of lime putties were determined by rotational rheometer with the solids characterized by laser particle size analyzer.

## 1. Introduction

Lime based mortars have been used in building construction for thousands of years. The non-hydraulic lime, in the form of lime putty, was one of the most common form of this binder. The characteristics of both historic and new mortars are known to be affected by the properties of the lime putty in addition to other factors [1].

After the development of Portland cement, it rapidly became the binder component of choice in mortar for new masonry works and, unfortunately, for use in conservation and rehabilitation of architectonic heritage, too. However, the rapid hardening and high mechanical strengths of cement mortars are incompatible with old building materials [2]. With the increased use of Portland cement, and other binders, in mortars, the use of air lime-based mortars fell into gradual disuse and, along with this, the knowledge and skills of craftsmen almost disappeared. Over time, it became apparent that the replacement of non-hydraulic lime mortars in the conservation and restoration of historic buildings and structures was a serious error. Subsequently, a new phase in the use of air lime, along with new research into the understanding of the properties of air lime and their mortars commenced [3, 4].

The aged, wet slaked, air lime (in the form of lime putty) is preferred for lime mortars preparation in certain situations. An alternative method, sometimes employed in the preparation of lime putty, is the mixing of powdered hydrated lime with water and aging the putty formed.

There is a great deal of skill and expertise required in good putty preparation and storing (optimal slaking temperature, water/quicklime ratio, aging of lime putty, avoidance of drying and inhibiting contact with air) and, last but not least, the knowledge gained by practitioners that lime putty cannot be allowed to freeze [5, 6]. The need to protect lime putties from frost is also highlighted by commercial suppliers in their instructions for use [7, 8] as lime putties lose the properties for which they are known to impart to mortars, such as plasticity, workability, cohesion, etc. when frozen. This has also led to the discarding of quantities of lime putty affected by frost, both in storage and on sites, and, therefore, any procedure that would permit the continued use of frost affected materials would be of significant value.

A satisfactory objective explanation of the risk of damage due to frost is still unknown. A possible explanation is (as in the cases of polymeric dispersions) that the freezing of the system causes a certain coagulation of slightly dispersed particles thereby resulting in the deterioration of its rheological properties [9].

Rheological properties of lime putties are very important, if the putty is used in mortar mix design. Lime putty is a non-Newtonian fluid, usually a Bingham type of fluid, i.e. in absence of external forces, it behaves as a solid. If the external forces are present, the putty starts to change its form and behaves as a liquid and starts to flow. The minimal stress necessary for putty to start to flow is called its yield stress. This parameter together with its plasticity is important for mortar applications. Other properties, called "time-dependent properties" i.e. thixotropy and rheopexy, are significant for the non-Newtonian fluids, too. When thixotropic fluid is sheared, its apparent viscosity decreases, whereas a shear of the rheopectic fluid results in an increase in its apparent viscosity. In order to determine the general flow behaviour of a material, a graph of the shear stress against the shear rate is plotted. This graph is called its flow curve. For fluids with time-dependency, the flow curve for increasing and decreasing shear rate are not identical and the curve forms a hysteresis loop. The area of the loop determines thixotropy value (the "falling" curve lies under the "rising" curve) or rheopexy value (the "falling" curve lies above the "rising" curve).

In the past, experiments were made to improve putty properties damaged by frost using the mechanical activation of lime putty [9, 10]. The aim of our study was to compare the time dependent behaviors of two aged putties, before and after freezing and after mechanical activation.

## 2. Experimental Procedure

### *2.1. The lime putty preparation*

Two types of lime putty were prepared from the same high calcium quicklime (Hydrated lime EN 459-1 CL 90-Q, producer Vápenka Čertovy schody a. s., Czech Republic):

- Lime putty prepared from technical dry slaking calcium hydrate by mixing the powder with water and aged for 3 years in plastic container (sample LP-D).
- Lime putty prepared by wet slaking of quicklime and aged for 3 years in a lime pit (sample LP-W).

The samples of lime putty (~ 2 kg) were homogenized for short time by employing a helical stirrer, fitted to a hand-operated electrically powered drill before freezing. Prior to use, the content of dry matter was determined gravimetrically. To allow comparison of rheological properties of the lime putties, the content of dry matter was corrected to 30 wt. % by adding a calculated quantity of demineralized water. Initial and corrected values are given in Table 1. Adjusted samples were used in the experiments and analyzes.

**Table 1.** Solid content in tested putties (wt. %) before freezing.

| | Solid content (original samples) | Solid content (adjusted samples) |
|---|---|---|
| LP-D | 48.10 | 29.43 |
| LP-W | 45.15 | 29.55 |

### *2.2. Frost treatment of samples*

Approximately 1 kg of lime putty, in plastic vessels, were placed in a temperature controlled environmental cabinet KD20 and the samples exposed to 5 freeze/thaw cycles according to the following schedule:

- Freezing from ambient temperature to -10°C (linear temperature decrease, 6 hours),
- Maintained at a constant temperature -10°C (6 hours),

- Thawing to ambient temperature (linear temperature increase, 6 hours),
- Maintained at a constant ambient temperature (6 hours).

The freezing device (environmental cabinet) was operated under fully automatic control for 5 days. After the freeze/thaw cycling treatment, samples were aged for 1 month, at ambient temperature (lab temperature ~24°C). After aging, the mechanical activation was applied by using a laboratory stirrer, model IKA RW 20, which has a digital display; this permitted all samples to be activated under the same conditions. The stirrer was fitted with a helical paddle. Each putty was reactivated by using the stirrer at speeds of 450 rpm for 10, 20, 30 and 40 minutes.

### *2.3. Measurement methods*

Particle size distribution was determined by laser particle size analyzer CILAS 920L in range 0.3 to 400 μm in isopropyl alcohol medium. The measurements of particle size of the putty samples was determined before freezing, after freezing, after 1 month aging and following each of the mechanical activation (stirring) periods. Particles coarser than 0.315 mm were removed by sieve to prevent clogging of apparatus measuring cell.

Rheometer TA Instruments DRH-1 was used to obtain flow curves measurement. Geometry of coaxial cylinders according to DIN/ISO 3219 was used in the measurement, and the range from 0 to 150 $s^{-1}$ for shear rate was investigated. Data was evaluated by firmware TRIOS and the yield stress and time-dependent rheological properties determined.

## **3. Results and Discussion**

The experiments have confirmed the crucial role of frost on the rheological properties of lime putty. Frozen putty loses the properties necessary for mortar preparation – plasticity (that corresponds with yield stress value) and optimal particle size distribution. The putties following freeze/thaw cycling were liquid coagulate solids and their rheological parameters were non-measurable by the rheometer employed (out of measurement range). An increase in the number of coarser particles was also found after freezing. Both these effects most likely cause the loss of plasticity and the rapid sedimentation of particles in suspension. This is probably related to changes in surface charge (analogous effect is known for organic material dispersions) along with the aggregation of portlandite crystals [12, 14, 15, 16].

The increase in the proportion of larger particles after 1 month of aging can be explained by portlandite crystals growth and recrystallization (see Figure 2 and 3 for LP-D and 6 and 7 for LP-W).

It was confirmed that the properties of frozen lime putties were improved by intensive remixing, by using the laboratory stirrer at 450 rpm for 10, 20, 30 and 40 minutes. Mechanical activation by external forces appears to have caused the breakage of crystal aggregates in the suspension. The quantity of larger particles, after activation, became lower, as is shown by the particle size analyses. The particle size distribution returned to approximately original values after activation (see Figure 1– 8 and Table 2). The medians of the particle sizes determined after activation are practically identical to those of the original putties, before treatment.

In general, colloidal suspensions show a variety of nonlinear rheological properties (yield stress, shear thinning, shear thickening, thixotropy, and/or rheopexy) that depend on interparticle forces, Brownian motion of the particles, and hydrodynamic interactions as well as on the shape, size, and volume fraction of particles, on the degree of aggregation, and on the fraction of fluid immobilized by the particles [11].

For the suspension of particles below 1 μm in size, the influence of Brownian motion is important. In this case, the probability of sedimentation can be predicted from the gravitational and Brown strength ratios using equation (1) [13]:

$$p = \frac{a^4.\Delta\rho.g}{k_B.T} \quad (1)$$

where

*a* radius of particle (m)

*Δρ* difference between particle and liquid density ($kg.m^{-3}$)

IOP Conf. Series: Materials Science and Engineering **379** (2018) 012009 doi:10.1088/1757-899X/379/1/012009

$g$ gravitational acceleration ($m.s^{-2}$)
$k_B$ Boltzmann constant $1.38064852 \times 10^{-23}$ ($m^2.kg\ s^{-2}.K^{-1}$)
$T$ absolute temperature (K)

If this ratio is greater than one, a degree of sedimentation can be expected, while a ratio of less than one is likely to indicate a stable system. However, this equation does not take into account the potential interactions between particles. Due to Brown's motion, the particles will continually collide with each other, and as a result, the particles can collapse due to Van der Waals's attractive forces. This may lead to the formation of much larger particles, or secondary particles (flakes) and, therefore, experience a larger contribution of gravity than that indicated in equation (1).

Using the equation (1), $\Delta\rho$ difference between calcium hydrate and water density = (2240 – 1000) $kg.m^{-3}$, $g = 9.81$ $m.s^{-2}$ and $T = 295$ K the critical value ($p = 1$) of particle radius is about 0.75 μm (i.e. diameter 1.5μm). The higher content of larger particles is the reason for the rapid particle sedimentation in freeze/thaw lime putties. Results of tests are shown in Figure 1 to 8, with the changes of median particle size, during test, shown in Table 2.

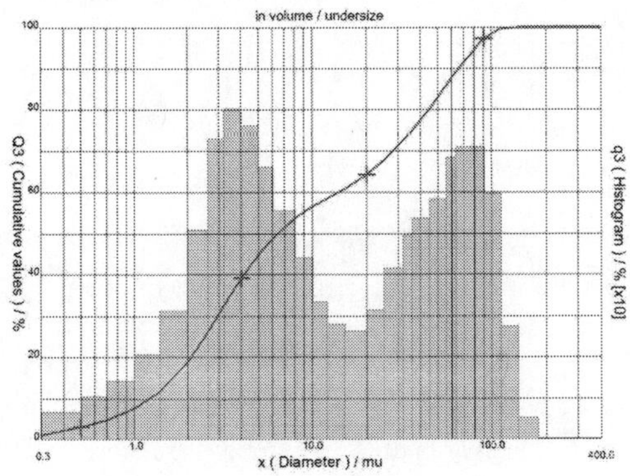


**Figure 1.** LP-D before freezing.

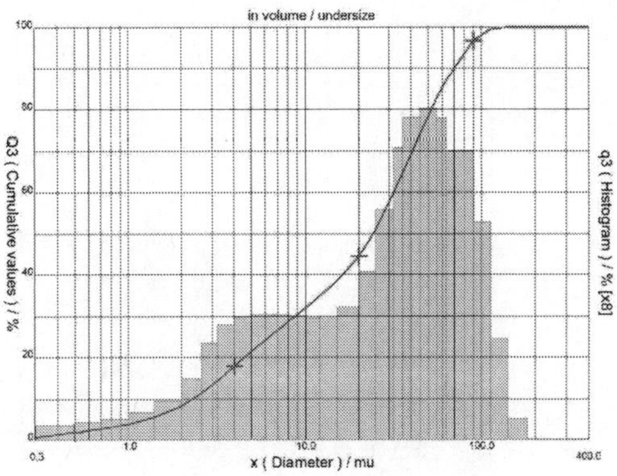


**Figure 2.** LP-D after freezing.

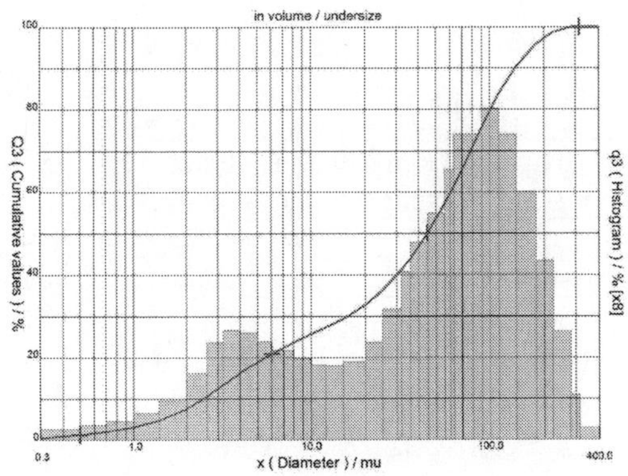


**Figure 3.** LP-D after freezing + 1 month of spontaneous aging.

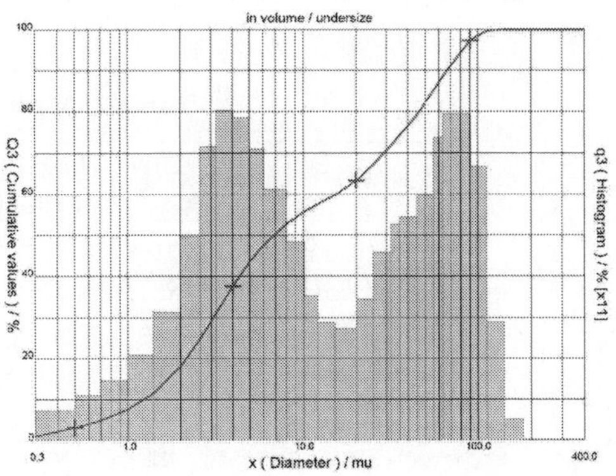


**Figure 4.** LP-D after 40 min of activation.

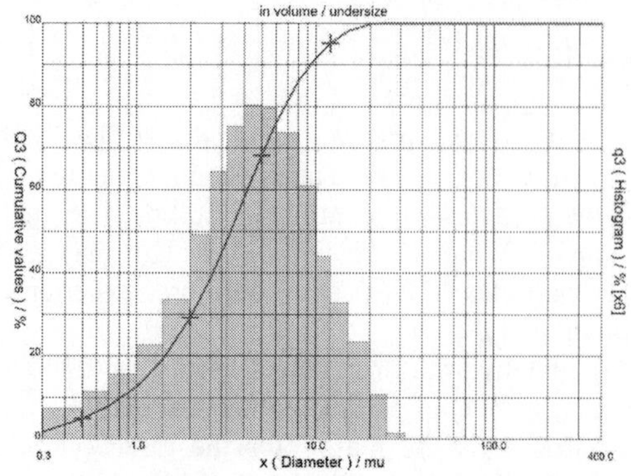


**Figure 5.** LP-W before freezing.

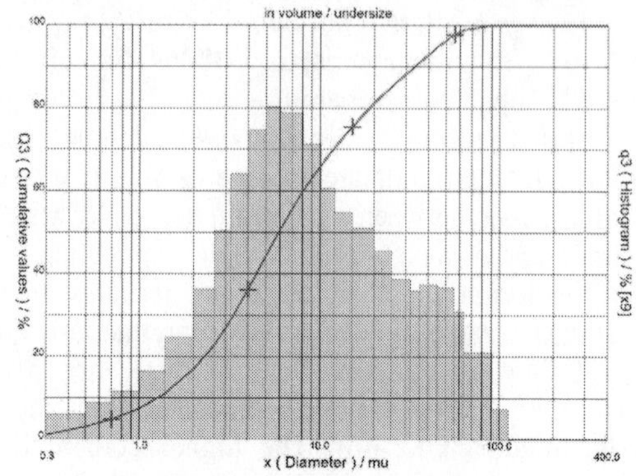


**Figure 6.** LP-W after freezing.

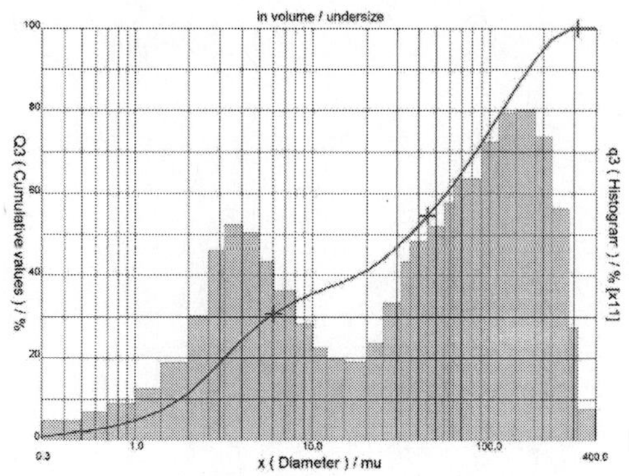


**Figure 7.** LP-W after freezing + 1 month of spontaneous aging.

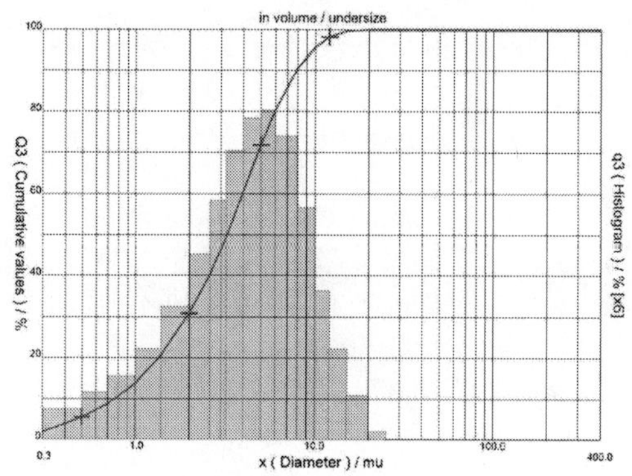


**Figure 8.** LP-W after 40 min of activation.

**Table 2.** Medians of particle size (μm).

| | Before freezing | After freezing | Frozen putty + 1 month aging | 10 min mixing | 20 min mixing | 30 min mixing | 40 min mixing |
|---|---|---|---|---|---|---|---|
| LP-D | 6.44 | 24.20 | 44.93 | 13.29 | 8.99 | 7.69 | 6.40 |
| LP-W | 3.36 | 5.89 | 35.67 | 4.51 | 3.05 | 3.06 | 3.27 |

For better clarity, the overlay curves of particle size distribution are shown in Figures 9–12.

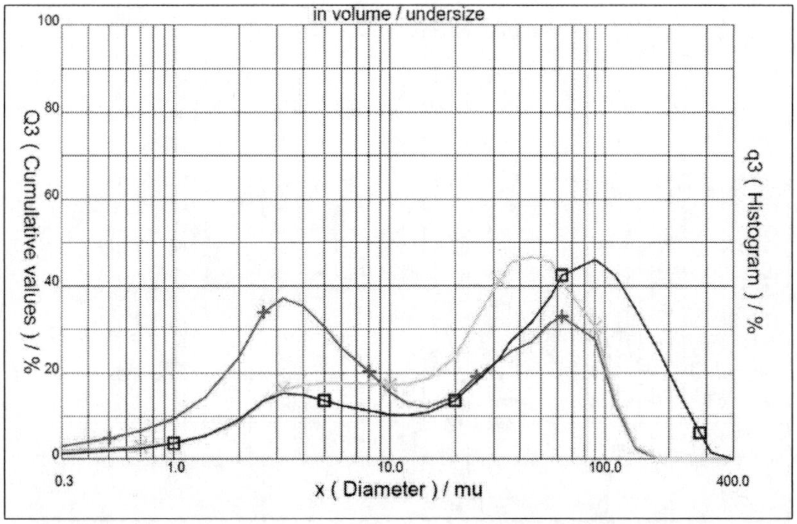


**Figure 9.** Overlay curves of particle size distribution of LP-D sample:
+ before freeze/thaw cycling; × after freeze/thaw cycling; □ after 1 month of aging.

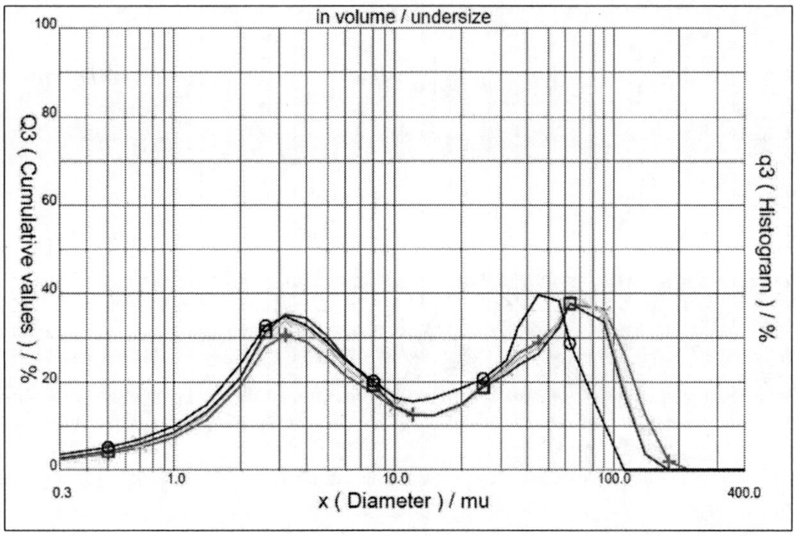


**Figure 10.** Overlay curves of particle size distribution of LP-D sample:
+ 10 min. of activation; × 20 min. of activation; □ 30 min. of activation; o 40 min. of activation.

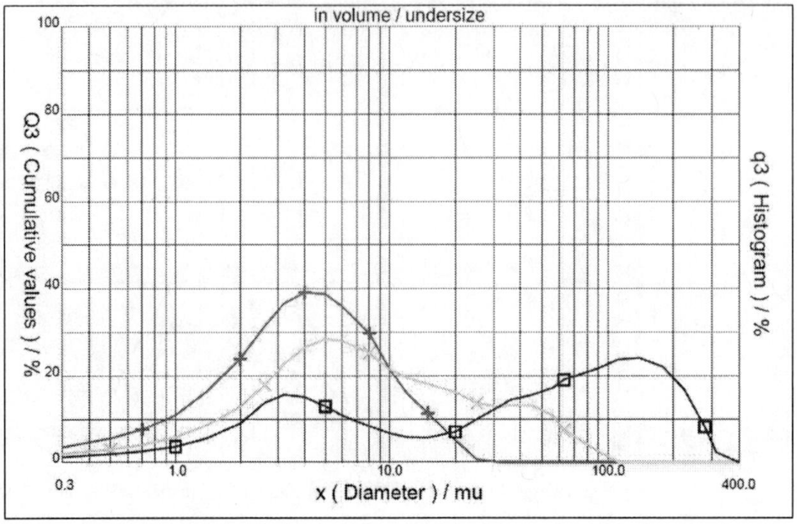


**Figure 11.** Overlay curves of particle size distribution of LP-W sample:
+ before freeze/thaw cycling; × after freeze/thaw cycling; □ after 1 month of aging.

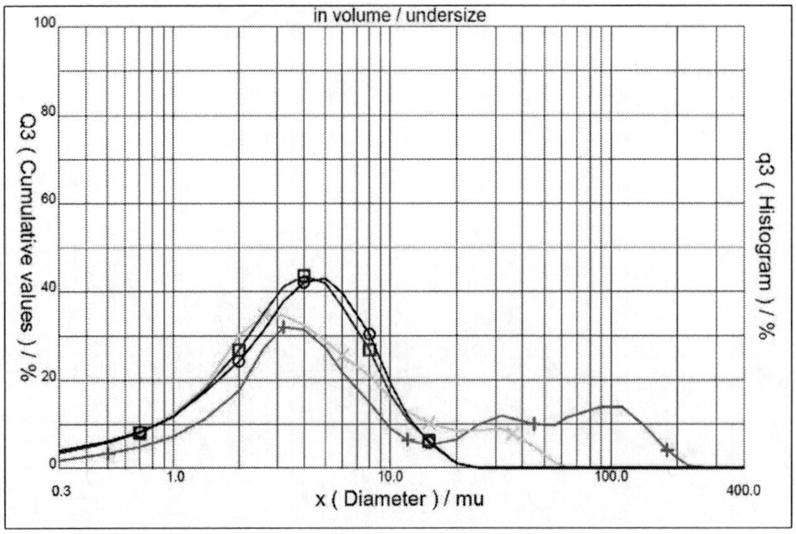


**Figure 12.** Overlay curves of particle size distribution of LP-W sample:
+ 10 min. of activation; × 20 min. of activation; □ 30 min. of activation; o 40 min. of activation.

## 4. Rheological properties

The rheological properties of putty samples before freezing and after 40 min of activation were measured only, as the samples after freezing, and after 1 month of aging, were not measurable by means of the rheometer. More precisely, the samples had the appearance and properties of a water suspension, and were very runny, displaying extremely fast sedimentation of solid matter, and this made the measurement impossible.

Based on the rheological measurements made, LP-D after reactivation showed a larger yield stress after 40 minutes activation, compared to the putty before freezing. Activation of LP-W sample was stopped after 40 minutes, because median of particles size didn't decrease in last twenty minutes of mixing. However, a lower yield stress was found in the case of LP-W. All measured samples of lime putty exhibited rheopectic behavior. At high shear rates, hydrodynamic stresses can act upon the lose clusters and fragment them. These stresses are manifested as velocity differences across the structure of the aggregates that produce several fragments of similar size and fluid drag forces that strip primary particles or small flocules from the surface of the original aggregates. As a consequence, both smaller aggregates and individual $Ca(OH)_2$ nanoparticles are released from the original clusters. This caused an increase in the viscosity of the suspension upon relaxation of the flowing system. This behavior may also help to explain the results by Vávrová and Kotlík [9] who indicate that, after prolonged shearing, fresh lime putties showed much higher values of viscosity then nonsheared pastes [11].

Values of rheopexy are lower (approximately half the value) after treatment by frost and activation. The flow properties of the lime putties are shown in Figures 13–16 and in Table 3.

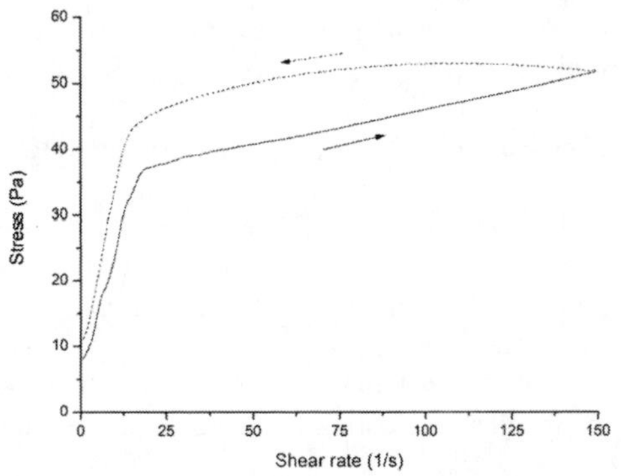


**Figure 13.** LP-D before freezing.

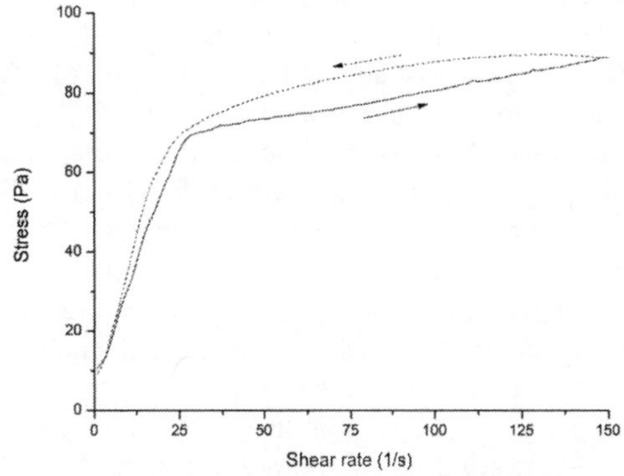


**Figure 14.** LP-D after freezing and 40 min of activation.

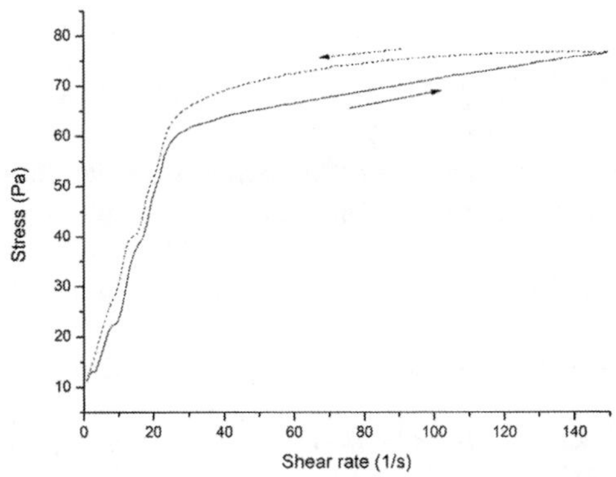


**Figure 15.** LP-W before freezing and activation.

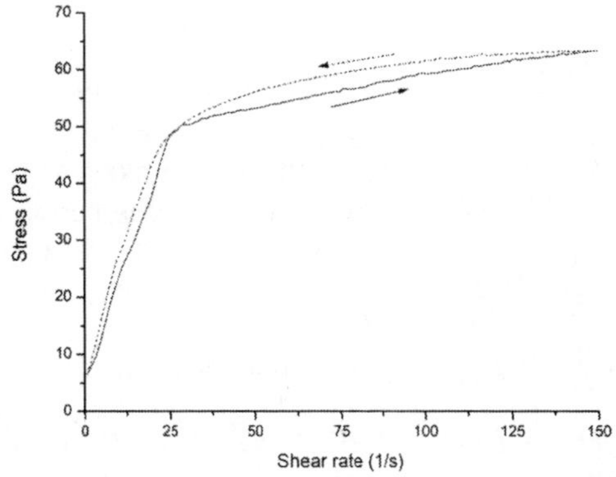


**Figure 16.** LP-W after freezing and 40 min of activation.

**Table 3.** Rheological properties of lime putties.

| Sample | Yield stress (Pa) | Thixotropy (+) or rheopexy (-) ($Pa.s^{-1}$) |
|---|---|---|
| LP-D (before freezing) | 35.3 | -1021 |
| LP-D (mixing 40 min) | 65.6 | -661 |
| LP-W (before freezing) | 59.7 | -622 |
| LP-W (mixing 40 min) | 48.5 | -347 |

## 5. Conclusion

There are minor differences in the behavior of lime putties prepared by wet slaking quicklime and by mixing of dry slaked calcium hydroxide with water, even when the original quicklime was the same for both putties, and the storage condition were comparable.

The action of frost had a negative effect on the rheological properties of both lime putties.

Frozen putties lose their plasticity due to significant changes in their particle size distribution. Decrease of plasticity and sedimentation of coagulated particles is related to changes in surface charge and aggregation of portlandite crystals.

The content of larger particles increased spontaneously after 1 month of aging, this is considered to be due to growth and recrystallization of portlandite crystals.

Properties of frozen lime putties can be improved by intensive mixing. The reactivation of lime putties can be achieved by mechanical operation, by applying external forces that cause the deagglomeration of aggregates in the suspension, and it has been shown that the particle size distributions can reach almost original values, after intensive activation.

Based on the results of these experiments, it can be said that the behavior of lime putties damaged by frost can be re-established by intensive mixing – by mechanical activation – to approximately the same condition, as before freezing.

**Acknowledgements**

The Czech Ministry of Culture and The National Heritage Institute financed the research under "Long-term Conceptual Development of Research Organizations" project (99H30PP110).

**References**

[1] Elert K, Rodriguez-Navarro C, Pardo E S, Hansen E and Cazalla O 2002 Lime mortars for the conservation of historic Buildings *Studies in Conservation* **47** pp 62–75
[2] Klemm A J and Wiggins D E 2015 Microstructural Characteristics of Modern Repair Materials vs. The Performance Requirements of Historic Lime Mortar *In International Symposium on Brittle Matrix Composites* RILEM pp 33–42
[3] Torraca G 1982 Porous Building Matrials *Materials Science for Architectural Conservation ICCROM*
[4] Wiggins D 2017 Traditional lime mortars and masonry preservation *The Journal of the Building Limes Forum* **24** pp 28–37
[5] Gibbons P 1995 *Preparation and Use of Lime Mortars, An Introduction to the principles of using lime mortars* 2sted. (Edinburg: Historic Scotland )
[6] Holmes S and Wintage M 2002 *Building with lime* 3nd ed. (London: ITDG Publishing)
[7] *Superfine lime putty* 2017 (*Online* http://www.roseofjericho.co.uk/product/superfine-lime-putty)
[8] *Heritage Restoration - Slaked Lime Putty* 2017 (*Online* http://www.stratacote.co.nz/heritage-restoration-slaked-lime-mortar/)
[9] Vávrová P and Kotlík P 2016 Rheological properties of lime putty (Online http://www.schleibinger.com/k2004/vavrova04/Prezentace_Regensburg.pdf)
[10] *What Happens When Lime Putty Freezes?* 2017 *(Online* http://www.preservationscience.com/materials/lime/FL.html)
[11] Vávrová P and Kotlík P 2005 Rheological properties of lime putty *Conference paper of 36th*

*International Conference on CoatingsTechnology CCT′ Seč, Czech Republic*

[12] Ruiz-Agudo E and Rodriguez-Navarro C 2010 Microstructure and rheology of lime putty *Langmuir* **26** pp 3868–3877

[13] Barnes H A 2000 A Handbook of Elementary Rheology *University of Wales, Institute of Non-Newtonian Fluid Mechanics*

[14] Atzeni C, Farci A, Floris D and Meloni P 2004 Effect of aging on rheological properties of lime putty *J. Am. Ceram. Soc.* **87** pp 1764–1766

[15] Boháč M and Nečas R 2016 The role of aging on rheological properties of lime putty *Procedia Engineering ICEBMP 2016*

[16] Lopes M G and Margalha B 2013 Microstructural Changes of Lime Putty during Aging *Journal of Materials in Civil Engineering* **25**(10) pp 1524–1532

# Properties of mixtures of cement with various raw materials

**T Staněk[1], R Nečas[1], M Boháč[1], A Zezulová[1] and A Rybová[1]**

[1] Research Institute for Building Materials, Hněvkovského 65, 617 00 Brno, Czech Republic

E-mail: stanek@vustah.cz

**Abstract.** This paper presents possibility of utilization of selected Czech and Slovak industrial and natural raw materials in blended cement production. The cement blends were prepared using 75 wt. % of cement and 25 wt. % of a substituent. The paper is focused only on test results of the laboratory-prepared blended cements. The influence of admixture on hydration process was observed by calorimetric measurement of heat evolution during hydration process within 24 hours. The hydration products of water-cured samples for 28 days were determined by thermal analysis. Further properties, such as short and long-term compressive strengths and setting period, were determined.

## 1. Introduction

The possibility of utilization of wastes or non-traditional raw materials in cement production is discussed all around the world. Portland cement industry is responsible for approximately 5% of global $CO_2$ emission [1]. Partial replacement of Portland cement by one or more additives to obtain blended cements is generally associated with significant savings in energy and cost and it also supplies more durable cementitious systems to construction industry. Nowadays, the range of building materials is extended and various blended cements are produced using fly ash, slag or mineral admixtures, which is documented in the EN 197-1 standard. By replacement of Portland cement or clinker by such secondary raw materials, a satisfactory quality of cement can be still reached, moreover, some advantageous properties can be obtained, such as a decrease of hydration heat, specific rheological and setting properties etc. [2, 3].

Recently in Turkey, a complete research on natural pozzolans such as zeolites and volcanic tuffs was conducted [4]. Natural pozzolans show excellent ability to reduce the alkali-silica expansion and they provide many beneficial properties of both fresh and hardened concrete [5]. Whereas natural pozzolans tend to increase the water ratio requirement for concrete, their use in blended Portland cement is generally limited by 30%. In Greece, the effect of other natural pozzolan – diatomite – on blended cement properties and on hydration process was studied as well. Blended cements, having up to 10% of diatomite content, develop the same compressive strength as corresponding Portland cement [6, 7]. It was proven that an effective mineral additive can be manufactured by thermal activation of raw kaolin with at least 41.6% of kaolinite [8], the thermally activated kaolin is very effective admixture in blended cements based on high strength. A long-term research of fly ash, slag and silica fume utilization has confirmed their suitability for the blended cement production.

This work is focused on properties of 8 secondary and non-traditional materials (2 samples of fly ash, 2 samples of slag, 2 samples of rock and 2 samples of other material) selected from a group of 59 monitored materials from the Czech Republic and Slovakia. Their suitability of utilization in composite cements is observed.

## 2. Materials and methods

Cement blends were prepared from Portland cement CEM I 42.5 R and selected industrial and natural raw materials in the ratio 3:1 by mass. Chosen raw materials were: fly ash, fluidized ash, ground granulated blast-furnace slag (GGBS), foundry slag, marlite, zeolite and two silica fumes. Before mixing with cement, every material except silica fumes was milled in laboratory ball mill so that the residue on 0.09 mm sieve was 2–3%. The basic characteristics of these raw materials are given in Table 1.

**Table 1.** Basic monitored components in raw materials in wt. % determined by X-ray diffraction and their specific surface.

| Raw material | Specific surface $m^2/kg$ | Amorphous content wt. % | Free CaO wt. % | Other parameters |
|---|---|---|---|---|
| Fly ash | 649 | 73.4 | - | - |
| Fluidized ash | 1256 | 55.2 | 6.6 | - |
| GGBS | 195 | 83.9 | - | - |
| Foundry slag | | 0.0 | - | high content of belite |
| Marlite | 681 | 22.9 | - | contains opal |
| Zeolite | 1211 | 43.0 | - | clinoptilolite, heulandite, opal |
| Silica fume I | 1187 | 100.0 | - | - |
| Silica fume II | 4041 | 100.0 | - | - |

X-ray diffraction of raw materials was performed on Bruker D8 Advance apparatus with Cu anode ($\lambda K\alpha$ = 1.54184 Å) and variable divergence slits at Θ-Θ Bragg-Brentano reflective geometry. Quantitative phase analysis was done by Rietveld method using software Topas. Amorphous phase was quantified by addition of internal standard (20 wt. % of $CaF_2$).

The influence of admixture on hydration process was observed by calorimetric measurement. Heat released during hydration process was measured within 24 hours. The hydration products of water-cured samples for 28 days were determined by thermal analysis. Further properties, such as short and long-term compressive and flexural strengths and setting process, were determined.

Observation of the course of the temperature during hydration process of prepared blended cement pastes was performed on isoperibolic calorimeter [9]. The measurements were carried out for 24 hours at 35°C.

The setting process of prepared cement blends, containing industrial and natural raw materials, was observed by a Tussenbrock method. Tussenbrock method is performed on a device with a needle similar to Vicats. Depth of submersion of six cylindrical probes of decreasing diameter in the cement paste is measured in time dependence. The depth of indentation of particular probe indicates a threshold of shearing stress (in kPa) which corresponds to consistency achieved at the moment of measurement. This method allows to capture several anomalies during the setting process such as false setting.

Compressive strengths were measured in mortars from blended cements containing industrial and natural raw materials prepared according to EN 196-1 standard, water-cured for 2, 7, 28, 56 and 90 days. Mortars were prepared according to standard, i.e. with binder to sand ratio 1:3 and water to cement ratio w/c = 0.5.

The hydration products of blended cement pastes were determined by means of the differential thermal analysis. The samples from calorimetric measurement, water-cured for 28 days, were crushed to grain size of 1–5 mm, rinsed by ethyl alcohol and acetone and dried. The samples were then ground to grain size less than 0.09 mm and subjected to DTA-TG measurement on the Netzsch Perseus STA 449 apparatus in temperature range of 20–1000°C. This method in some cases enables to quantify the processes related to thermal decomposition of hardened cement (when peaks do not overlap each other).

## 3. Results and discussion

### *3.1. Technological properties*

Figure 1 compares the hydration process of samples containing admixtures to the reference sample of ordinary Portland cement. Except the fluidized ash, all replacements of cement by different admixtures result in a reduction of hydration heat and temperature maximum of hydration. A shorter time needed for the maximal temperature is achieved when fluidized ash, marlite, zeolite and silica fume were used. Other blends show a slower hydration in the comparison with the reference sample. A very slow development of hydration heat was observed in the case of foundry slag blend. The fast increase and high temperature of fluidized ash hydration process is probably caused by a content of free lime. Similar results are also observed by other authors [10]. Silica fume hydration process is influenced mainly by its fineness and the amount of used water [11]. In our experiment, amount of water is the same for all samples, therefore finer silica fume shows a delay in hydration because of not sufficient water content.

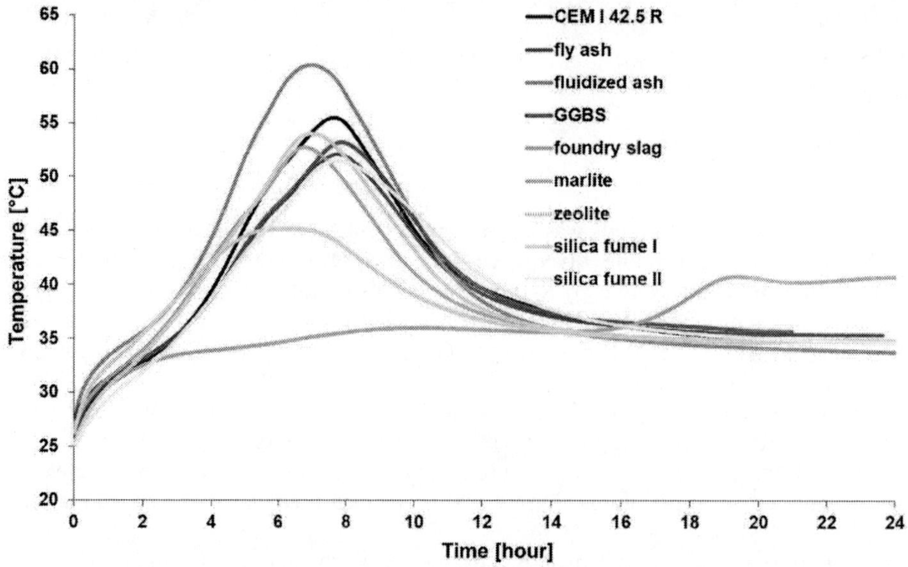


**Figure 1.** Hydration process of cement blends containing industrial and natural raw materials.

Figure 2 shows results of the blended cement setting process observed by means of the Tussenbrock method. The setting process of cement blends with fluidized ash, marlite, zeolite and silica fume is faster in comparison with the reference sample. This fact is in accordance with results of hydration properties mentioned above. The longest setting process was recorded for the blend with foundry slag. No signs of the false setting were recorded among prepared cement blends.

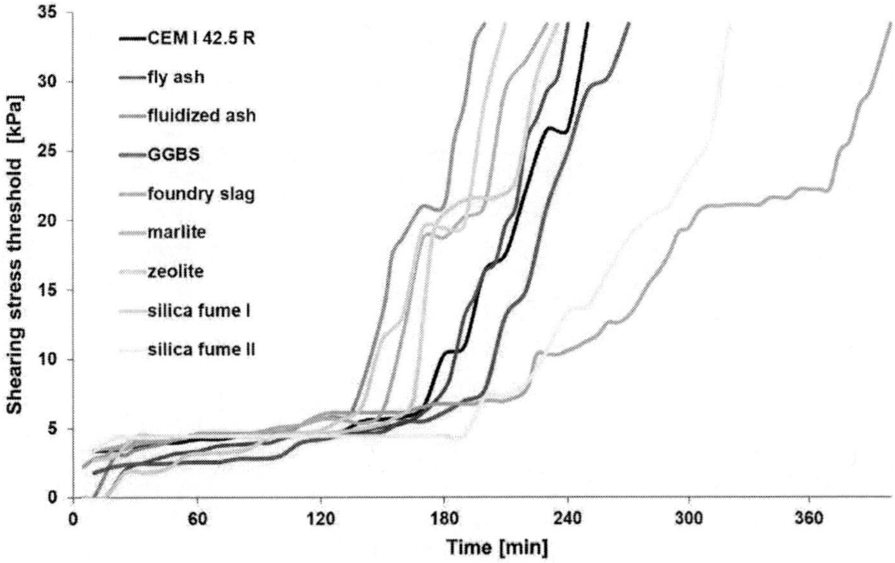


**Figure 2.** Setting process of cement blends containing industrial and natural raw materials.

The compressive strength data for 2, 7, 28, 56 and 90 days of hydration for blended cements and reference cement are plotted in Figure 3. All blended cements exhibit lower early strengths after 2 and 7 days. The reduction in early strengths was considerable especially in the case of foundry slag and silica fume II samples. After 28 days of aging, blends with fluidized ash and zeolite possessed a little higher compressive strength than reference CEM I. Admixtures as fly ash, silica fumes show the positive effect on the cement hydration process (good activity) after 56 days of aging. Finally, after 90 days also sample with fly ash shows comparable strengths. The lowest values of strengths were recorded for cement blend with foundry slag. By the amount of constituent in prepared cements, this cement blend can be categorized as a type CEM II according to EN 197-1. Based on the strength development of cements with fly ash, fluidized ash and silica fume I, these cements remain in strength class 42.5R as reference CEM I. Cement containing granulated blast furnace slag, foundry slag, marlite, zeolite and silica fume II belong to strength class 42.5N. From this point of view, the worst cement with foundry slag is better than for example CEM II prepared from natural pozzolan Turkish trachyandesite that belongs to strength class 32.5R [12].

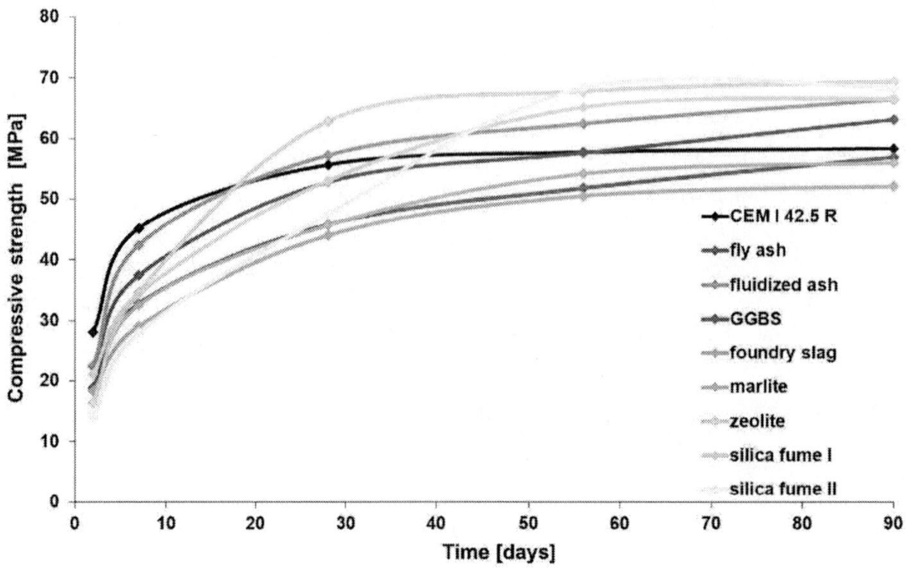


**Figure 3.** Compressive strengths of cement blends containing industrial and natural raw materials.

*3.2. Hydration products*

Four basic endothermal processes were observed in thermograms of hydrated blended cement samples. The first endotherm with minimum at 150–180°C covers several consequential and overlaying processes which include physically bonded water release and calciumhydrosilcates, ettringite or gypsum dehydration. The smaller and the least endotherm follow with the minimum at 390–430°C, which belongs to dehydration of calciumhydroaluminates. Here the $3CaO.Al_2O_3.6H_2O$ phase was selected for quantity evaluation. The next significant endotherm around 540 °C is related to portlandite $Ca(OH)_2$ dehydroxylation. The last endotherm with the minimum at 800–900°C defines the calcite $CaCO_3$ decarbonatation. The quantitative evaluation of hydration products is given in Table 2.

**Table 2.** Quantitative evaluation of cement pastes, water-cured for 28 days, by DTA in wt. % (*overlapping of processes, ** Loss on ignition).

| Cement blend with: | Hydrate $H_2O$ | $C_3AH_6$ | $Ca(OH)_2$ | $CaCO_3$ | LOI** |
|---|---|---|---|---|---|
| CEM I 42.5R | 10.1 | 4.9 | 12.6 | 12.5 | 20.0 |
| Fly ash | 10.0 | 5.0 | 2.7* | 14.9* | 18.6 |
| Fluidized ash | 9.8 | 3.4 | 7.5 | 10.2 | 17.2 |
| GGBS | 11.5 | 5.2 | 8.8 | 10.6 | 19.8 |
| Foundry slag | 10.2 | 4.3 | 10.5 | 10.8 | 18.8 |
| Marlite | 11.5 | 4.3 | 6.6* | 19.1 | 22.8 |
| Zeolite | 13.8 | 4.7 | 8.1 | 9.5 | 21.3 |
| Silica fume I | 12.3 | 5.0 | 4.7* | 12.0 | 20.0 |
| Silica fume II | 16.9 | 5.7* | 2.4* | 10.2* | 23.5 |

## 4. Conclusions

The results show that the hydration reactions within observed blended cements proceed in various ways and do not necessarily indicate the final strengths. The phase composition and microstructure of raw materials could be the reason for this phenomenon. The most significant factors are fineness of raw materials and content of amorphous phase, in case of fluidized ash also the amount of free CaO.

Compared to the reference sample, the best values of long-term compressive strengths along with comparable hydration process were obtained in cement blends with silica fume II and zeolite as admixtures.

Zeolite decreases the hydration heat development, keeping the compressive strengths still quite high due to its pozzolan character and it seems to be a suitable admixture in the amount of 25 wt.% for a blended cement production. It is necessary to pay attention to a possible increased mixing water demand. The mixing water demand can differ from type to type and may be the cause of lower short-term strength rate.

Samples with fly ash and fluidized ash showed increased heat development during first 24 hours, and reached similar, relatively high long-term strengths.

Strong delay of initial setting and low hydration heat, yet satisfactory long-term strength can be observed in the sample containing foundry slag.

**Acknowledgements**
This paper was elaborated with the Institutional support for long-term development of research organizations by the Ministry of Industry and Trade of the Czech Republic.

**References**

[1] *Online* 2018 http://www.wbcsdcement.org/index.php/key-issues/climate-protection
[2] Li B, Liang W and He Z 2002 *Cem. Concr. Res.* **32** 1341
[3] Fu X, Wang Y, Huang S, Hou X and Hou W 2003 *Cem. Concr. Res.* **33** 851
[4] Uzal B and Turanli L 2012 *Cem. Concr. Comp.* **34** 101
[5] Uzal B and Turanli L 2003 *Cem. Concr. Res.* **33** 1777
[6] Kastis D, Kakali G, Tsivilis S and Stamatakis M G 2006 *Cem. Concr. Res.* **36** 1821
[7] Staněk T 2016 *Mater. Sci. Forum* **865** 1
[8] Arikan M, Sobolev K, Ertűn T and Yeginobali A 2009 *Constr. Build. Mater.* **23** 62
[9] Brandštetr J, Polcer J, Krátký J. Holešinský R and Havlica J 2001 *Cem. Concr. Res.* **31** 941
[10] Šiler P, Bayer P, Sehnal T, Kolářová I, Opravil T and Šoukal F 2015 *Cons. Build. Mat.* **78** 181
[11] Šiler P, Krátký J, Kolářová I, Havlica J and Brandštetr J 2013 *Chem. Pap.* **67** 213
[12] Alp I, Deveci H, Süngün Y H, Oilmaz A O, Kesimal A and Yilmaz E 2009 *Iran. J. Sci. Tech. Trans. B Eng.* **33** B4 291

IOP Conf. Series: Materials Science and Engineering **379** (2018) 012011 doi:10.1088/1757-899X/379/1/012011

# Rheological and mechanical properties of alkali-activated brick powder based pastes: effect of amount of alkali activator

**M Vyšvařil[1], E Vejmelková[2] and P Rovnaníková[1]**

[1] Faculty of Civil Engineering, Brno University of Technology, Veveří 331/95, Brno 602 00, Czech Republic

[2] Czech Technical University in Prague, Faculty of Civil Engineering, Thákurova 7, 166 29 Praha 6, Czech Republic

E-mail: vysvaril.m@fce.vutbr.cz

**Abstract.** The rheological behaviour and mechanical properties of alkali-activated materials prepared by activation of a brick powder by alkaline solution (alkali + water glass) is presented in this paper. The effect of quantity of alkaline activator on the rheological and mechanical properties of brick pastes prepared from two types of brick dust was monitored. The rheological investigation included the characterization of flow properties and viscoelastic properties of the fresh mixtures. Compression and bending strengths were determined at 7, 28, 56 and 90 days. These materials have a low yield stress and a high viscosity strongly influenced by the amount of alkali activator. The fresh brick pastes become liquid more easily with increasing amount of alkali activator. The rising amount of alkali activator leads to a decrease of the material cohesion. The results of rheological and mechanical properties investigation showed considerable differences in behaviour of particular types of brick dust.

## 1. Introduction

Recently, the aspiration to exploit the waste brick dust as a precursor for alkali-activated materials has been developed [1–6]. Brick dust as a waste material is available from the development of building elements for the construction of low-energy houses, or from a demolition waste. Bricks waste is usually rich of quartz and fired clays, consisting in dehydrated alumino-silicates not only in crystalline phase but also in amorphous state which is important for alkaline reaction. After an alkaline activation of aluminosilicate precursors, partially or fully amorphous inorganic polymers are formed with Si–O–Al bonds arranged through $SiO_4$ and $AlO_4$ tetrahedra into 3D-network (geopolymerization reaction). As activators, a wide range of alkaline solutions is applied: NaOH, $Na_2CO_3$, sodium silicate (water glass), sodium aluminate, or their potassic or lithic equivalents. The elevated temperature promotes the alkali reaction, mostly used is temperature of 60 to 90°C, although even at ambient temperature an aluminosilicate gel is formed by brick powder activation [3].

The effect of using different amounts of sodium silicate for brick powder activation has been superficially monitored [3]. Pastes obtained from brick powder activation at room temperature, cured for 7 days, exhibited bulk density of 1.5–2.0 $g\cdot cm^{-3}$ and water absorption of 20–27%; high content of sodium silicate solution in the mix design promoted more compact microstructures.

In recent years, an academic research on the binders has considerably extended the understanding of the geopolymerization reaction and rheological measurements greatly contributed to it. Furthermore, most of them are carried out on metakaolin, fly ashes or slags based geopolymers [7–12].

In this study, the influence of quantity of alkaline activator on the rheological and mechanical properties of brick pastes prepared from two types of brick dust (differing by content of Ca and amorphous phase) was monitored. From the point of view of the rheology, it can assume a similar behaviour of the brick powder paste system and the metakaolin paste system. Both of them are suspensions of particles (brick powder, metakaolin) in a continuous fluid (sodium silicate solution) and their rheological behaviour can be comprehended by the physics of concentrated suspensions of rigid particles and therefore the results obtained in this study are compared with the results of similar metakaolin based geopolymers.

## 2. Materials

Two types of waste material from skiving of heat-insulating bricks from the HELUZ Brickworks factory, v.o.s., plant in Libochovice (L) and Hevlín (H) were the basic raw materials for alkali-activated pastes. The powder were used as-received, just the particles larger than 1 mm were removed by sieving. The chemical composition of the brick powder is shown in Table 1. The phase composition of the brick powder was studied via XRD using Rietveld analysis (Table 2) and the particle size distribution was determined using a laser diffraction analyzer (Figure 1). Fundamental physical parameters of the brick powders are summarized in Figure 1 as well.

The brick pastes were prepared by homogenization of brick powder with alkaline activator prepared from sodium silicate solution (water glass, $SiO_2/Na_2O$ = 1.6, Vodní sklo, a.s., Czech Republic), sodium hydroxide p. a. (Penta Chemicals, a.s.), water and silicone defoamer (Lukosan S, Lučební závody a.s. Kolín). The $SiO_2/Na_2O$ ratio of prepared alkali activator was 1.0 and it was added in three different dosages (Table 3). For decreasing amount of sodium silicate solution a higher content of water in mixture was necessary to reach an acceptable consistency.

Fresh mixtures were cast into prismatic moulds of size 20 × 20 × 100 mm. Standard conditions of sample storage were 21 ± 2°C and relative humidity of 50 ± 5%. After 7, 28, 56 and 90 days the flexural and compressive strengths were determined.

**Table 1**. Chemical composition of brick powders examined by XRF spectroscopy (mass %).

| | $SiO_2$ | $Al_2O_3$ | $Fe_2O_3$ | CaO | MgO | $K_2O$ | $Na_2O$ | $P_2O_5$ | $TiO_2$ | $SO_3$ | sum |
|---|---|---|---|---|---|---|---|---|---|---|---|
| L | 54.2 | 17.2 | 4.4 | 14.1 | 1.7 | 2.8 | 0.3 | 0.2 | 1.0 | 1.1 | 97.0 |
| H | 55.5 | 17.0 | 5.8 | 10.5 | 2.4 | 2.8 | 0.7 | 0.2 | 0.8 | 1.9 | 97.6 |

**Table 2**. Phase composition of brick powders (mass %).

| Mineral | Formula | L | H |
|---|---|---|---|
| Quartz | $SiO_2$ | 20.0 | 26.2 |
| Hematite | $Fe_2O_3$ | 2.0 | 2.3 |
| Albite | $NaAlSi_3O_8$ | 3.5 | 13.0 |
| Microcline | $KAlSi_3O_8$ | 4.2 | 3.6 |
| Orthoclase | $KAlSi_3O_8$ | 6.5 | 3.5 |
| Muscovite | $KAl_2(AlSi_3O_{10})(OH)_2$ | 4.4 | 12.5 |
| Illite | $K_{0.65}Al_2(Al_{0.65}Si_{3.35}O_{10})(OH)_2$ | 3.1 | 3.8 |
| Diopside | $CaMgSi_2O_6$ | 1.6 | 4.4 |
| Akermanite | $Ca_2MgSi_2O_7$ | 2.9 | 2.8 |
| Mullite | $Al_6Si_2O_{13}$ | 2.9 | – |
| Anhydrite | $CaSO_4$ | 2.0 | – |
| Amorphous | – | 47.0 | 27.8 |

IOP Conf. Series: Materials Science and Engineering **379** (2018) 012011 doi:10.1088/1757-899X/379/1/012011

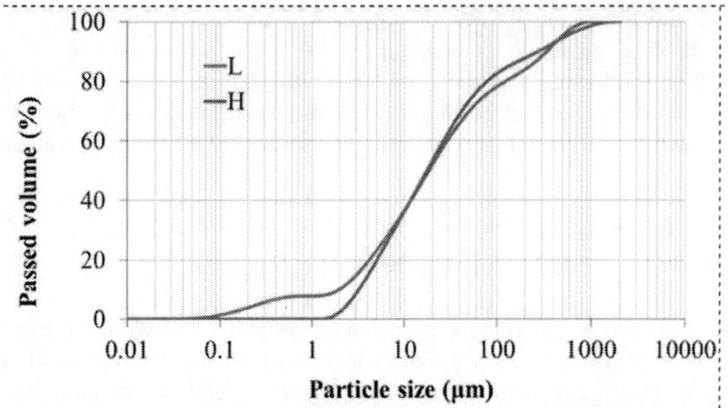


| | L | H |
|---|---|---|
| ρ (g/cm$^3$) | 2.628 | 2.654 |
| $d_{50}$ (µm) | 41 | 42 |
| $S_{g\ Blaine}$ (m$^2$/g) | 0.2 | 0.2 |
| $S_{g\ BET}$ (m$^2$/g) | 3.7 | 3.9 |
| $V_{BJH}$ (cm$^3$/g) | 0.049 | 0.030 |
| $P_{BJH}$ (%) | 13 | 7.9 |

**Figure 1**. Particle size distribution and fundamental physical parameters of brick powders.

## 3. Experimental procedure

### *3.1. Rheological properties*

Rheological tests were carried out on the hybrid rheometer Discovery HR-1 (TA Instruments) in the Peltier Concentric Cylinder system with a DIN rotor and TRIOS 4.0.2.30774 software was used for a data evaluation. All the measurements have been done at 25°C.

*3.1.1. Flow properties.* The brick suspension was introduced into the measurement system at the end of the mixing cycle. The rheological measurements were started after a period of rest of 60 s. The testing routine comprised a shear rate increase (from 0.1 to 100 s$^{-1}$) applied through 30 steps with 15 s of measuring time at each shear rate followed by a decrease of shear rate on the same conditions. The results were expressed as shear rate vs. shear stress (flow curves) and the Herschel-Bulkley model (1) was applied to descending branches of flow curves to fit the experimental data and used to describe the pastes rheological behaviour [13]:

$$\tau = \tau_0 + k\dot{\gamma}^n \tag{1}$$

where $\tau$ is the shear stress, $\dot{\gamma}$ is the shear rate, $\tau_0$ corresponds to the yield stress, $k$ is the consistency coefficient and $n$ is the fluidity index which characterizes shear-thinning ($n < 1$) or shear-thickening ($n > 1$) behaviour of a material.

*3.1.2. Viscoelastic properties.* Time evolution of the viscoelastic properties of pastes were evaluated by small amplitude (0.01%) oscillation tests measuring dynamic moduli, $G'$ and $G''$, represent the elastic and viscous behaviour of a material. If the elastic behaviour dominates the viscous one ($G' > G''$), the sample exhibits a certain rigidity. Conversely, if $G'' > G'$, the sample shows the character of a liquid without a consistent chemical or physical network-of-forces [13]. Since one of the purposes of the this study is to determine the variation over time of the viscoelastic parameters according to the geopolymer composition, it is important to determine the linear viscoelasticity region (LVR) in order to prevent structural decomposition of the geopolymerized network. The LVR is the region in which measurements are nondestructive to the microstructure of a material and it ends at a critical strain $\gamma_c$. The strain of 0.01% used in the oscillation tests is lower than the critical strain for all brick pastes studied. The rheological characterization was completed by frequency sweep tests with increasing the frequency from 0.1 Hz to 10 Hz. Experimental data were reported in terms of complex modulus, $G^*$, and loss tangent, tan($\delta$), defined as follows (2–3):

$$G^* = \sqrt{(G')^2 + (G'')^2} \tag{2}$$

$$\tan(\delta) = G''/G' \tag{3}$$

**Table 3**. Composition of mixtures.

| Mixture | Brick powder (g) | Sodium silicate, $M_s = 1.0$ (g) | $H_2O$ (ml) | Lukosan S 1% (ml) |
|---|---|---|---|---|
| L-1 | 100 | 65.5 | – | 2 |
| L-2 | 100 | 57.3 | 5 | 2 |
| L-3 | 100 | 49.1 | 10 | 2 |
| H-1 | 100 | 65.5 | – | 2 |
| H-2 | 100 | 57.3 | 5 | 2 |
| H-3 | 100 | 49.1 | 10 | 2 |

## 4. Results and discussion

### *4.1. Rheological properties*

*4.1.1. Flow properties.* Figure 2 illustrates the descending branches of flow curves of the fresh brick powder suspensions. The shape of curves indicates nearly Newtonian behaviour of brick pastes as confirmed by the flow index values, *n*, varying around 1 (Figure 3). It is apparent that the yield stress of pastes increase with decreasing dosage of sodium silicate solution (Figure 3). This is expected because the lower silicate solution content in paste limits the solubility of the precursor particles and the paste is less fluid. Likewise, low yield stresses has been observed at metakaolin based geopolymers [10, 11, 14]. Lower value of the yield stress for sample H-1 (versus L-1) may be caused by lower content of calcium in this brick powder. The presence of divalent cations can result in attractive ion correlation forces (as in C-S-H particles). Brick powder H responded to the decreasing amount of activator by a significant increase in the yield stress.

The consistency coefficient (which can be compared to a plastic viscosity of the material) increased with decreasing dose of alkali activator (Figure 3) which was caused by decrescent amount of mixing solution. H powder exhibited a higher consistency coefficient. This is due to the different content of the amorphous phase in the precursors, where mixtures with higher crystalline content are more reluctant to flow. The consistency coefficients of the pastes were higher than of cement pastes or metakaolin-based geopolymer pastes despite of bigger particle sizes of brick powders than of common cement or metakaolin [14, 15]. This is due to the viscosity of sodium silicate solution which is 10 to 100 times higher than the viscosity of water. As a consequence the viscosity of these fresh pastes is expected to be 10 to 100 times higher than of standard cement pastes.

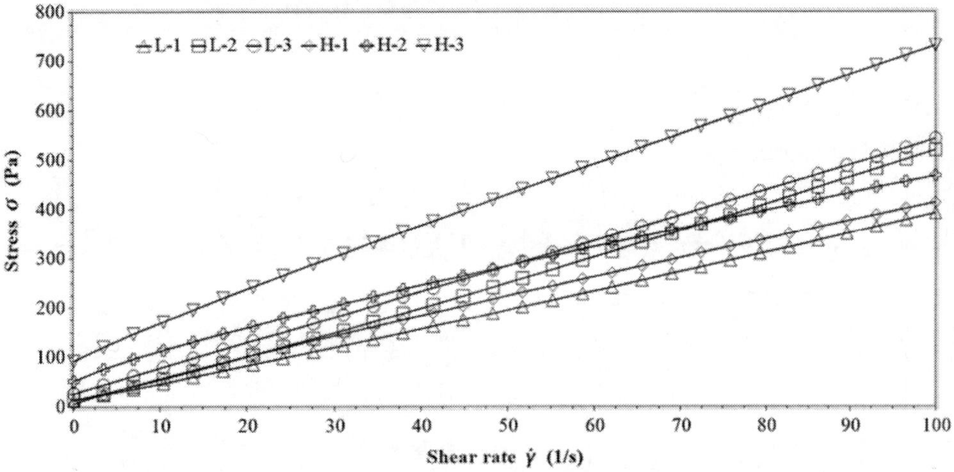


**Figure 2**. Descending branches of flow curves of brick powder suspensions.

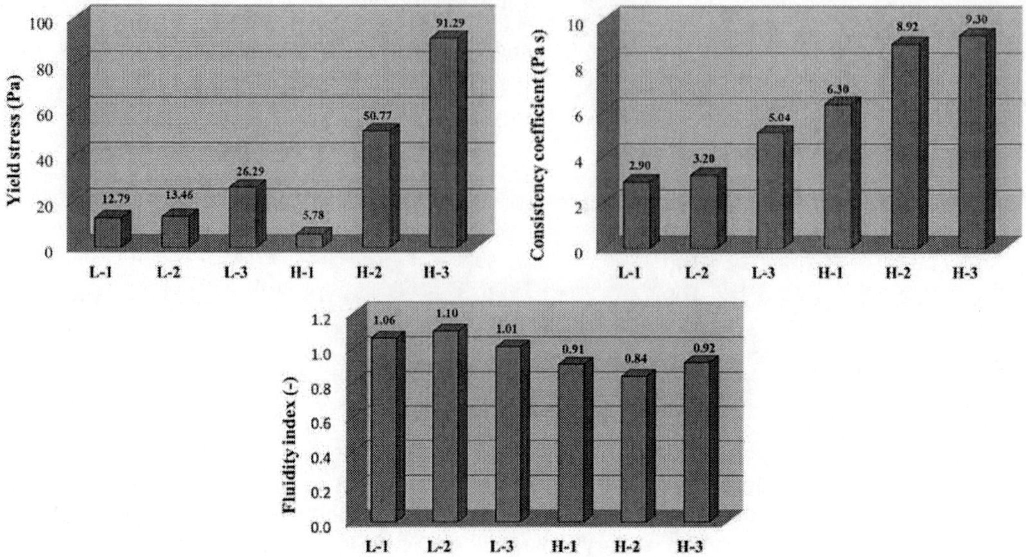


**Figure 3**. Comparison of yield stresses ($\tau_0$), consistency coefficients ($k$) and fluidity indexes ($n$) of brick powder suspensions.

*4.1.2. Viscoelastic properties*. Figure 4 shows the time evolution of the viscoelastic parameters (G′ and G″) during a small amplitude (0.01%) oscillation test. Geopolymer gel formation was monitored by a crossover point G′ = G′′ (gel point). Before the gel point, samples show the behaviour of a liquid with G′′ > G′, after the gel point, the samples show gel-like or solid behaviour with the existence of interactions between constituents [13]. The results showed that the geopolymer gel formation is faster in the brick powder paste with higher content of calcium and amorphous phase (L). It confirmed the enhanced reactivity of amorphous precursors for alkali activated materials. Decreasing amount of alkali activator in the mixtures leaded to an acceleration of the gel formation probably due to stronger interactions between particles in more concentrated aluminosilicate solution. L-3 mixture exhibited the most rigid structure.

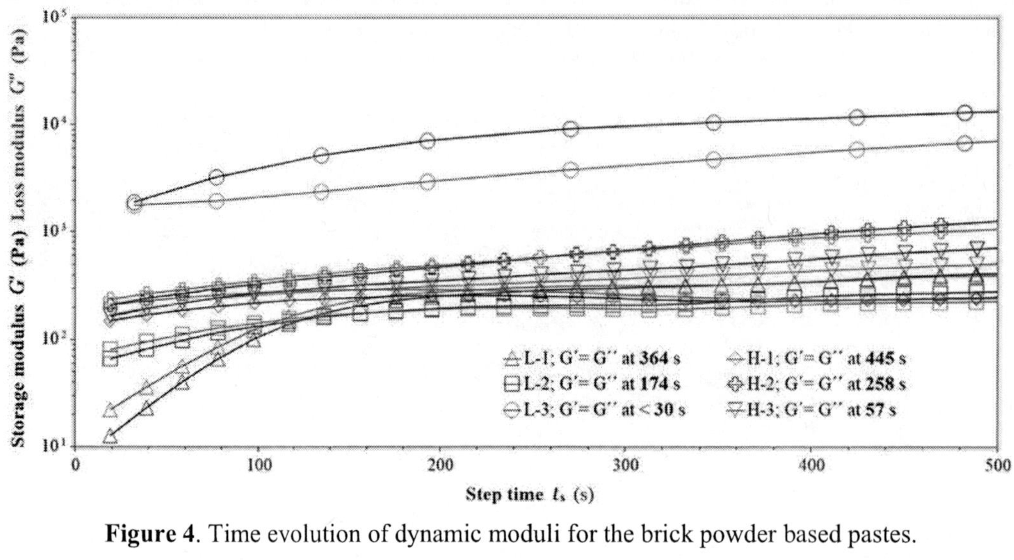


**Figure 4**. Time evolution of dynamic moduli for the brick powder based pastes.

The results of frequency sweep tests proved the connection between the amount of alkali activator and the stiffness of brick pastes; the ratio between dissipation and elasticity increased (growing $G^*$) with decreasing amount of alkali activator and the pastes became more rigid and more viscous (Figure 5). Since the tan($\delta$) of the pastes vary around 1, the brick paste showed the viscoelastic behaviour with almost the same ratio of the viscous and elastic portions. This does not apply to mixtures L-3 and H-3 which exhibited lower tan($\delta$) values of about 0.5, indicating a higher proportion of the elastic component ($G'$), suggesting that these pastes have a stronger structure more resistant to external interference.

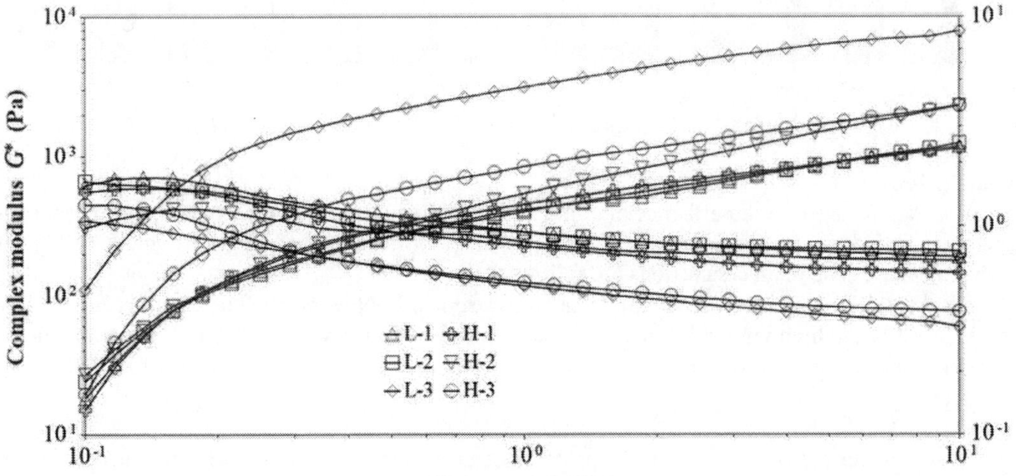


**Figure 5**. Evolution of G* and tan(δ) as a function of frequency for the brick pastes.

*4.2. Mechanical properties*

Obviously, because of binder hardening, the mechanical strengths of pastes increased with curing time (Figure 6). It can be stated that the pastes made of the brick powder with higher content of calcium and amorphous phase (L) have better strength properties than the other pastes. This again confirms the advantage of using a precursor with high content of amorphous phase for the production of alkali-activated materials. H mixtures showed a gradual increase in strengths with decreasing amount of alkali activator. A different trend can be observed for L mixtures where the best strength results were obtained for L-2 sample. When comparing the strength characteristics of the pastes in a fresh and hardened state, it can be concluded that already in the fresh state, the L brick powder has better properties than the H brick powder. However, the sample with the strongest structure in the fresh state, did not achieve the highest strength after hardening.

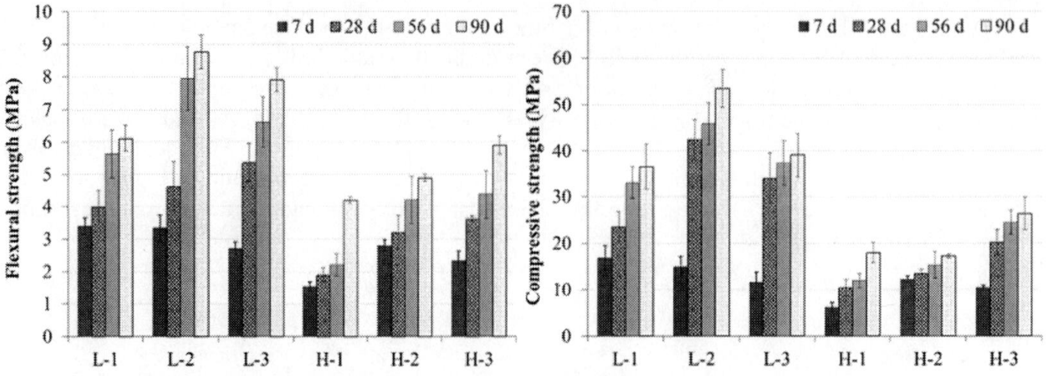


**Figure 6**. Flexural and compressive strengths of brick powder based pastes.

## 5. Conclusions

This paper has focused on the effect of alkaline activator quantity on the rheological and mechanical properties of brick pastes prepared from two types of brick powder differing by content of Ca and amorphous phase. The main conclusions are follow:

- Brick powder based pastes prepared in this study are nearly Newtonian fluids with relatively low yield stresses and high viscosities that increase with decreasing dosage of sodium silicate solution.
- Increased values of yield stress and consistency coefficient of H brick powder are caused by higher content of crystalline phase in this precursor, since mixtures with higher crystalline content are more reluctant to flow.
- The geopolymer gel formation in brick paste is faster if higher content of calcium and amorphous phase is present in a brick powder. Decreasing amount of alkali activator in the mixtures leads to an acceleration of the gel formation.
- The pastes made of the brick powder with higher content of calcium and amorphous phase (L) have better strength properties than the H pastes. This confirms the advantage of using brick powder precursors with high content of calcium and amorphous phase for the production of alkali-activated materials.

**Acknowledgments**

This work has been financially supported by the GA ČR research project 16-02862S and by Ministry of Education, Youth and Sports under the „National Sustainability Programme I“ – the project No. LO1408 „AdMaS UP – Advanced Materials, Structures and Technologies“.

**References**

[1] Reig L, Tashima M M, Soriano L, Borrachero M V, Monzó J and Payá J 2013 *Waste Biomass. Valoriz.* **4** pp 729–736
[2] Reig L, Tashima M M, Borrachero M V, Monzó J, Cheeseman C R and Paya J 2013 *Constr. Build. Mater.* **43** pp 98–106
[3] Bignozzi M C, Fusco O, Fregni A, Guardigli L and Gulli R 2014 *Adv. Sci. Tech.* **92** pp 26–31
[4] Sassoni E, Pahlavan P, Franzoni E and Bignozzi M C 2016 *Ceram. Int.* **42** pp 14685–14694
[5] Rovnaník P, Řezník B and Rovnaníková P 2016 *Procedia Eng.* **151** pp 108–113
[6] Reig L, Sanz M A, Borrachero M V, Monzó J, Soriano L and Payá J 2017 *Ceram. Int.* **43** pp 13622–13634
[7] Chindaprasirt P, Chareerat T and Sirivivatnanon V 2007 *Cem. Concr. Compos.* **29** pp 224–229
[8] Palacios P, Banfill P F G and Puertas F 2008 *ACI Materials Journal.* **105** pp 140–148
[9] Palomo A, Banfill P F G, Fernández-Jiménez A and Swift D S 2005 *Adv. Cem. Res.* **17** pp 143–151
[10] Poulesquen A, Frizon F and Lambertin D 2011 *J. Non-Cryst. Solids.* **357** pp 3565–3571

[11] Romagnoli M, Leonelli C, Kamse E and Lassinantti Gualtieri M 2012 *Constr. Build. Mater.* **36** pp 251–258
[12] Steins P, Poulesquen A, Diat O and Frizon F 2012 *Langmuir* **28** pp 8502–8510
[13] Mezger T G 2017 *The Rheology Handbook* (Hanover: Vincentz Network)
[14] Favier A, Hot J, Habert G, Roussel N and d'Espinose de Lacailleriecd J-B 2014 *Soft Matter.* **10** pp 1134–1141
[15] Aboulayt A, Riahi M, Anis S, Ouazzani Touhami M and Moussa R 2014 *IJIAS* **7** pp 1170–1177

IOP Conf. Series: Materials Science and Engineering **379** (2018) 012012 doi:10.1088/1757-899X/379/1/012012

# Potential use of blast furnace slag for filtration membranes preparation: A pilot study

**V Bílek Jr.[1], P Bulejko[2], P Kejík[2], J Hajzler[1], J Másilko[1], J Bednárek[1], J Tkacz[1] and L Kalina[1]**

[1] Brno University of Technology, Faculty of Chemistry, Materials Research Centre, Purkyňova 118, 612 00 Brno, Czech Republic
[2] Brno University of Technology, Faculty of Mechanical Engineering, Heat Transfer and Fluid Flow Laboratory, Technická 2896/2, 616 69 Brno, Czech Republic

Email: bilek@fch.vut.cz

**Abstract**. This paper investigates the possibilities of the use of sodium hydroxide activated blast furnace slag for preparation of filtration barriers. The motivation for this research is searching for cheaper and more eco-friendly alternative to sintering, which is commonly used for filtration membranes preparation. Such an alternative could be alkaline activation, which is able to bind slag particles together. Low amount of activator, low water to slag (w/s) ratio together with pressure compaction was used to obtain microstructure with preserved continuous porosity. To confirm this presumption scanning electron microscopy (SEM) and mercury intrusion porosimetry (MIP) analyses were carried out.

## 1. Introduction

Blast-furnace slag (BFS) is an industrial by-product from production of raw iron. It is used in blended cements and has been a subject of research focused on alkali-activated materials with an application in civil engineering [1, 2]. A lot of attempts to utilize secondary materials from energy (fly ash) and iron production (BFS) industry in advanced applications have been done. Several chemical engineering applications such as adsorption, membrane separations or hot gas cleaning have recently been of growing interest [3–6].

Membrane separations are important chemical engineering operations involving membrane, an active layer/barrier with selective properties. They have widely been used in air/wastewater treatment in tubular [7], flat [8] or hollow fibre geometry [9]. The membrane materials including polymers, ceramics or metals are chosen depending on the process conditions. Ceramic membranes are used mainly for harsh conditions where high temperatures and chemically aggressive environment is expected. Most of them are prepared of expensive materials via several steps one of which includes sintering at temperature around 1,000°C. Some of them may follow a route via hydrothermal reaction [10].

This work is an attempt to prepare a filtration membrane based on alkali-activated BFS without the additional sintering or hydrothermal treatment and thus via easier, cheaper and more eco-friendly process. Two different membranes differing in the initial BFS particle-size distribution were prepared and characterized in terms of pore-size distribution and observed using SEM.

## 2. Materials and methods

### *2.1. Materials and their characteristics*

Filtration membranes were prepared via alkaline activation of ground granulated blast furnace slag (GGBFS) from Czech production (Kotouč Štramberk, Ltd.) by sodium hydroxide solution. Two series of specimens with different particle size distribution of slag were prepared. The first one contained GGBFS without any modifications, i.e. in the same form as it was obtained from supplier, while for the second series one narrower and coarser GGBFS fraction separated using an Alpine Windsichter ATP50 cyclone classifier was chosen.

Particle size distribution and morphology (Figure 1) of both types of GGBFS were determined using a Sympatec Helos KR laser diffraction analyzer and a ZEISS EVO LS 10 scanning electron microscope, respectively. Blaine fineness of unclassified GGBFS was about 400 $m^2/kg$ while its selected fraction had Blaine fineness of about 250 $m^2/kg$. The phase composition was characterized using the X-ray diffraction method (XRD). Both types of GGBFS had the same phase composition containing about 88% of amorphous phase, 9% of akermanite and 3% of calcite. Activating solution consisted of 50% sodium hydroxide solution and demineralized water. Water to slag ratio was 0.095 including water from NaOH and $Na_2O$ to slag ratio was 0.02. Such composition of the mixture resulted in a consistency like that of a wet powder.

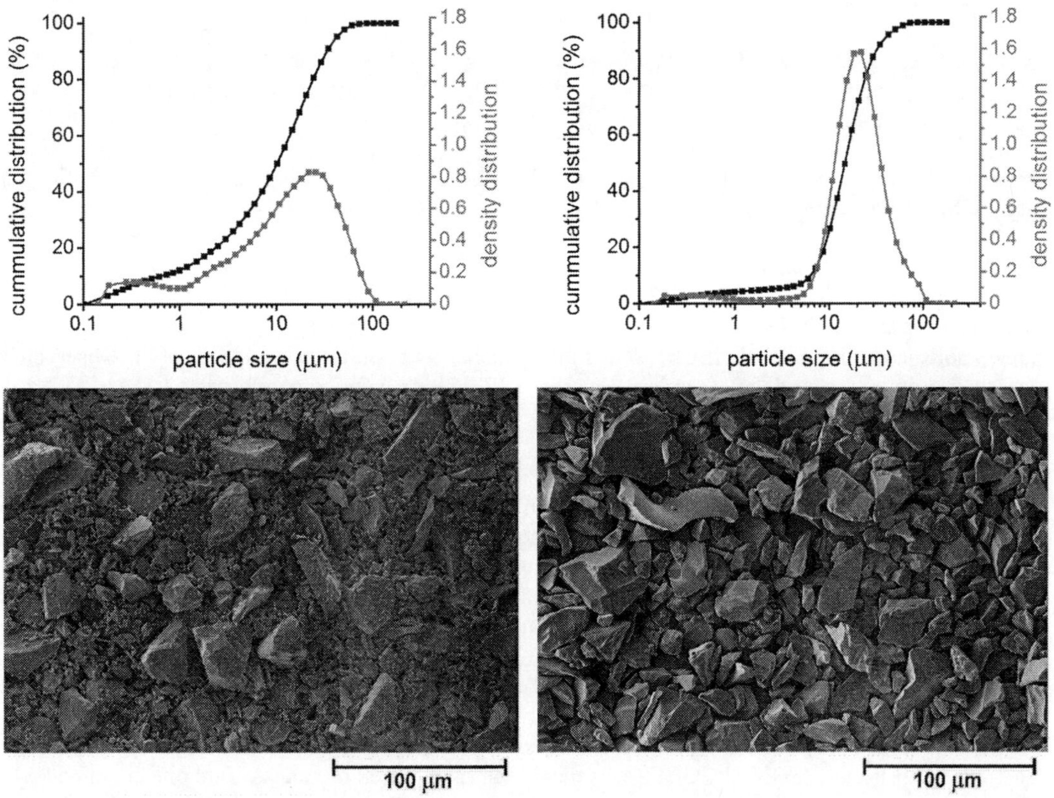


**Figure 1.** Particle size distribution and particle shape of the used unclassified (left) and classified (right) GGBFS; magnitude 1,000×.

### *2.2. Specimen preparation*
Mixture of GGBFS and activating solution was homogenized using a Thinky ARV-310 planetary centrifugal vacuum mixer for 30 seconds at a frequency of 2,000 rpm. Homogenized mixture was cast into round steel molds with a diameter of 25 mm and pressed at a pressure of 40 MPa. Thickness of prepared samples was 2 mm. The fresh samples were then moist cured at 60°C for 24 hours and dried at laboratory conditions. Examples of the final membranes are given in Figure 2.

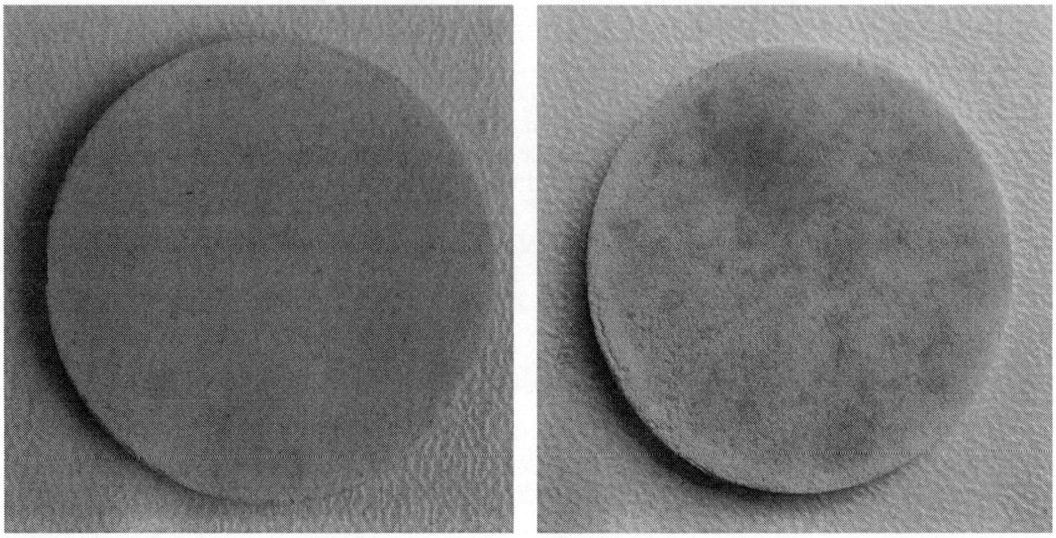

**Figure 2.** Examples of the membranes prepared from GGBFS before classification (left) and after that (right). Membrane diameter is 25 mm.

### *2.3. Scanning electron microscopy*
Microstructure of the prepared membranes was investigated using a ZEISS EVO LS 10 scanning electron microscope. Before the observation cross-section of a specimen was smoothed by a sand paper, polished using a JEOL IB09010CP ionic polisher and sputter coated with gold. Observation was carried out in the mode of secondary electrons (SE) at an accelerating voltage of 10 kV.

### *2.4. Mercury intrusion porosimetry*
Pore structure of the prepared membranes was investigated using MIP. For these measurements specimens were abraded by sand paper to obtain small bars with a cross-section of approximately 5 × 2 mm and a length of 15 mm. Intrusion of mercury into the samples was performed at different pressures up to 225 MPa. Surface tension of mercury was assumed to be 480 mN/m and wetting angle 140° and thus according to Washburn equation the finest pores which were possible to detect had a diameter of 6.5 nm. Four measurements of each mixture composition were done.

## 3. Results and their discussion
Microstructure overview of the prepared membranes is given in Figure 3. It can be seen that membranes with both GGBFS types were successfully prepared regarding their preserved porosity. The reason for use of classified GGBFS was that membranes based on unclassified one contained a lot of places where potential larger pores were stuck by fine GGBFS particles and thus number of desired pores decreased. Moreover, microstructure of the membrane prepared from unclassified GGBFS is less homogenous compared to classified because many areas with both very low porosity and higher porosity can be found.

On the other hand increase in porosity is generally connected with decrease in strength and despite strength of membranes was not determined those prepared from classified GGBFS were more prone to

IOP Conf. Series: Materials Science and Engineering **379** (2018) 012012 doi:10.1088/1757-899X/379/1/012012

cracking during the sample preparation and easier to abrade. Therefore, further study will investigate the effect of increasing activator dose to prepare stronger membranes from various fractions of GGBFS.

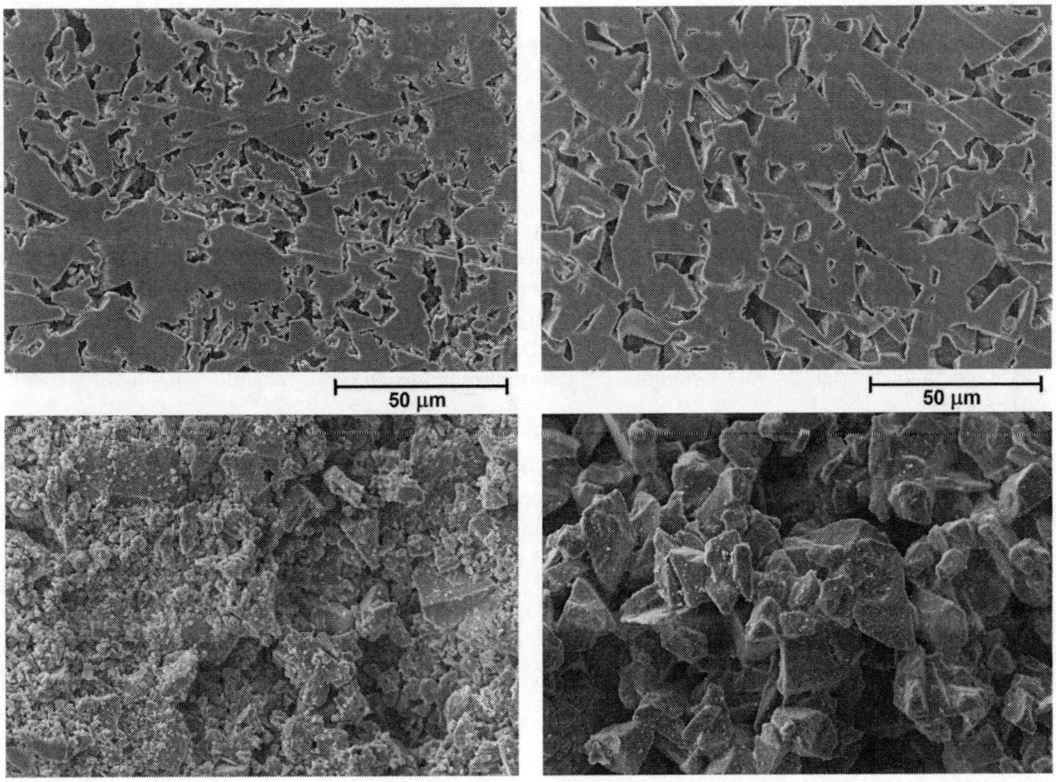


**Figure 3.** Microstructure of polished (up) and unpolished (down) parts of the membrane from unclassified (left) and classified (right) GGBFS; magnitude 2,000×.

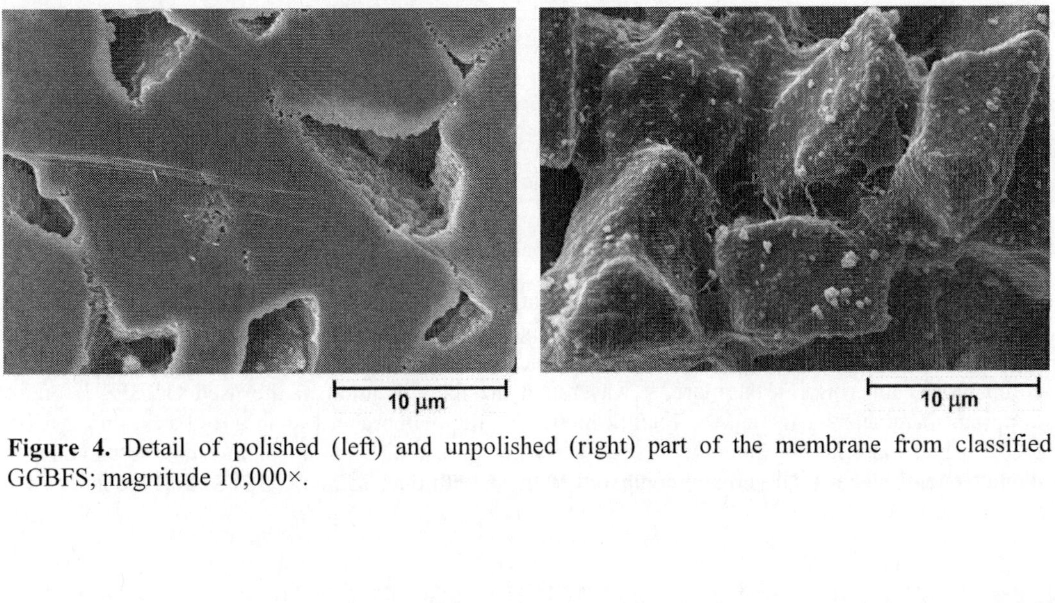


**Figure 4.** Detail of polished (left) and unpolished (right) part of the membrane from classified GGBFS; magnitude 10,000×.

The main binding phase formed during alkaline activation of slag is calcium-aluminium-silicate-hydrate (C-A-S-H) [11, 12]. Thank to this phase, GGBFS grains are connected together and form filtration membrane (Figure 4). Its amount is sufficiently high to facilitate compactness of the membrane but low to determine its composition, e.g. using energy dispersive analysis.

Impact of GGBFS classification on pore structure of the prepared membranes studied using MIP is shown in Figure 5. Both unclassified and classified GGBFS had significant intrusions in the range of pore diameter between 0.2 and 5–10 μm. While the curves of unclassified GGBFS were flat with very diffuse maximum in the range of 0.6–3 μm, for classified GGBFS new very intense sharp peak appeared, whose maximum corresponded to the pore diameter of 6 μm. Classification of GGBFS also resulted in significant total porosity increase from 23.8 ± 0.4% for unclassified slag to 37.8 ± 2.0% for the selected coarser fraction.

The main reason for such observations is that unclassified GGBFS contain particles of various sizes and thus finer particles can fill the free space between the coarser ones, which was also confirmed by SEM. General complication of GGBFS for membrane preparation is that its particles are angular. If they are in contact with each other along their edges or walls, continuous porosity can be reduced. More favorable particle shape compared to GGBFS has fly ash which may be an alternative. On the other hand, it contains lower amount of amorphous phase and has variable and inhomogeneous composition.

Due to the lack of results in literature, the determined properties of filtration membranes can hardly be compared to some others. Probably the most similar in terms of the used material and preparation process (but still noticeably different) were membranes prepared by Mohammadi [13]. Their porosity was determined to 45–65% which seems to be higher compared to our membranes from classified GGBFS, but it should be noted that the methods of porosity evaluation were different.

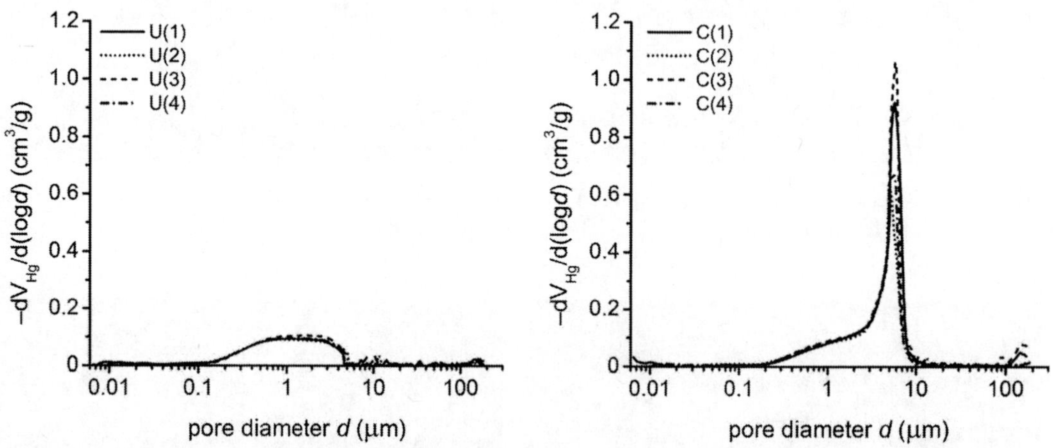


**Figure 5.** Pore structure of the prepared membranes with unclassified (U, left) and classified (C, right) GGBFS as determined using MIP.

## 4. Conclusion

This work investigated the possibilities of preparation of alkali-activated slag systems with preserved porosity which can be potentially used for preparation of filtration membranes. The results confirmed that alkaline activation is a possible way to bind the slag particles together at a low temperature compared to sintering. Furthermore, it was found that classification of the used GGBFS is one of important parameters affecting the final properties of the membrane. Membranes based on classified GGBFS had more homogenous microstructure with narrow distribution of the most abundant pore diameters and higher total porosity compared to those with unclassified GGBFS. Further studies will be focused on the effect of activator dose, different fractions of classified GGBFS and leaching of soluble components on properties of the prepared membranes.

**Acknowledgement**
This outcome was achieved with the financial support of the junior specific research program at the Brno University of Technology, project No. FCH/FSI-J-17-4606, Filtration membranes based on alkali-activated materials: preparation, characterization and properties; with financial support from the National Programme for Sustainability I (Ministry of Education, Youth and Sports) by the project NETME Centre Plus, Project LO1202; and with financial support of the Czech Science Foundation under project No 18-12289Y.

**References**

[1] Provis J L and van Deventer J S J (eds.) 2014 *Alkali Activated Materials: State-of-the-Art Report, RILEM TC 224-AAM* (Dordrecht: Springer Netherlands) 1st ed (doi: 10.1007/978-94-007-7672-2)

[2] Shi C, Krivenko P and Roy D 2006 *Alkali-Activated Cements and Concretes* (New York: Taylor & Francis) 1st ed

[3] Namburath M, Joshi G, Cholemari M, Shet C, Sreekrishnan T R and Veeravalli S 2015 Feasibility Study of Indigenously Developed Fly Ash Membrane in Municipal Wastewater Treatment, *Aquat. Procedia.* **4** pp 1492–99 (doi:10.1016/j.aqpro.2015.02.193)

[4] Lee N K, Khalid H R and Lee H K 2017 Adsorption characteristics of cesium onto mesoporous geopolymers containing nano-crystalline zeolites *Microporous Mesoporous Mater.* **242** pp 238–44 (doi:10.1016/j.micromeso.2017.01.030)

[5] Runtti H, Luukkonen T, Niskanen M, Tuomikoski S, Kangas T, Tynjälä P, Tolonen E T, Sarkkinen M, Kemppianen K, Rämö J and Lassi U 2016 Sulphate removal over barium-modified blast-furnace-slag geopolymer *J. Hazard. Mater.* **317** pp 373–84 (doi:10.1016/j.jhazmat.2016.06.001)

[6] Jo Y M, Hutchison R B and Raper J A 1997 Characterization of ceramic composite membrane filters for hot gas cleaning *Powder Technol.* 91 (1997) pp 55–62 (doi:10.1016/S0032-5910(96)03246-9)

[7] Fang J, Qin G, Wei W and Zhao X 2011 Preparation and characterization of tubular supported ceramic microfiltration membranes from fly ash *Separ. Purif. Technol.* **80** pp 585–91 (doi:10.1016/j.seppur.2011.06.014)

[8] Baraka N E, Saffaj N, Mamouni R, Laknifli A, Younssi S A, Albizane A and El Haddad M 2014 Elaboration of a new flat membrane support from Moroccan clay *Desalin. Water Treat.* **52** pp 1357–61 (doi:10.1080/19443994.2013.797542)

[9] Sverak T, Bulejko P, Ostrezi J, Kristof O, Kalivoda J, Kejik P, Mayerova K and Adamcik M Separation of gaseous air pollutants using membrane contactors *IOP Conf. Ser.: Earth And Environ. Sci.* **92** 012061 (doi:10.1088/1755-1315/92/1/012061)

[10] Azarshab M, Mohammadi F, Maghsoodloorad H and Mohammadi T 2016 Ceramic membrane synthesis based on alkali activated blast furnace slag for separation of water from ethanol *Ceram. Int.* **42** pp 15568–74 (doi:10.1016/j.ceramint.2016.07.005)

[11] Myers R J, Bernal S A, Nicolas R S, Provis J L, Ivanova E and Vilchevskaya E 2013 Generalized structural description of calcium–sodium aluminosilicate hydrate gels: The cross-linked substituted tobermorite model *Langmuir* **29** 17 (doi: 10.1021/la4000473)

[12] Puertas F, Palacios M, Manzano H, Dolado J S, Rico A and Rodríguez J 2011 A model for the C-A-S-H gel formed in alkali-activated slag cements *J. Ceram. Soc.* **31** 12 (doi:10.1016/j.jeurceramsoc.2011.04.036)

[13] Mohammadi F and Mohammadi T 2017 Optimal conditions of porous ceramic membrane synthesis based on alkali activated blast furnace slag using Taguchi method *Ceram. Int.* **43** pp 14369–79 (doi:10.1016/j.ceramint.2016.07.005)

# Effect of waste glass addition on mechanical properties of slag based geopolymers

**M Marcin**[1], **M Sisol**[1] **and I Brezani**[1]

[1] Technical University of Košice, Faculty of Mining, Ecology, Process Control and Geotechnologies, Institute of earth resources, Department of mineral processing, Park Komenského 19, 042 00, Košice, Slovak Republic

Email: michal.marcin@tuke.sk

**Abstract.** Slag is one of the by-products of energy industry which is suitable for secondary industrial processing. Although the slag has been successfully used in industrial production for several decades its use does not achieve the level of its potential. The Slags can be used as alkali activated materials and utilized in synthesis of geopolymers. This work describes an improvement of mechanical properties of alkali activated binders – geopolymers made of blast furnace slag. The effect of addition of waste glass on mechanical properties of geopolymers was examined. The Waste glass was added in a form of broken and crushed glass particles. Samples were tested after 7, 28 and 90 days.

## 1. Introduction

The Ground granulated blast-furnace slag (GGBFS) is a by-product of the steel industry by adding limestone to ore for the purpose of removing non-ferrous contaminants, and it consists of four major chemical components: CaO, $SiO_2$, $Al_2O_3$ and MgO [1]. As one of the main by-products during the process of iron and steel making, the blast furnace slag is discharged in a super-high temperature of 1450–1650°C [2]. These components are highly reactive with alkaline solutions to form alkali-activated GGBS [3].

The slag can be used as an alkali activated material and utilized in synthesis of geopolymers. The term geopolymer was first used by Joseph Davidovits. He defined the material that is formed in inorganic polycondensation called geopolymerization. In the geopolymerization reaction, three-dimensional structures of $AlO_4$ and $SiO_4$ tetrahedra are created. Later the term geopolymer was used for all alkali activated aluminosilicates [4, 5].

The geopolymers now represent a new group of inorganic substances, because they have a significant environmental and energy potential. They belong to a group of the inorganic polymer covalently bound macromolecules with the chain consisting of -Si-O-Al-O-. Geopolymers are obtained from the chemical reaction of alumino-silicate oxides with sodium silicate solutions in a highly alkaline environment. As an alkali activating solution, a strongly alkaline aqueous solution of sodium or potassium hydroxide is most commonly used [5, 6].

The geopolymers find a broad range of applications in the field of transportation, emergency repairs, metallurgy, coating, membrane materials, and nuclear waste disposal. Despite a significant commercial and technological potential geopolymers' easy-brittle character limits their extensive applications where great efforts are made to overcome such a disadvantage. Numerous studies are dedicated to optimize the strength of geopolymer products and to understand the mechanism of the geopolymerization [7, 8, 9, 10].

In the last years, different studies have investigated the possibility of using different types of wastes mixed with fly ash or slag as raw materials. The selection of the materials to create the geopolymers depends on influences such as availability, disposal urgency, difficulty for recycling and final applications [11].

The primary aim of this paper is to demonstrate that a mechanical activation of GGBFS improve mechanical properties of all geopolymers made in our research.

## 2. Materials and methods

The material used for the alkali activation was the blast furnace slag (BFS) after the mechanical activation in a ball mill. Milling was performed in the laboratory ball mill with steel balls with different diameters. A chemical analysis of the material is in Table 1. After grinding stage d80 was 120 µm. The material was homogenized before the alkali activation. No other treatment was applied to the material. SEM pictures of the materials after grinding are in Figure 1. As shown, slag after mechanical activation in ball mill has plate-like and needle-like structures. SEM images have been taken at different magnification. The SEM images of GGBFS display rough and angular shape.

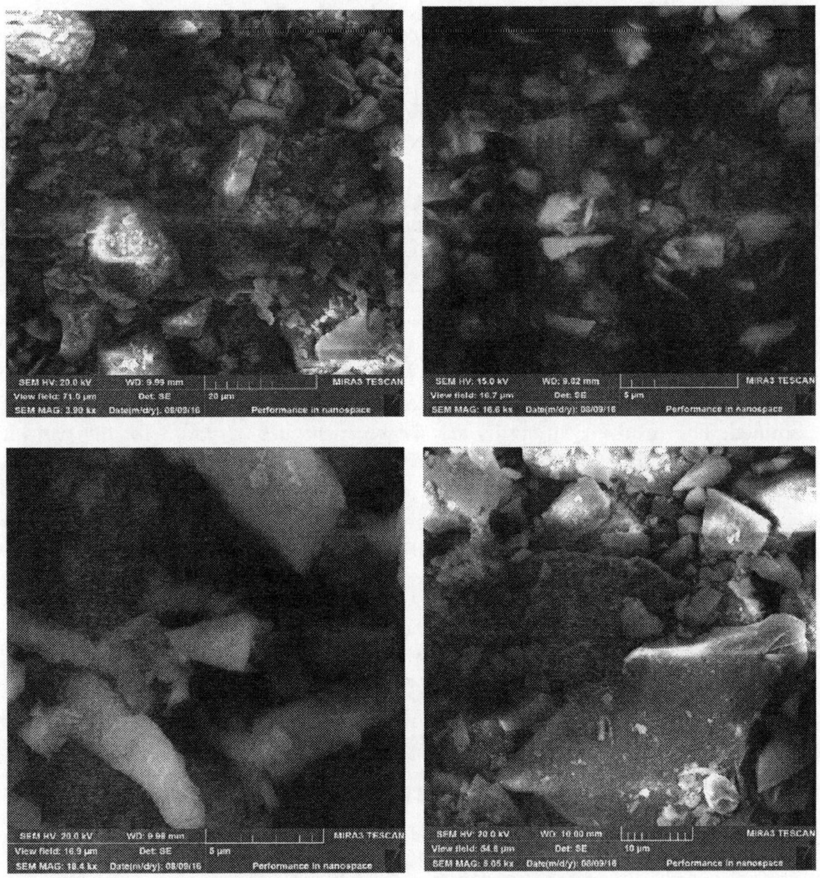

**Figure 1***:* SEM pictures of GGBFS.

The activation solution was prepared by mixing solid NaOH pellets with Na-water glass and water. The sodium water glass from the Kittfort Praha Co. with the density of 1.328–1.378 g/cm$^3$ was used. It contains 36–38% $Na_2SiO_3$ and the molar ratio of $SiO_2/Na_2O$ is 3.2–3.5. The solid NaOH with the density of 2.13 g/cm$^3$ was obtained from Kittfort Praha Co. containing at least 97% – 99.5% of NaOH.

**Table 1.** Chemical composition of GGBFS.

| Material | $SiO_2$ | CaO | MgO | $Al_2O_3$ | $Fe_2O_3$ | Other |
|---|---|---|---|---|---|---|
| **BFS [%]** | 40.3 | 37.01 | 12.1 | 8.51 | 0.3 | 1.78 |
| **WG [%]** | 71.3 | 9.2 | 1.2 | 1.9 | | |

The slag mixture samples with different amount of waste glass (WG) were prepared. WG came from old colorless glassy waste such as old bottles. The samples with 10%, 20%, and 30% by weight and also a reference sample which was created only from GGBFS were prepared. Two different particle sizes of the waste glass were used. In the first set of experiments the particle size of the waste glass was below 1mm. In the second set it was between 1–4 mm.

GGBFS was first mixed with WG and after their homogenization, the activation solution was added. The value of the Ms modulus in the activation solution was adjusted to 1.3. An overall concentration of the alkaline activation agent was adjusted to 7% $Na_2O$ in the binder mass. The water-to-fly ash ratio was adjusted to 0.23. GGBFS and WG mixture was stirred with the activation solution for 10 minutes, until a creation of a homogenous mixture. The mixture was then filled into prismatic molds with the dimensions 40 × 40 × 160 mm and compacted on the vibration table VSB-40. The pastes were cured in a hot air drying chamber at 80°C for 6 hours. Thereafter, the samples were removed from the molds, marked, and stored in laboratory conditions till the moment of the strength test. The values of compressive strength were determined after 7, 28 and 90 days according to the Slovak Standard STN EN 12390-3. A part of the samples was kept for 28 days at laboratory temperature, then a water absorption test according to the Slovak Standard STN 73 1316 was performed.

The compressive strength of the hardened samples was determined after 7, 28 and 90 days using the hydraulic machine Form+Test MEGA 100-200-10D.

## 3. Results and discussion

Properties and a composition of material play an important role in the alkali activation. The samples were synthetized from GGBFS with addition of WG by several proportions and in different particle sizes. The effect of WG addition on final compressive strength and flexural strength as a function of time was examined. The resulting strengths are shown in Figures 2 and 3 for the particle size below 1 mm.

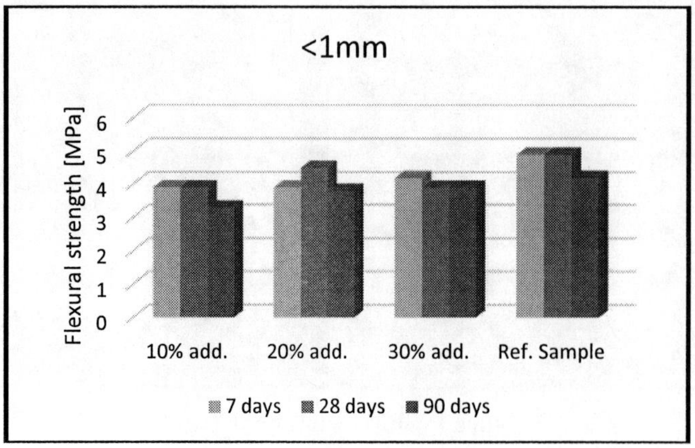


**Figure 2.** Effect of WG addition on flexural strength.

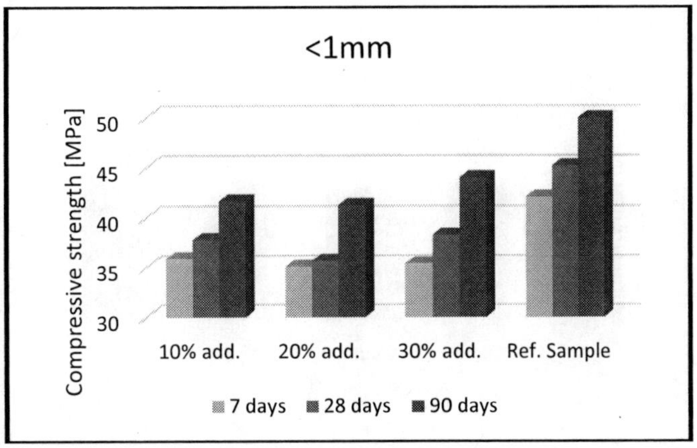


**Figure 3**. Effect of WG addition on compressive strength.

The flexural strength results shows that maximum strength of all samples is not achieved on 90th day, but the slag addition in the samples proves that these strengths were lower than the flexural strength of the reference sample. The flexural strength of all samples was decreasing over the time.

The same tendency showed results of the compressive strength tests. The addition of WG with particle size below 1mm did not improve the compressive strengths of the mixtures over the reference sample. The highest strengths were achieved for the samples with 30% addition of WG. Results also showed that on 90th day, all the strengths were about the same, which means that the added portion of WG have no significant effect on the compressive strength of the hardened material, but when compared to the reference sample, the compressive strength of all the samples is lower.

The water absorption was examined over 24 hours and was measured after 10, 40, 90, 360 and 1440 minutes. The measured water absorption in the prepared geopolymeric materials is shown in Figure 4. The results show that 20% addition of WG in the sample decreases the water absorption. The highest water absorption was measured for the sample with 10% addition of WG.

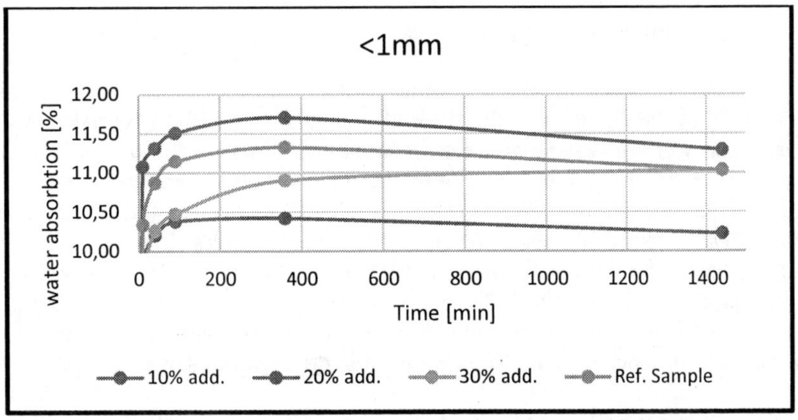


**Figure 4**. Water absorption of GGBFS geopolymers with <1mm WG addition.

The second set of experiments had the same parameters, but the particle size of WG was different. Results are in Figures 5 and 6.

IOP Conf. Series: Materials Science and Engineering **379** (2018) 012013 doi:10.1088/1757-899X/379/1/012013

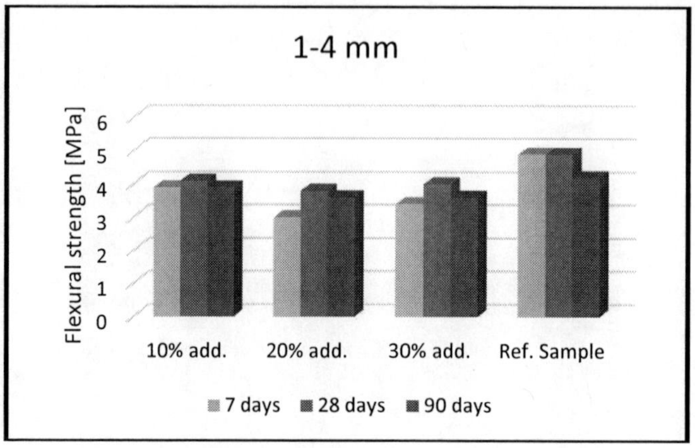


**Figure 5**. Effect of WG addition on flexural strength.

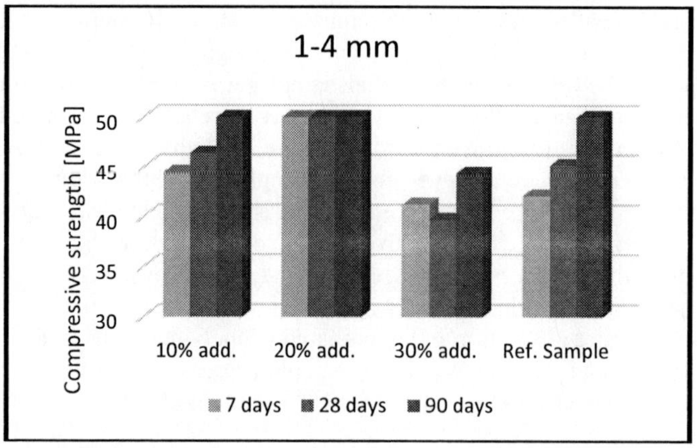


**Figure 6**. Effect of WG addition on compressive strength.

The flexural strength results show very similar results than results with WG with smaller particle size. Also not only they did not improve the flexural strengths but strengths were decreasing over time. The highest strengths were achieved by the reference sample.

The compressive strengths results with 1–4 mm particle size WG finally showed an improvement. 10 and 20% of WG in the mixture showed a little bit higher strengths than the reference sample in each day of testing. 20% addition of WG also showed that the strengths were almost the same from $7^{th}$ to $90^{th}$ day of measuring. On the contrary 30% addition of WG on the contrary showed decreasing in strengths.

WG shows similar strengths as in [12, 13], we can compare the high of compressive strengths. Compressive strength values are comparable to Portland cement, close to 45 and 50 MPa, were achieved for samples produced with 7 M solution in contrary to [12] where highest strengths were achieved with 6 M solution. Similar results were also obtained in [13], where highest strengths were achieved in mixture containing 20% addition of ground container glass. Optimum addition levels of WG are in 10 to 20% range.

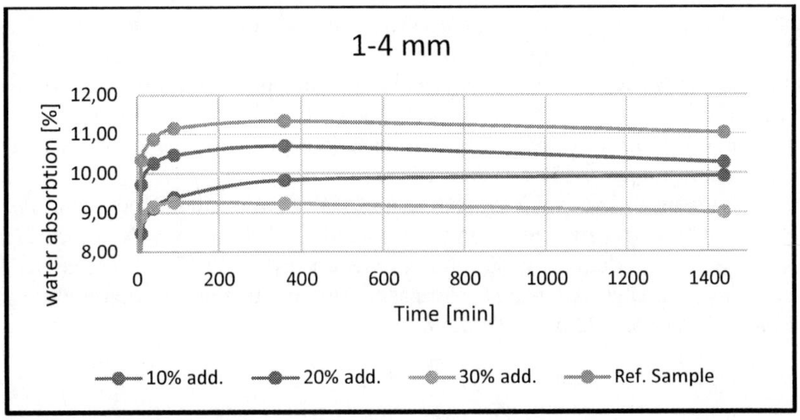


**Figure 7.** Water absorption of GGBFS geopolymers with 1–4 mm WG addition.

The water absorption test with these geopolymers mixtures showed that an increase in the portion of WG in the sample decreases the water absorption. The highest water absorption was measured for the reference sample, the lowest was measured for the sample with 30% addition of WG. WG has a positive effect to reduce the water absorption in these geopolymers.

## 4. Conclusions

The alkali activated materials – geopolymers, are a new generation of inorganic binders. Any aluminosilicate materials can be used to prepare the geopolymers, including the slag. This paper describes possibilities of preparation of the geopolymers from GGBFS with the addition of a different amount of the waste glass. The results showed, that WG has not always a positive effect on strengths of the produced material. With smaller particle size there is very small positive effect and only in the water absorption. Bigger particle size has achieved an improvement in the compressive strengths and also in the water absorption.

**Acknowledgments**

This work was supported by the research grant project VEGA, no. 1/0472/18.

**References**

[1] Snellings R., Mertens G and Elsen J 2012 Supplementary cementitious materials Rev. Mineral. *Geochem.* **74** pp 211–278

[2] Wang H, Ding B, Liu X, Zhu X, He X and Liao Q 2017 Solidification behaviors of a molten blast furnace slag droplet cooled by air *Applied Thermal Engineering* **127** pp 915–924

[3] Arai Y, Powell B A and Kaplan D I 2017 Sulfur speciation in untreated and alkali treated ground-granulated blast furnace slag *Science of The Total Environment* **589** pp 117–121

[4] Davidovits J 1991 Geopolymers: Inorganic Polymeric new materials *Journal of thermal analysis* **37** pp 1633–1656

[5] Skvara F 2007 Alkali activated materials or geopolymers *Ceramics – Silikáty* 51/3 pp 173–177

[6] Xu H and van Deventer J S J 2000 The geopolymerisation of alumina-silicate materials. *Int. J. Miner. Process* **59** pp 257–266

[7] van Deventer J S J, Provis J L, Duxson P and Lukey G C 2007 Reaction mechanisms in the geopolymeric conversion of inorganic waste to useful products *Journal of Hazardous Materials* **139** pp 506–513

[8] Zhang H Y, Kodur V, Qi S L and Wu B 2015 Characterizing the bond strength of geopolymers at ambient and elevated temperatures *Cement and Concrete Composites* **58** pp 40–49

[9] He J, Jie Y, Zhang J, Yu Y and Zhang G 2013 Synthesis and characterization of red mud and rice

husk ash-based geopolymer composites *Cement and Concrete Composites* **37** pp 108–118
[10] Zhang J, He Y, Wang Y, Mao J and Cui X 2014 Synthesis of a self-supporting faujasite zeolite membrane using geopolymer gel for separation of alcohol/water mixture *Materials Letters* **116** pp 167–170
[11] Toniolo N and Boccaccini A R 2017 Fly ash-based geopolymers containing added silicate waste *Ceramics International* **43** pp 14545–14551
[12] Toniolo N, Rincón A, Avadhut Y S, Hartmann M, Bernardo E and Boccaccini A R 2018 Novel geopolymers incorporating red mud and waste glass cullet *Materials Letters* **219** pp 152–154
[13] Tho-In T, Sata V, Boonserm K and Chindaprasirt P 2018 Compressive strength and microstructure analysis of geopolymer paste using waste glass powder and fly ash *Journal of Cleaner Production* **172** pp 2892–2898

IOP Conf. Series: Materials Science and Engineering **379** (2018) 012014 doi:10.1088/1757-899X/379/1/012014

# Application of fly ashes from fluid bed coal combustion within refractory materials

**J Kotrla[1], F Šoukal[1], J Másilko[1], J Švec[1] and M Janča[1]**

[1] Brno university of technology, Faculty of Chemistry, Materials Research Centre, Purkyňova 464/118, Brno Cz-612 00, Czech Republic

Email: xckotrla@fch.vut.cz

**Abstract.** The study deals with the possibility of utilization of fly ashes from coal combustion in refractory materials and its influence on resulting properties of the material. The effect of bed and filter ash addition on the properties of the droplet and the resulting mechanical properties, bulk density, shrinkage and firing temperature of stove fireclay. The subject of this study is also reduction of $SO_2$ emissions evolved burning of fireclay containing fluid ash rich of anhydrite. Emissions of $SO_2$ were observed by TG-DTA method. Characterization of prepared samples and study of heat treatment processes were based on XRD and TG-DTA.

## 1. Introduction

With technological advances of mankind, the demand for energy is growing. Today, the main source of energy is coal. In recent years, the trend of energy-saving and non-waste economy has grown. For this reason, it is important to look for a new and efficient use of energy industry byproducts [1].

Chemical, physical and mineralogical properties of the energy products, depends on a type of coal and a type of combustion process [2], [3]. A main type of the combustion process today is high-temperature and fluidized bed combustion [4]. Fluidized bed combustion is increasingly being applied, the advantage of it is that the desulfurization process is situated in a boiler and desulphurization stations are not needed [5]. Limestone is added directly to the boiler, where it reacts with $SO_2$ to form $CaSO_4$. The main difference between high-temperature fly ash and fluid fly ash is a higher content of $SO_3$ and CaO in the fluid fly ash [6]. Fluid bed combustion takes place at a lower temperature and therefore it differs from high-temperature fly ash by chemical properties and content of amorphous phase [5].

Because of the high content of CaO, fluid fly ashes should not be used as an additive to concrete. There are some studies that deal with the possible use of fluid fly ash. Mostly fluid fly ash was used in smaller quantities as a replacement for high-temperature fly ash [7]–[9].

Fluid fly ash also contains a certain amount of anhydrite produced in the reactor during the desulfurization process. The presence of anhydrite in the fly ash is undesirable from an ecological point of view because during the firing process $SO_2$ is emerging to the atmosphere. One way to prevent $SO_2$ escaping to the atmosphere is to bond it to thermostable compounds such as hauyne. Hauyne is a tectosilicate mineral with endmember formula $Na_3Ca(Si_3Al_3)O_{12}(SO_4)$ and it is formed at higher temperature and pressure, so naturally it could be found in volcanic rocks [10]. Bo Wei et al. [11] dealt with the processes occurring between $SiO_2$, $Al_2O_3$, $Na_2O$, and $CaSO_4$ at high temperature. Bo Wei used fly ash to remove sodium ions which induces severe fouling and slagging in pulverized coal furnaces and found out that optimal temperature for hauyne formation is 1 050°C.

## 2. Experimental

### *2.1. Material*

Fireclay samples were prepared for the measurements where the primary raw materials were partially replaced by fluid fly ash. For the preparation of samples was used fluid fly ash from the power plant Tisová and stove fireclay materials (BX shale, fireclay, and clay TPU) comes from P-D refractories CZ a.s. As a source of sodium ions, sodium water glass ($Na_2O$ 16.15%) was used, which is important for the formation of thermally stable phases. This premise was based on the work of Vodová [10].

The chemical composition of the used ashes was determined by ICP-OES and the values are in Table 1. The content of free lime in fly ashes was determined by the saccharate method and values are given in Table 2.

**Table 1.** The chemical composition of used ashes.

| **Major oxides [%]** | **Tisová** | |
|---|---|---|
| | **Filther ash** | **Bed ash** |
| $SiO_2$ | 32.2 | 24.5 |
| $Al_2O_3$ | 25.4 | 16.1 |
| CaO | 10.5 | 24.5 |
| $K_2O$ | 5.1 | 5.8 |
| MgO | 0.4 | 0.4 |
| $SO_3$ | 7.7 | 14.8 |
| $TiO_2$ | 5.7 | 3.4 |
| $Fe_2O_3$ | 7.0 | 7.5 |

**Table 2.** The content of free lime in used ashes.

| | **Tisová** | |
|---|---|---|
| | **Filter ash** | **Bed ash** |
| Content of free lime [wt.%] | 4.29 | 18.20 |

### *2.2. Sample preparation*

Sample preparation consisted of mixing the ceramic mixture, pressing, drying and firing. The homogenization of the mixture was carried out in a planetary mixer. The pressing was carried out by one-sided compression in a 55 mm diameter steel cylindrical form on a hydraulic press BetonSystem DESTTEST 3310. Compression pressure was 25 MPa, acting on the body for 50 seconds. After pressing, the samples were dried at 105°C to a constant weight. Thereafter, the samples were placed in the Classic 7013L muffle furnace. The temperature mode consisted of a temperature ramp of 2 °C/min to 1 100 °C/min, a two hours isotherm and a ramp 2 °C/min to 500°C. The weights of raw materials for preparation of test samples are shown in Table 3.

**Table 3.** The composition of samples.

| **Filter ash Sample identification** | | | | **[g]** | | | |
|---|---|---|---|---|---|---|---|
| | **0%** | **1%** | **2%** | **5%** | **10%** | **20%** | **30%** |
| Fireclay | 34.6 | 41.3 | 40.9 | 39.7 | 37.6 | 33.4 | 29.3 |
| BX shale | 103.3 | 123.9 | 122.5 | 118.8 | 112.5 | 100 | 87.5 |
| TPU clay | 68.6 | 82.4 | 81.6 | 79.1 | 74.9 | 66.6 | 58.3 |
| Ash | 0 | 2.5 | 5 | 12.5 | 25 | 50 | 75 |
| Water | 25 | 25 | 25 | 27 | 30 | 34 | 40 |
| **Bed ash Sample identification** | | | | **[g]** | | | |
| | **0%** | **1%** | **2%** | **5%** | **10%** | **20%** | **30%** |
| Fireclay | 34.6 | 41.3 | 40.9 | 39.7 | 37.6 | 33.4 | 29.3 |
| BX shale | 103.3 | 123.9 | 122.5 | 118.8 | 112.5 | 100 | 87.5 |
| TPU clay | 68.6 | 82.4 | 81.6 | 79.1 | 74.9 | 66.6 | 58.3 |
| Ash | 0 | 2.5 | 5 | 12.5 | 25 | 50 | 75 |
| Water | 25 | 25 | 25 | 27 | 30 | 30 | 30 |

### *2.3. Methods*

Mechanical properties, compressive strength, were measured on the complex device for strength test on building materials Betonsystem DESTTEST 3310. Compressive strength was measured on each testing body, which dimensions were 55 × 50 mm. Bulk density and shrinkage of samples were determined by dimensional and weight measurements.

Saccharate method was used to determine a content of free lime in ashes, according to ČSN P 72 2080 standard. The content of free lime (Table 2) was calculated according to the formula:

$$\% \, CaO = \frac{c \cdot V \cdot M}{m \cdot v} \tag{1}$$

where $c$ is a concentration of hydrochloric acid solution, $V$ is the volume of the hydrochloric acid solution, $M$ is the molar weight of CaO, $m$ is the weight of the sample of ash and $v$ is the stoichiometric ratio of reaction.

The TG-DTA was used to determine the behavior of the ceramic mixture. Measurements were performed on the TA INSTRUMENTS SDT Q600. The measurement started at 30°C, the heating rate was 10°C per minute to 1 350°C in the air atmosphere.

The XRD method was used to find out the occurrence of the hauyne crystalline phase. For observation, XRD Empyrea from Panalytical was used measured by referencing the intensity ratio method. For the measurement, the copper anode was used, voltage was 40 kV and the current was 30 mA.

## 3. Results and discussion

The influence of filter and bed fluid fly ash on compressive strength, bulk density and shrinkage are shown in Figure 1, 2 and 3.

IOP Conf. Series: Materials Science and Engineering **379** (2018) 012014 doi:10.1088/1757-899X/379/1/012014

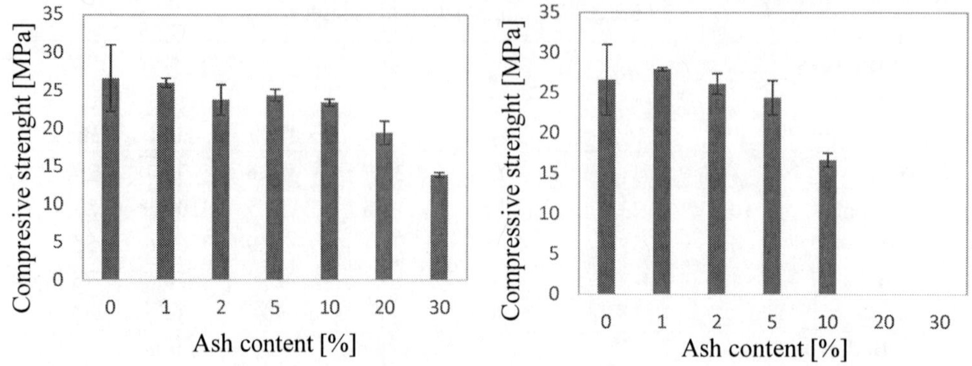


**Figure 1.** Influence of filter (left) and bed (right) ash on compressive strength.

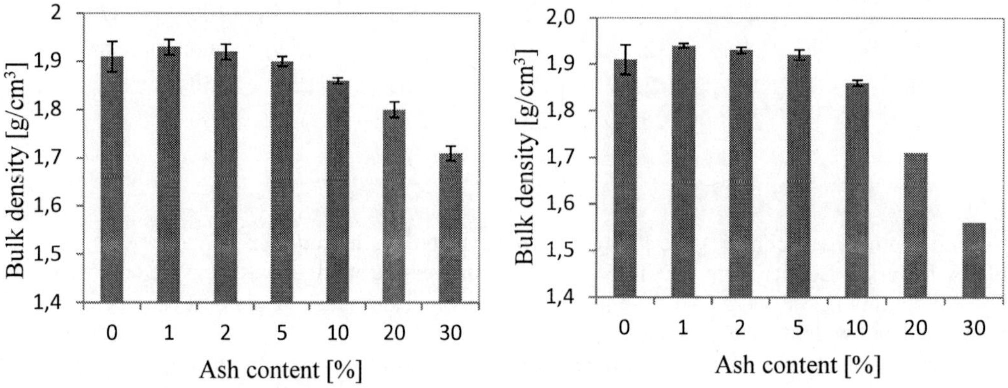


**Figure 2.** Influence of filter (left) and bed (right) ash on bulk density.

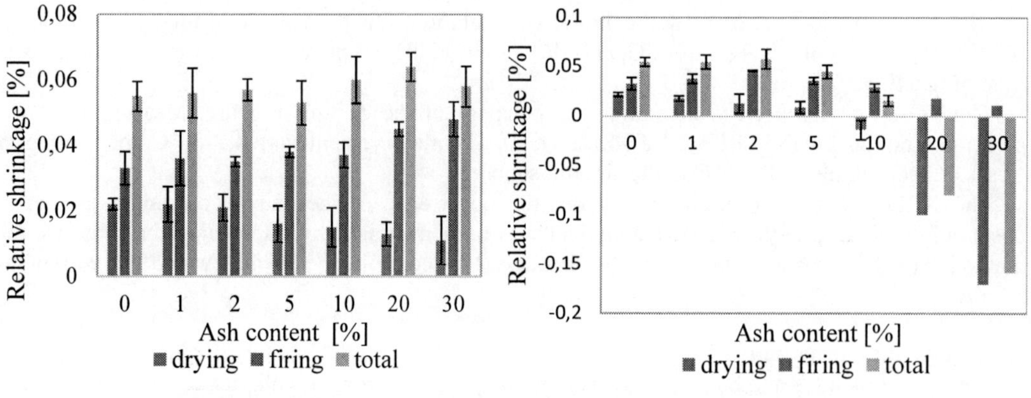


**Figure 3.** Influence of filter (left) and bed (right) ash on shrinkage.

From the measured values, it is evident, that the addition of filter ash, up to 10 wt. %, and bed ash, up to 5 wt. %, does not have a significant effect on the resulting mechanical properties. Higher addition of fluid fly ashes, cause a decrease in compressive strength due to hydration of CaO and anhydrite. When there is a change in volume and the formation of micro-cracks. For samples with higher bed fly ash content, a large number of cracks occurred and it was not possible to determine their compressive strength.

Keršnerová et al. [12] found that bulk density decreases with the increasing addition of fluid fly ash. This trend has been also observed, but only with higher additions (over 2–5 wt. %). The addition of fluid fly ash up to 2–5 wt. % increased the bulk density, due to filling the space between coarse fractions in the sample.

There was also observed shrinkage which occurred during drying and firing. From Figure 3 it is apparent that the addition of fluid filter ash reduces shrinkage during drying. However, this contraction is compensated by higher shrinkage during firing, so the addition of filter ash does not have the effect to the overall shrinkage. Higher fluid bed ash additives did not cause shrinkage during drying, but on the contrary to their expansion and cracking, due to CaO hydration. Cracks were evident in all samples containing bed ash (Figure 4).

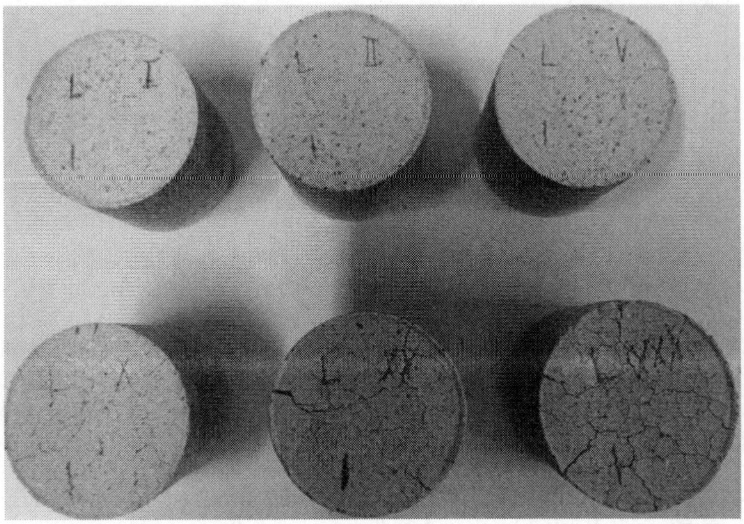

**Figure 4.** Samples with a fluid bed fly ash (top left: 1, 2 and 5 wt. %, bottom left: 10, 20 and 30 wt. %).

TG-DTA analysis was used to determine the chemical processes occurring during firing. Figure 5 shows the TG-DTA curve of the ceramic mixture with the addition of filter ash.

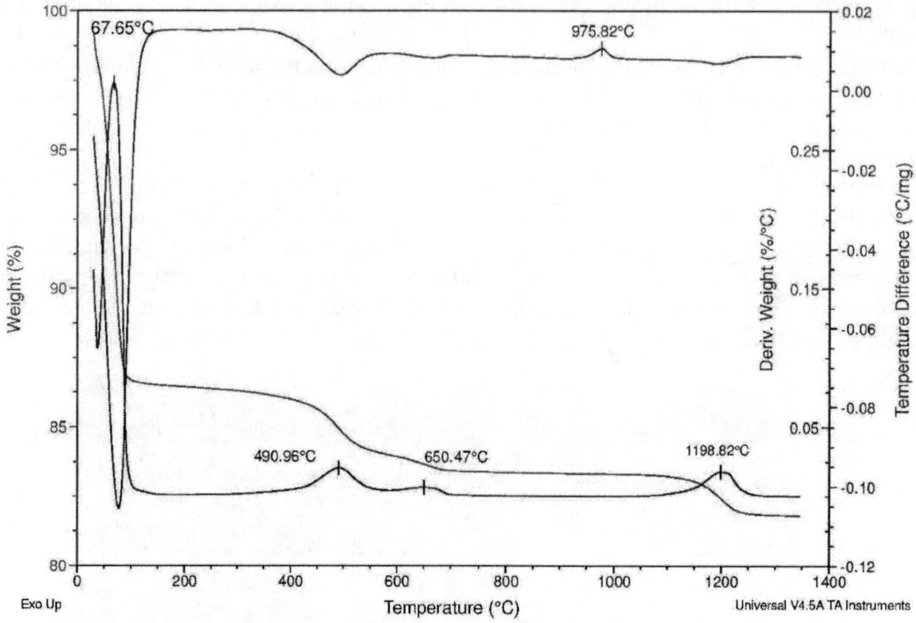


**Figure 5.** TG-DTA curve of ceramic mixture with filter ash (30 wt.%).

During the firing of the ceramic mixture with filter ash addition, the evaporation of water occurs to 68°C. Next weight loss is due to the dehydroxylation of clay minerals at about 491°C contained in TPU and decarbonization of carbonates in fly ash at about 650°C. At 976°C there is an exothermic reaction, which is the crystallization of the defective spinel [13]. Furthermore, from 1 070°C, mass loss occurs, which is associated with the decomposition of anhydrite and released $SO_2$.

To reduce $SO_2$ emissions sodium water glass was added to the ceramic mixture with ash, which, according to dissertation [10], should provide sodium ions to react with anhydrite and form a heat-resistant compound (hauyne). In the Figure 6 are DTG curves of two samples with 30 wt. % filter ash, where one of the sample being enriched with 12 wt. % of sodium water glass (water glass was calculated on the weight of fly ash). The addition of water glass has caused that $SO_2$ release in two steps. The first release was from anhydrite decomposition and the second was from heat-resistant hauyne. It has been demonstrated [10] that the use of 12 wt. % of water glass reduce the $SO_2$ content in flue gas. This quantity proved to be inadequate because part of anhydrite remained in the sample.

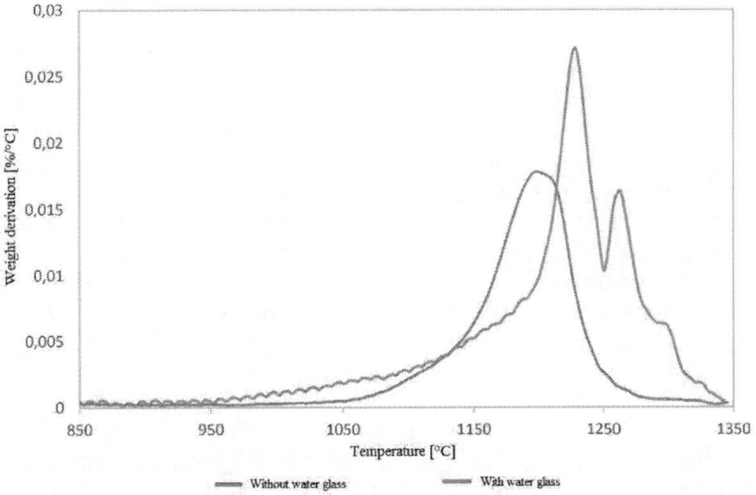


**Figure 6.** DTG curves of samples with and without sodium water glass.

Created hauyne was observed by XRD analysis. However, demonstrable presence was less than in the dissertation of Vodová [10]. Samples with 30 wt. % of water glass were also observed (Figure 7). Anhydrite was still present in the sample, which signifies that the firing time was insufficient.

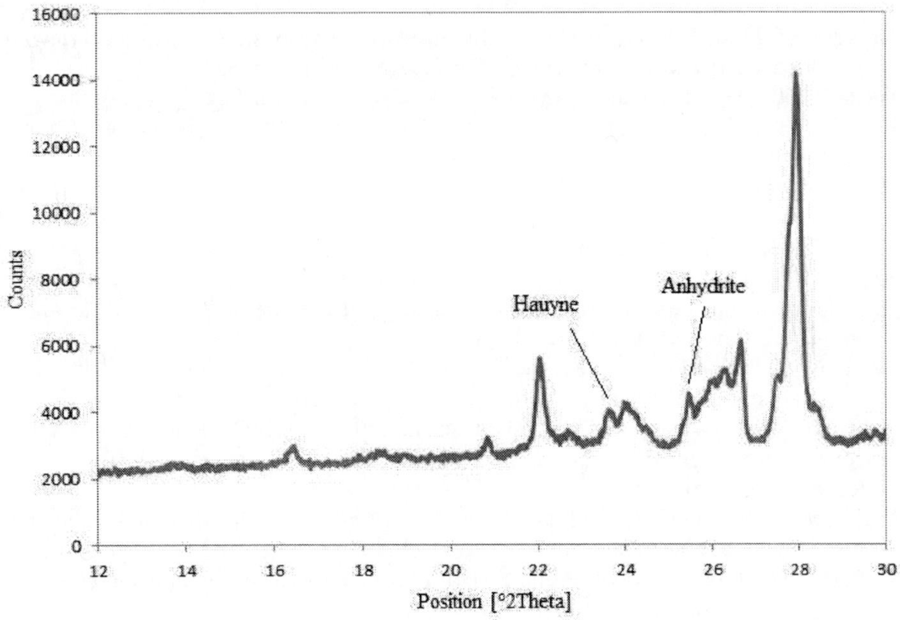


**Figure 7.** XRD analysis of sample with 30 wt. % of filter ash and 30 wt. % of water glass.

## 4. Conclusion

The work was focused on the utilization of filter and bed fluid fly ash in refractory materials. From experimental observations, the highest values of pressure strength were measured on samples containing 1 to 10 wt. % of filter fly ash and 1 to 5 wt. % bed fly ash. However, samples with bed fly ash contained cracks. This has shown that ash with a high content of CaO is unsuitable for the preparation of refractory materials.

The use of filter ash reduces shrinkage during drying, but this is compensated by bigger firing shrinkage and resulting shrinkage corresponds with the reference sample. The use of bed fly ash significantly influences the dimensions of the prepared samples due to the high content of CaO.

The use of fluidized fly ash in low additions (up to 10 wt. %) increases the bulk density, but the higher fly ash additions decrease the bulk density.

In the mixture, with the addition of sodium water glass, the heat-resistant compound was identified by XRD after firing. However, the firing time or addition of the water glass was not sufficient to react with all the present anhydrite.

The study has shown that fluid fly ash with a lower content of free lime could be serving as a partial replacement of primary raw materials.

**Acknowledgements**
This work was supported by the project Materials Research Centre at FCH BUT - Sustainability and Development, REG LO1211, with financial support from National Programme for Sustainability I (Ministry of Education, Youth and Sports).

**References**
[1] 2012 Těžíme uhlí *OKD* (*Online* http://www.okd.cz/cs/tezime-uhli/soucasnost-u-nas-i-ve-svete)
[2] Ahmaruzzaman M 2010 A review on the utilization of fly ash *Prog. Energy Combust. Sci.* **36** pp 327 363
[3] Siddique R 2010 Utilization of coal combustion by-products in sustainable construction materials *Resour. Conserv. Recycl.* **54** pp 1060–1066
[4] Kooernneef J 2007 Development of fluidized bed combustion – An overview of trends, performance and cost *Prog. Energy Combust. Sci.* **33** pp 19–55
[5] Fečko P 2003 Popílky (Ostrava: Vysoká škola báňská – Technická univerzita)
[6] EN 450-1:2005 European standard: Fly ash for concrete (European Committee for Standardization)
[7] Havlica J, Brandštetr J and Odler I 1998 Possibilities of utilizing solid residues from pressured fluidized bed coal combustion (PSBC) for the production of blended cements *Cement Concrete Res.* **28** pp 242–260
[8] Keprdová Š and Bydžovský J 2013 Vliv přídavku fluidních popílků na fyzikálne-mechanické a tepelně-technické vlastnosti cementových kompozitů s organickou výplní *Sborník přednášek Popílky ve stavebnictví 2013* pp 111–116
[9] Temuujin J, Rickard W and Riessen A Van 2013 Characterization of various fly ashes for preparation of geopolymers with advanced applications *Adcanced Powder Technology (Online* https://www.sciencedirect.com/science/article/pii/S092188311300023X)
[10] Vodová L 2015 Možnosti využití fluidních popílků v keramické technologii (Brno: Vysoké učení technické v Brně)
[11] Wei B, Wang X, Tan H, Zhang L and Wang Y 2016 Effect of silicon-aluminum additives on ash fusion and ash mineral conversion of Xijnjiang high-sodium coal *Fuel* **181** pp 1224–1229
[12] Keršnerová L, Lang K, Kovář P and Šoukal F 2017 Experiences with using class C fly ash in refractory materials *EuroCoalAsh 2017* pp 39–44
[13] Ptáček P, Křečková M, Šoukal F, Opravil T and Havlica J 2012 The kinetics and mechanism of kaolin powder sintering I. The dilatometric CRH study of sinter-crystallization of mullite and cristobalite *Powder Technology* **232** pp 24–30

IOP Conf. Series: Materials Science and Engineering **379** (2018) 012015 doi:10.1088/1757-899X/379/1/012015

# Dynamic moduli of cement paste with different kinds of recycled concrete admixtures

**Z Prosek[1,2], J Trejbal[1], P Tesárek[1] and J Ďureje[1]**

[1] Czech Technical University in Prague, Faculty of Civil Engineering, Prague, 166 29, Czech Republic
[2] Czech Technical University in Prague, University Centre for Energy Efficient Buildings, Buštěhrad, 273 43, Czech Republic

E-mail: zdenek.prosek@fsv.cvut.cz

**Abstract**. This article deals with using recycled concrete powder as a partial replacement for cement. In particular, we focus on mechanical properties of cement pastes containing recycled concrete powder. We used three different recycled materials. The first recycled concrete powder was obtained from the PB2 and the SB8 railway sleepers, the second one was from the drainage gutter and the third one was from the monolithic construction of a pillar. All recycled materials were milled by LAVARIS Ltd. at high-speed milling process. Investigated mechanical properties were the dynamic modulus of elasticity and the dynamic shear modulus and were obtained by a non-destructive method: the resonance method. The dynamic modulus of elasticity and the dynamic shear modulus were examined and compared after 7, 14, 21 and 28 days of mixture curing. Results obtained from these materials were compared with the reference material.

## 1. Introduction

Nowadays, recycled concrete is used for backfilling and landscaping or underlying road layers [1]. Another possibility is to use recycled concrete as a new aggregate. This option is possible if precise sorting and recycling of old concrete are used [2]. Only the fraction larger than 1 mm is used in this recycling. However, the problem arises when processing a very fine fraction (<1 mm) of recycled concrete. The processing of this very fine fraction brings many complications and pitfalls with it and, unfortunately, other financial costs. One option is thermal activation of an old cement paste. The utilization of recycled concrete by thermal activation have been spelled out clarified within various researchers, as showed e.g. in [3, 4]. Due to high demands on energy consumption and $CO_2$ production, it is particularly suitable for sustainable and environmentally friendly methods where the very fine fraction of recycled concrete is used in its raw form [5, 6].

Unlike the mentioned researchers, we focused directly on mechanical activation of old cement pastes. The article deals directly with the use of possible hydration potential of micronized concrete. The micronizing of the old cement paste into fine particles results in uncovered unhydrated clinker minerals, which can further hydrate upon contact with water [7]. For this mode of activation, a high-speed grinding method was used. The high-speed milling process has been successfully used for micronizing of marble sludge [8].

The possibility of replacing the binder with a fine fraction of recycled concrete was dealt with in several works [9, 10], directly works deals with replacing the binder in mortar mixes up to 25 wt. %. As a result, the compressive strength was reduced 15 to 20%, the difference in the values may be due to

IOP Conf. Series: Materials Science and Engineering **379** (2018) 012015 doi:10.1088/1757-899X/379/1/012015

the type of old concrete. The dynamic module was not solved directly, but its value is directly dependent on the compressive strength of the cement composite [11].

## 2. Materials and specimens

Cement pastes were composed of Portland cement CEM I 42.5R (Radotín, Czech Republic) and micronized recycled concrete. Cement was replaced by the micronized recycled concrete by the amount of 50 wt. %, (Table 1). The selected ratio of cement and recycled material (1:1) was chosen to highlight the differences between single-use recycled materials. Three different types of recycled concrete were used. The first recycled concrete was obtained from the PB2 and the SB8 railway sleepers, the second one was from the drainage gutter and the third one was from the monolithic construction of a pillar. All recycled materials were in first pre-crushed to do fraction 0–1 and then milled. Recycled concrete was micronized using a high-speed mill from Lavaris Ltd. (Libčice, Czech Republic) for its mechanical activation of recycled concrete. The result was micronized material from railway sleepers (MRS), the drainage gutter (MDG) and the construction of a pillar (MCP). The water ratio is described as water to cement and the micronized recycled concrete weight ratio. The water ratio was equal to 0.35. The workability of the fresh mixture was measured within the production of specimen. The consistency of fresh mixtures was determined by means of a flow expansion test after 10 ($d_{10}$) and 20 ($d_{20}$) impulses (Table 1).

Mixtures made according to Table 1 were molded to rectangular forms having inner dimensions equal to 40 × 40 × 160 mm. These specimens were kept in the forms for 24 h at a room temperature of 22°C. All testing mixtures included 6 specimens. After demolding, these samples were stored in a water bath for 28 days.

**Table 1.** Composition of the mixtures.

| Set/ Material | Cement (g) | Micronized concrete (g) | Water ratio (-) | Flow expansion test $d_{10}$ (mm) | Flow expansion test $d_{20}$ (mm) | Bulk density (kg·m$^{-3}$) |
|---|---|---|---|---|---|---|
| **Ref. CEM** | 3,000 | 0 | 0.35 | 160 | 190 | 2,052 ± 4 |
| **MRS** | 1,500 | 1,500 | 0.35 | 110 | 120 | 1,957 ± 3 |
| **MDG** | 1,500 | 1,500 | 0.35 | 110 | 120 | 1,951 ± 11 |
| **MCP** | 1,500 | 1,500 | 0.35 | 105 | 115 | 1,981 ± 2 |

## 3. Experimental methods

The characterization of recycled concrete was carried out with a number of experimental approaches in order to investigate all relevant features. In the initial stage of the research, the micronized recycled concrete was examined from the perspective of material composition by X-ray fluorescence technique (XRF). The result of XRF was chemical composition of recycled concrete. It was determined 48 elements among which were also measured common harmful heavy metals.

The pH value was determined from aqueous extracts (in a weight ratio of 1: 100) after 24 hours at a room temperature (25 ± 1°C) by using pH meter (from Monokrystaly a.s.). This type of pH meter is suitable for fast measurement with a deviation of a value about 0.2.

The particle size distribution was determined by devices Laser Particle Sizer Fritsch ANALYSETTE 22 NanoTec. The device was capable of measuring sizes ranging from 0.01 mm up to 2.1 mm and used the combination of two lasers. The first laser was infrared and was used for large particles. Small particles were measured by green laser. The particle size distribution was performed using wet dispersion.

Values of the dynamic shear modulus and the dynamic modulus of elasticity of the samples were determined after 7, 14, 21 and 28 days. The resonance method was applied for the detection of these moduli. Six samples with dimensions 40 × 40 × 160 mm were examined. Measurements of the dynamic moduli were carried out by the instrument from Brüel & Kjær. Dynamic moduli were evaluated from basic natural frequencies by using standard equations for the prismatic sample. The dynamic modulus of elasticity was achieved by using the basic longitudinal natural frequency of samples and basic flexural natural frequency of samples. An acceleration transducer was placed on the surface of the sample in the center of the vertical side perpendicular to the length of the sample and the hammer hits were performed on the center of the opposite surface, in the case of the longitudinal natural frequency. The acceleration transducer was placed on the edge of the top side of the surface sample and hammer hits were performed on the edge of the same side of the sample, in the case of the flexural natural frequency. Samples were supported by a smooth and a flexible block at the midpoint of its length, which is the major longitudinal nodal location, in the case of the longitudinal natural frequency. Samples were supported by two smooth and flexible blocks at the major bending nodal lines, which were located at a distance of 0.224 × length of the sample from each end of the sample, in the case of the flexural natural frequency. The dynamic shear modulus was achieved by using the basic torsional natural frequency of the sample. The acceleration transducer was placed on the surface of the sample in the corner of the vertical side parallel to the length of the sample and hammer hits were performed diagonally on the same side. Samples were supported in the same way as in the case of measuring of the longitudinal natural frequency of samples [8].

## 4. Results and discussion

The result of chemical composition is shown in Table 2. Table 2 shows percent weight of individual oxides with a content greater than 0.5%. Results showed that the micronized drainage gutter (MDG) had a high volume of silica, from the reason of the high volume of silica aggregate in fraction 0-1 mm and the low volume of the cement matrix with unhydrated clinker minerals. The other two recycled concrete, the micronized railway sleepers (MRS) and the micronized construction of a pillar (MCP), exhibit the same amount of the cement matrix, making it more suitable for possible using of recycled concrete powder as a partial replacement for cement. None of recycled materials does contain an excessive amount of harmful components for future application in concrete. According to composition, waste micronized recycled concrete can be used as a partial replacement for cement.

**Table 2.** Chemical composition of tested materials.

| | Percent weight of individual oxides [%] | | | |
|---|---|---|---|---|
| Oxides | CEM I 42.5R | MRS | MDG | MCP |
| $CaO$ | 65.76 | 26.85 | 16.89 | 19.90 |
| $SiO_2$ | 16.85 | 51.63 | 50.71 | 54.23 |
| $SO_3$ | 4.85 | 1.52 | 0.79 | 1.37 |
| $Al_2O_3$ | 3.61 | 14.95 | 16.82 | 12.92 |
| $Fe_2O_3$ | 3.34 | - | 4.85 | 4.36 |
| $Na_2O$ | 2.88 | 1.92 | 3.26 | 2.20 |
| $MgO$ | 2.31 | 2.92 | 3.00 | 2.43 |
| $K_2O$ | - | - | 3.45 | 2.02 |

The particle size distribution was shown in Figure 1. Portland cement with fraction 0–200 microns had mean grain size of 12.8 microns and 90 wt. % of grains were smaller than 44.1 microns. The micronized railway sleepers (MRS) with fractions 0–130 μm had mean grain size of 12.4 microns and 90 wt. % of grains were smaller than 45.4 microns. The micronized drainage gutter (MDG) with fractions 0–80 μm had mean grain size of 7.4 microns and 90 wt. % of grains were smaller than 22.3 microns. The last recycled concrete micronized construction of a pillar (MCP) with fractions 0–40 μm had mean grain size of 3.1 microns and 90 wt. % of grains were smaller than 11.5 microns. The all three materials were milled by the same disintegrator and had different mean values of grains size, this is due to the different composition of the recycled concrete (size and amount of silica aggregate, water-cement ratio, kind of cement) and thus also different strength and resistance to the milling process.

Results showed that all recycled concrete is finer or same as the reference Portland cement. The high speed milling process could cause crushing of cement grains, which uncovered unhydrated clinker minerals. It is possible at the same time, that micronized concrete can be a binder and a microfiller.

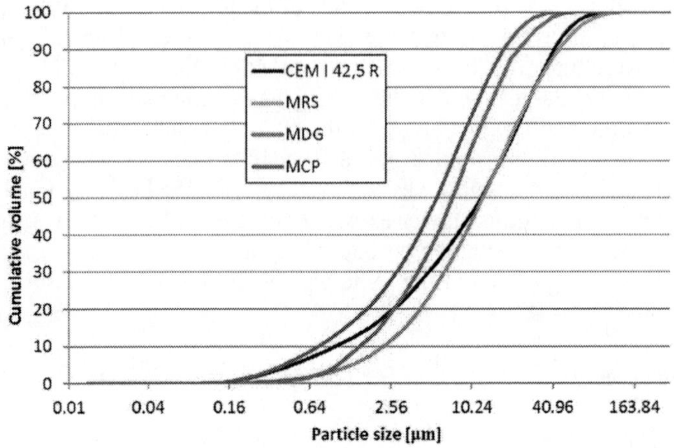


**Figure 1.** Particle size distribution of tested materials.

Values of pH have been monitored for the research to provide hydration properties. The pH values of recycled concrete were compared with the pH value of Portland cement (12.54). The value of pH was 11.68 in the case of micronized railway sleepers (MRS), 11.99 in the case of the micronized drainage gutter (MDG) and 12.52 in the case of the micronized construction of a pillar (MCP). Recycled concrete has a similar pH value, so it does not seem necessary to solve it in this case.

Results from the resonance non-destructive method are shown in Figure 2 and Figure 3 and showed differences between values of the moduli at the 7, 14, 21 and 28 days. Differences between the values after 21 and 28 days were in the size of standard deviations. It can be stated that after 21 days, a minimal change of the dynamic modulus of elasticity and the dynamic shear modulus occurred. The average value of moduli of the reference samples after 28 days were 25.31 ± 0.08 GPa in the case of the modulus of elasticity (Figure 3) and 9.90 ± 0.09 GPa in the case of the shear modulus (Figure 2). The samples with the micronized construction of a pillar (MCP) had the largest value among samples with the recycled material of the dynamic shear modulus, namely 7.82 ± 0.13 GPa (Figure 2) and the dynamic modulus of elasticity, namely 19.19 ± 0.23 GPa (Figure 3). The samples with the micronized drainage gutter (MDG) had the lowest value of the dynamic shear modulus, namely 7.30 ± 0.01 GPa (Figure 2) and the dynamic modulus of elasticity, namely 18.55 ± 0.08 GPa (Figure 3).

The results obtained are in accordance with the results of the authors [9, 10, 11] The type of used old concrete influences the resulting mechanical properties of the cement composite. The concrete with high strength (pillar, sleepers) has a positive effect on the mechanical properties of the resulting cement composite.

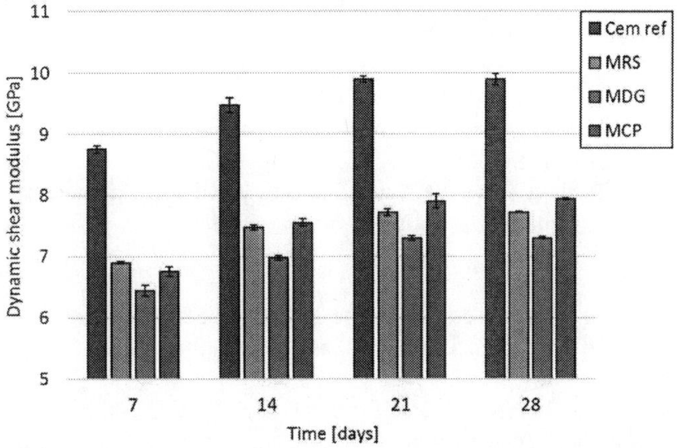


**Figure 2.** Comparison of the dynamic shear modulus (with standard deviation).

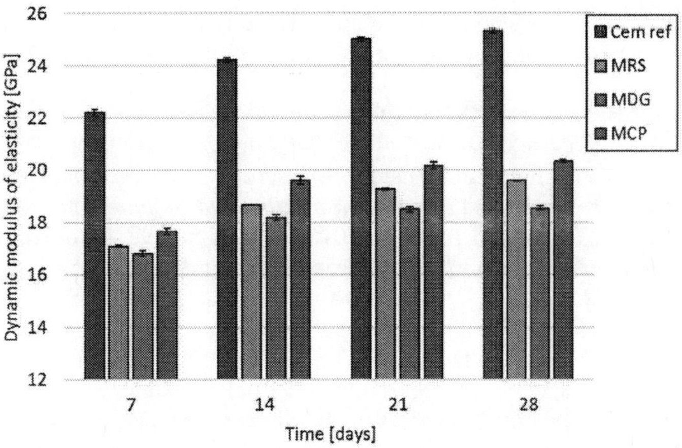


**Figure 3.** Comparison of the dynamic modulus of elasticity (with standard deviation).

## 5. Conclusion

This article deals with the influence of micronized recycled concrete on mechanical properties of resulting cement pastes. Researched cement pastes were composed of Portland cement and three different recycled materials in a ratio of 50/50. The first recycled concrete powder was obtained from the PB2 and the SB8 railway sleepers, the second one was from the drainage gutter and the third one was from the monolithic construction of a pillar. Based on results, it can be concluded that:

- None of recycled materials does contain excessive amount of harmful components for future application in concrete.
- The high speed milling process cause crushing of cement grains, which uncovered unhydrated clinker minerals.
- The average value of the dynamic modulus of elasticity and the dynamic shear modulus were unchanged between 21 and 28 days.
- For 50 wt. % of recycled concrete, the dynamic modulus of elasticity decreased by approximately 20%. The largest value of the dynamic modulus of elasticity had samples with recycled concrete from the micronized construction of a pillar (MCP) and the lowest from

the micronized drainage gutter (MDG). The difference between these values was approximately 10%.

**Acknowledgements**
This work was financially supported by the Czech Technical University in Prague – the project SGS16/201/OHK1/3T/11, by the Ministry of Industry and Trade the research project FV20503. The authors also thank to the Centre for Nanotechnology in Civil Engineering at the Faculty of Civil Engineering, Czech Technical University in Prague.

**References**
[1] Barbudo A, Agrela F, Ayuso J and Jiménez J R 2012 Statistical analysis of recycled aggregates derived from different sources for sub base applications *Construction and Building Materials* **28** pp 129–38
[2] Vegas I, Ibanez J A, Lisbona A and Cortazar A S 2011 Pre-normative research on the use of mixed recycled aggregates in unbound road sections *Construction and Building Materials* **25** pp 2674–82
[3] Shui Z, Xuan D, Wan H and Cao B 2008 Rehydration reactivity of recycled mortar from concrete waste experienced to thermal treatment *Construction and Building Materials* **22** pp 1723–29.
[4] Shoon J, Heyden L and Eloy P 2012 Waste fibrecement: An interesting alternative raw material for a sustainable Portland clinker production *Construction and Building Materials* **36** pp 391–403.
[5] Corinaldesi V and Moriconi G 2009 Behaviour of cementious mortars containing different kinds of recycled aggregate *Construction and Building Materials* **23** pp 289–94
[6] Ahmari S, Ren X, Toufigh V and Zhang L 2012 Production of geopolymeric binder from blended waste concrete powder and fly ash *Construction and Building Materials* **35** pp 718–29.
[7] Prošek Z, Šeps K, Topič J and Tesárek P 2016 Effect of particle size distribution waste micronized marble powder on the mechanical properties of the cement composite *EAN 2016 - 54th International Conference on Experimental Stress Analysis (Srní)* (Plzeň: Department of Mechanics, Faculty of Applied Sciences, University of West Bohemia)
[8] Prošek Z, Trejbal J, Topič J, Plachý T and Tesárek P 2017 Utilization of the waste from the marble industry for application in transport infrastructure: mechanical properties of cement pastes *IOP Conference Series: Materials Science and Engineering* **236**
[9] Bílek V 2012 The use of concrete filler as a mineral admixture in concrete *Proc. of 8 th int. Conf. Concrete in the Low Carbon Era* pp 667–675
[10] Elaqra H A 2015 Valorization of Recycled Aggregate in Concrete and Mortar *IUG Journal of Natural and Engineering Studies* **23** pp 26–37
[11] Plachý T, Tesárek P, Hlubocký L, Hrůza J, Fládr J, Trejbal J and Prošek Z 2017 Effect of Microfillers on Selected Destructive and Nondestructive Mechanical Properties of Cement Mortars: Different Types of Recycled Materials *Key Engineering Materials* **722** pp 54–58

# Interaction assessment between fiber reinforcement and cement matrix containing finely ground recycled concrete

**J Trejbal[1], J Ďureje[1], R Hlůžek[1] and Z Prošek[1,2]**

[1] Czech Technical University in Prague, Faculty of Civil Engineering, Thákurova 7, 166 29 Prague, Czech Republic

[2] Czech Academy of Science, Institute of Physics, Cukrovarnická 10, 162 00 Prague, Czech Republic

Email: jan.trejbal@fsv.cvut.cz

**Abstract**. This work deals with adhesion enhancement between polymer macro-fibers and cement matrix (CEM I 42.5 R). Two types of fibers were used – polypropylene (PP) and polyethylene terephthalate (PET), both having approx. 0.3 mm in diameter. To achieve the increase between the two materials, fibers were surface roughened by means of an oxygen plasma treatment. Moreover, a concrete finely ground recyclate containing grains at the size of less than 0.25 mm was admixed into the cement matrix (the amount was equal to 30 wt. % of cement) to promote the required adhesion between the two materials by principle of adhesive locks. To assess the adhesion, individual fibers were pulled-out from the matrix, while a force needed to their movement was recorded. The fiber embedded length was 30 mm. It was shown, that plasma treated fibers exhibited increased adhesion with the matrix by ca. up to 20% (PP), resp. 5% (PET) than them with no surface treatment. If the concrete recyclate was used, the adhesion further increased by ca. 5–10% for both fiber types.

## 1. Introduction

It is obvious that production of waste is unsustainable on a global scale. More than elsewhere, the same assertion applies to the construction industry. It is reported that from 3 to 10 billion tons of construction waste is produced per year all over the word. In the Czech Republic, the amount of this waste is equal to 3.5 million tons per year. These facts justify creation of authority's and ecologist's requirements and regulations aiming to reduce production and to re-use of waste. The European Union has therefore established the prohibiting of the storage of waste from 2024 [1]. Although this ordinance has rather a political than a technical origin, it is necessary to adapt it and to re-use construction waste, concrete from demolished building including [2].

For example, Australia, Denmark, Germany, Japan, Holland and the Great Britain recycle more than 50% of waste concrete, some of them even more than 70%. On the contrary, developed countries as Italy or France less than 20%. The Czech Republic with its ca. 40% share of the concrete recycling is ranged side by side rather to worse countries from this point of view [3]. Therefore, it is necessary to find out new ways how to reuse the waste concrete and how to reach a rate of the recycling of the most progressive countries.

T. Pavlů et al. summarized that possibilities of concrete recycling rests in re-use of aggregate separated from waste concrete and in an utilization of dust emerging during this process. It was shown that recycled aggregate can be used for production of new concrete again [4]. Unfortunately, it is obvious that potential of this material is significantly limited due to presence of various contaminants, a random shape and a size of particles etc. [3, 5, 6].

On the contrary to the utilization of recycled aggregate, the finely ground powder seems to have good potential to be used for production of cement-based products, if processed properly from waste. As Plachý et al. remarked, the powder originated from the waste concrete contains more than 10% of un-hydrated grains [6]. Therefore, the powder can be used not only as micro-aggregate but also as an active binder. To exploit potential of this material, it must be processed by a high-speed milling. This leads to unification of a particles size and to stripping of un-hydrated grains [7]. As reported in many studies, thus obtained material can be used for production of mortars, lightweight concrete blocks and so on [6, 7, 8]. Throughout scientific and technical studies, there can be found many ways to use milled dust effectively. So far, in the field of fiber reinforced cement composites, an impact of this recycled material on an adhesion between fibers and the matrix has not been unambiguously researched. A few papers have been published, but their results are not conclusive [9, 10]. In can be assumed from what is obvious, that presence of finely milled dust in cement-based materials contributes to the adhesion between the matrix and the fiber reinforcement. Nevertheless, the rate of this contributing has not been clearly quantified.

To solve this unanswered question, a set of pull out tests of polymer (PP and PET, d = ca. 0.3 mm) macro-fibers was done from two types of cement matrix: (i) reference and (ii) containing 30 wt. % of recycled concrete powder. Two types of each fibers were used – as received and surface treated via a plasma treatment.

## 2. Materials

### *2.1. Fibers*

Two types of polymer fibers were used – PP (polypropylene) and PET (polyethylene terephthalate). Each type was distinguished by their surface treatment on untreated (marked PET and PP) and plasma treated (marked PET-T and PP-T). According to methodology published in our previous research [11], the treatment was done in order to roughen fiber surfaces and to increase their wettability by water (classified by a contact angle measurement). Their basic mechanical and geometric properties are summarized in the table below (Table 1). Both types were made by manufacture Spokar Inc. (Czech Republic). These fibers are primarily made for production of brushes, however, as it was confirmed earlier, they can be used as an alternative to common types of fibers designated as the reinforcement of fiber reinforced concrete [11].

**Table 1.** Basic mechanical and geometric parameters of fibers.

| | Diameter [μm] | Density [kg/m$^3$] | Young's modulus of elasticity [GPa] | Tensile strength [MPa] | Elongation [%] | Contact angle [°] |
|---|---|---|---|---|---|---|
| **PET / PET-T** | 335 | 950 | 5.8 | 238 | 51 | 85.0/26.2 |
| **PP / PP-T** | 305 | 900 | 6.1 | 440 | 8 | 88.3/31.1 |

### *2.2. Cement and finely ground recycled concrete*

For purposes of experiments described in following captures, two types of matrixes were made. The first was composed only from Portland cement CEM I 42.5 R with water to cement ratio (w/c) equal to 0.4 (in the chapter of results marked as "CEM"), while the second contained moreover finely ground recycled concrete in the amount of 30 wt. % of cement (marked as REC). In this case, w/c ratio was increased to 0.41 due to a high specific surface area of the recyclate. Recycled concrete comes from old drainage channels. It was ground using the high speed milling technology. The size of grains was up to 0.25 mm. Curing and hardening of mixtures took 28 days in standard laboratory conditions.

## 3. Experimental Methods

### *3.1. Pull-out tests*

For purposes of pull-out testing, samples from above mentioned mixtures were made using prismatic molds of dimensions equal to 25 × 20 × 25 mm. During samples preparation, the single fiber was placed into the each form (located in its center), while its embedded length was equal to the height of specimen, i.e. 25 mm. One of thus prepared samples is captured in the picture below (Figure 1). After 28 days of mixture curing and hardening, the pull-out tests were performed. Samples was fixed by means of self-locking chuck as a part of a loading frame Web Tiv Ravestein FP100. A free-end of the single fiber protruded from the specimen was fixed to a movable part of the device. Then, the fiber was pulled out from the specimen at the constant rate of 2 mm/min (the experiment was displacement controlled), while the response of the fiber during the pulling was recorded in the form of dependence between the fiber free-end displacement and a force resisting to the movement. The experiment was interrupted when displacement reached on 5 mm. To obtain statistically relevant data, ten measurements were realized for each type of fiber and matrix. At the whole, 40 pull-out tests were thus realized. The resulting data were evaluated using MatLab based software DiPro. In the field of fiber reinforced composites, it is in a habit to indicate the adhesion between the matrix and the fiber in the form of interface shear stress due to provide values normalized by the fiber surface area. Therefore, the shear stress was calculated from load-displacement curves at distinct stages (i) when the force reached its maximum (further marked by index "MAX") and (ii) when displacement reached value 3.5 mm (marked as "3.5"), according to calculations given below (1),

$$\tau_{\max} = \frac{F_{\max}}{C_f l_e} \quad \text{and} \quad \tau_{3.5} = \frac{F_{3.5}}{C_f l_e} \tag{1}$$

where $F_{\max}$ and $F_{3.5}$ represent the maximal force and the force necessary to reach 3.5 mm free-end displacement, respectively. $C_f$ is a fiber circumference and $l_e$ is the embedded fiber length.

**Figure 1.** Digital image of specimen specially made for pull-out testing.

## 4. Results

For clarity, it is appropriate to repeat marking of fibers, matrixes and pull-out stages. Collectively, "PP" and "PET" mark the fiber type, polypropylene and polyethylene terephthalate, respectively. If used "T", it means "surface treated". "CEM" and "REC" mark matrixes, cement and cement with recyclate, respectively. Finally, "MAX" and "3.5" refer to the stage during pull out tests in which was the force recorded and further used to a calculation of the shear stress, in more detail in the chapters above.

Figure 2 clearly shows that the strongest adhesion was attained in the case of treated PP fibers pulled-out from the matrix containing finely ground recycled concrete. In this case, the maximal interfacial shear stress between the two materials reached almost to 0.45 MPa and to ca. 0.38 at 3.5 mm displacement of the fiber free-end. When the cement matrix and the same fibers were used, results were following: ca. 0.4, resp. 0.34 MPa. It is also worth noting that the adhesion between PP fibers and the cement matrix was practically unchanged, compared to them and the matrix containing recyclate.

Similar results were obtained also in the case of PET fibers, as shown in the Figure 3. Contrary to our assumptions, we found out that the adhesion between reference fibers and the matrix containing recyclate was rather worse than between them and the cement-only matrix. The maximal shear strength decreased from ca. 0.31 (cement matrix) to 0.28 (matrix with recyclate) MPa, while the shear strength detected after 3.5 mm fiber free-end displacement decreased from 0.31 to 0.3 MPa. On the other hand, the adhesion was slightly increased, if treated fibers were used, especially in the case of samples made from recyclate containing matrix. Concretely, the use of the recyclate led to increase of the maximal shear stress by ca. 5–10% to ca. 0.33 MPa.

In summary, it can be concluded, that the use of finely ground concrete as an admixture to cement may improve adhesion between fibers and the matrix, but only in cases, when treated fibers are applied. If reference fibers having smooth surface were used, the utilization of the recyclate seems to be useless. Therefore, it is obvious, that the physical interaction between individual particles of the recyclate and roughened fiber surfaces plays an important role in the attainment of the required adhesion increase. Our finding corresponds with Sebiabi's et al. work [9], who confirmed that presence of very small particles of the silicate powder can improves the adhesion between the reinforcement and the matrix. Similarly, J. Zhou et al. [12] found out, that with increasing amount of limestone powder (up to 2 wt. %) in cement composite reinforced with PVA fibers, the tensile strain capacity of the material increased. It must be however added, that both cited authors did not provide an information about fibers morphology.

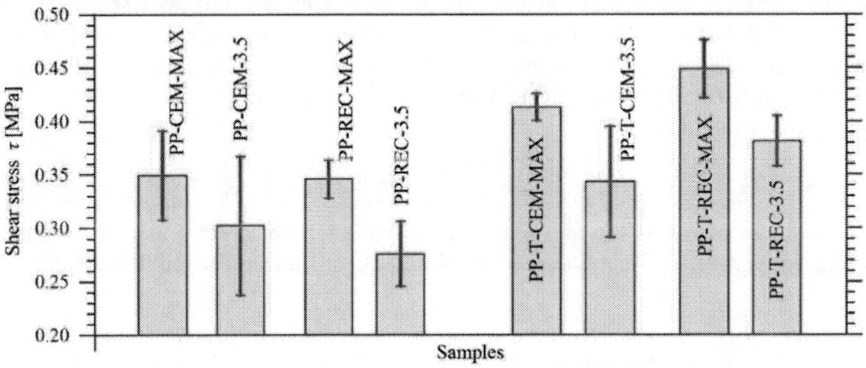


**Figure 2.** Interfacial shear stress between PP fibers and cement matrix obtained from pull-out tests.

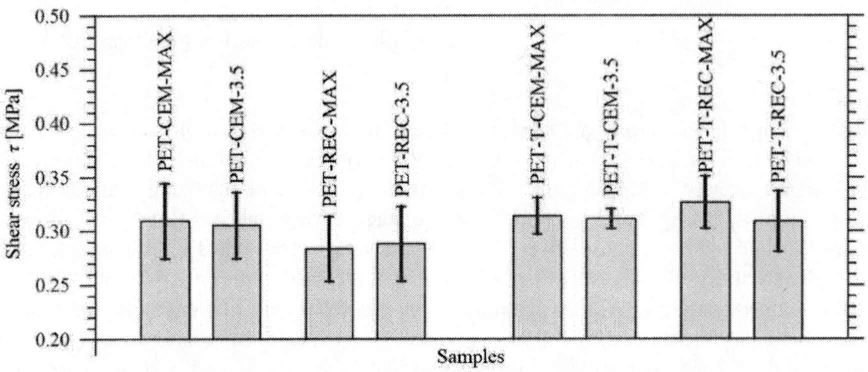


**Figure 3.** Interfacial shear stress between PET fibers and cement matrix obtained from pull-out tests.

## 5. Conclusion

The adhesion between two types of polymer fibers (polypropylene and polyethylene terephthalate) having ca. 0.3 mm in a dimeter and cement matrixes was researched. The first matrix was made from Portland cement only (CEM I 42.5 R, w/c 0.4), while the second one contained 30 wt. % of finely ground recycled concrete. The influence of the recyclate use on the interfacial interaction between the two materials was assessed. Apart of that, two variants of each type of fiber was used – reference and surface roughened via plasma treatment. The adhesion was assessed via single-fiber pull out tests from samples (25 × 20 × 25 mm). It was shown, that: (i) the adhesion between reference fibers (characterized by smooth surfaces) and the matrix was not increased if used concrete recyclate, (ii) the interfacial shear stress was increased by using of the recyclate, but only if surface treated fibers were used.

## Acknowledgments

This work was financially supported by Czech Technical University in Prague – SGS project SGS16/201/OHK1/3T/11 and Czech Ministry of Industry and Trade project FV20503. The authors would also like to thank to Tereza Horová for her help with this research.

## References

[1] Škopán M 2015 Analysis of the current state of C&DW recycling in the Czech republic and assessment of development options *Recycling 2015 – Možnosti a perspektivy recyklace stavebních odpadů jako zdroje plnohodnotných surovin* ed. M Škopán (Lomnička u Tišňova: VUT in Brno, ARSM) pp 19–27

[2] Schepper M, Snellings R, Buysser K, Driessche I and Belie N 2014 The hydration of cement regenerated from completely recyclable concrete *Construction and Building Materials* **60** pp 33–41

[3] Růžička V 2017 *Lightweight blocks based on recycled concrete* (Prague: Czech Technical University)

[4] Pavlů T and Šefflová M 2015 Environmental assessement of the structu ral elements with recycled aggregate *Recycling 2015 – Možnosti a perspektivy recyklace stavebních odpadů jako zdroje plnohodnotných surovin* ed. M Škopán (Lomnička u Tišňova: VUT in Brno, ARSM) pp 108–112

[5] Etxeberria M, Vázquez E, Marí A and Barra M 2007 Influence of amount of recycled coarse aggregates and production process on properties of recycled aggregate concrete *Cement and Concrete Research* **37** pp 735–42

[6] Jianzhuang X, Jiabin L and Zhang Ch 2005 Mechanical properties of recycled aggregate concrete under uniaxial loading *Cement and Concrete Research* **6** pp 1187–94

[7] Evangelista L and Brito J 2007 Mechanical behaviour of concrete made with fine recycled concrete aggregates *Cement and Concrete Composites* **29** pp 397–401

[8] Topič J, Trejbal J, Plachý T and Prošek Z 2017 Relationship between compressive strength and Young's modulus of cement paste with recycled concrete powder *Key Engineering Material* **722** pp 254–59

[9] Sebabi N, Benzerzour M, Abriak E and Binetruy Ch 2011 Mechanical properties of concrete-reinforced fibers and powders with crushed thermoset composites: The influence of fiber/matrix interaction *Construction and Building Materials* **29** pp 332–38

[10] Akca R K, Cakir Ö and Ipek M 2015 Properties of polypropylene fiber reinforced concrete using recycled aggregates *Construction and Building Materials* **98** pp 620–30

[11] Hlůžek R, Prošek Z, Trejbal J, Fládr J and Potocký Š 2017 Plasma treatment impact on physical and chemical properties of polymeric fibers *NMM 2017 – Nano & Macro Mechanics* ed. P. Padevět (Prague: Czech Technical University in Prague) pp 49–54

[12] Zhou J, Qian S, Beltran M G S, Ye G, Breugel K and Li V C 2009 Development of engineered cementitious composites with limestone powder and blast furnace slag *Materials and Structures* **43** pp 803–14

# Study on potential of fibre reinforced concrete with recycled aggregate for applications

**K Šeps[1], I Broukalová[1], J Fládr[1] and P Bílý[1]**

[1] Czech Technical University in Prague, Faculty of Civil Engineering, Department of Concrete and Masonry Structures, Thákurova 7, 166 29, Praha 6, Czech Republic

Email: karel.seps@fsv.cvut.cz

**Abstract**. The construction and demolition waste constitute a quarter of all types of waste production. Recovery of this waste in other areas of the economy is a step in the right direction to achieve sustainable development. Department of Concrete and Masonry Structures at the Faculty of Civil Engineering of the Czech Technical University is concerned in the problematic of sustainable constructions by research on recycled materials. One of the focus is on replacing of natural aggregate by rubble concrete recycled from demolished concrete structures and use of the recycled aggregate in fibre-reinforced concrete with synthetic fibres. The research focused also on possibilities of the structural design of structures from investigated material. The investigations proved usability of fibre-concrete with recycled crushed concrete in certain applications. The composite has satisfactory material characteristics and its application would consume a great deal of recycled concrete aggregate. The paper will present the material parameters and its efficiency in examples of applications.

## 1. Introduction

Application of recycled aggregate as a substitution for natural aggregate is well established and widespread. The use of recycled concrete aggregate (RCA) is principally same as the use of concrete with natural aggregate. Properties of the hardened RCA composite depend on the proportion of recycled and natural aggregate and on characteristics of the original concrete used for recycling.

The density does not change if the RCA content is up to 25%. The density of concrete with more than 25% of RCA is lower than density of common concrete with natural aggregate. Generally, the substitution of natural aggregate with RCA decreases the compressive strength (by 4–45%) [1–3]. Some investigations showed no changes of compressive strength [4] or even increase (by 9%) [1, 5–7], probably due to quality high-strength original concrete.

Tensile splitting strength if usually lower (by 33%) if more than 25% of natural aggregate is replaced by RCA [1, 7]. On the other hand, some investigations indicated increase of splitting tensile strength by 20% [8–10].

The article does not deal with the utilization of recycled concrete aggregate sorted to separate fractions. Therefore, it does not address potential use of recycled concrete aggregate as a partial or full replacement of natural aggregates in the production of dense concrete, classified by strength classes.

This paper shows a different, completely new way of utilizing of recycled concrete aggregate in practice. Porous fibre-reinforced concrete made entirely with RCA is applicable in appropriately selected constructions.

## 2. Examples of possible applications of the composite

Based on Inverse design approach and material characteristics determined in initial investigations, potential applications of the porous fibre reinforced composite with recycled concrete aggregate were proposed. The composite can be used in sub-layers, in earthworks, in earthen flood dams, noise barriers, walls and columns for low load-bearing structures.

### 2.1 *Fibre reinforced composite from RCA in sub-layers*

The composite can be efficiently used for stabilizing layers in pavements, cycle routes, roadways, parking areas and sports grounds. This application takes the advantage of permeability of the composite that will contribute to draining of sub-layers in these types of structures. Technology of the composite compacting complies with the road and pavements construction technology of roller compacting. Vast amounts of material are consumed in sub-layers, thus a big portion of waste material will be utilised and not deposited in the waste landfills.

### 2.2 *Fibre reinforced composite from RCA in earthworks*

Use of the composite as a strengthening layer in earthworks benefits from tensile strength, ductility and permeability of the composite. The pervious layer allows water from precipitation to pass through, prevents runoff and contributes to groundwater recharge.

The reduction of the embankment width by 2 x 10 m determined in the studied case would bring significant savings. Considering the purchasing price of farmland in the Czech Republic as 1 €/$m^2$, saving of 20 x 1000 = 20 000 € per one kilometre of the dam can be estimated. The savings can be even much higher in case that the structure is built on a more expensive estate such as forest or building land.

### 2.3 *Fibre reinforced composite from RCA in earthen flood dams*

Fibre reinforced composite with recycled aggregate can be utilized for stabilizing layers in the earthen flood dams. This application was proposed for recycled aggregate that incorporated higher amount of fine aggregate and the porosity of resulting composite was lower.

### 2.4 *Noise barriers*

Porous materials are generally exploited as sound absorbing materials. On the premise that sound energy will be dissipated in a series of interconnected cavities in the porous fibre reinforced composite with recycled aggregate it based the next proposed application in the noise barrier. The assumption must be verified by a laboratory test. Sound absorbing coefficient must be determined and reduction of sound pressure level must be verified for particular frequencies.

### 2.5 *Fibre reinforced concrete walls and columns*

Recycled aggregate can be used in a cement composite for application in walls and columns similarly to masonry walls and pillars. The walls and columns from the composite are applicable in low-storey houses with moderate loading. The composite has strength comparable to strength of clay masonry, comparable heat insulation performance and the advantage is lower labour intensity compared to clay masonry.

## 3. Materials and methods

In contrary to common concrete mix design, which begins by determining the requirements on concrete; setting of required concrete class, according to conditions that the concrete will be exposed to in service, a new approach was introduced, in which the procedure of the structural design and concrete mix design are inverted. In the initial phases a mixture is specified, subsequently material properties of hardened concrete are examined and identified, and application of the composite is proposed. The mixture composition is adjusted and modified for particular applications.

The composite consists of recycled concrete aggregate, which is not sorted into fractions. Maximum particles size 22 mm, gradation curve in Figure 1. Composite has a minimal dose of cement for structural application (260–300 kg/m$^3$), polymer fibres and water. The dosage of mixing water depends on water absorption of recycled aggregate and aggregate moisture state. The mixing water is dosed to reach the mixture consistency convenient for compaction by ramming or tamping.

A composite is intended for application in walls (columns). It consisted of 1,650 kg/m$^3$ of recycled concrete aggregate, 260 kg/m$^3$ of Portland cement (42,5 R Mokrá), 9.1 kg/m$^3$ of polypropylene fibres (length 55 mm) and 150 kg/m$^3$ of water. The composite properties were investigated in various tests. Parameters related to structural function of the composite were determined and, also thermal parameters were followed.

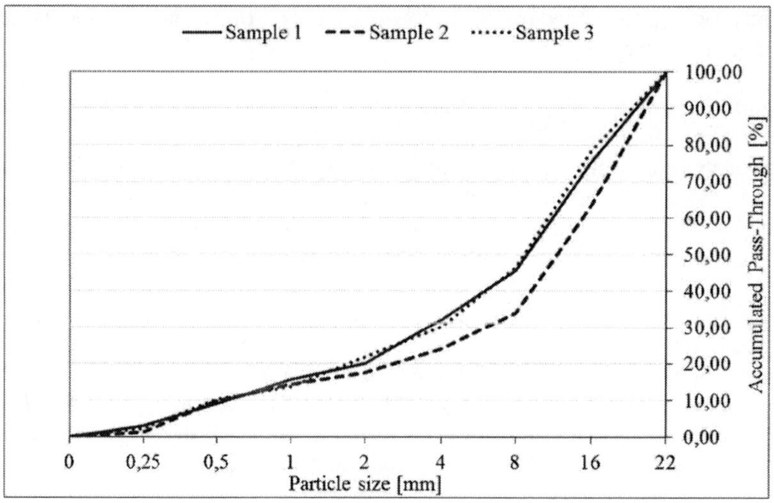


**Figure 1.** Gradation curve of recycled concrete aggregate.

## 4. Results and discussion

Common tests with cube specimens were performed to determine compressive strength (Table 1) and tensile splitting strength.

**Table 1.** Results of compression tests.

| Specimen | Dimensions Width (mm) | Length (mm) | Height (mm) | Weight (kg) | Density (kg/m$^3$) | Load (kN) | Compressive strength (MPa) |
|---|---|---|---|---|---|---|---|
| A1 | 149 | 149 | 149 | 6.088 | 1,840 | 276 | 12.43 |
| A2 | 149 | 147 | 149 | 5.946 | 1,822 | 216 | 9.86 |
| A3 | 149 | 145 | 148 | 6.032 | 1,887 | 350 | 16.20 |
| A4 | 149 | 149 | 149 | 6.034 | 1,824 | 254 | 11.44 |
| A5 | 149 | 150 | 148 | 6.302 | 1,903 | 282 | 12.62 |
| A6 | 148 | 149 | 151 | 5.980 | 1,796 | 214 | 9.70 |
| A7 | 149 | 149 | 149 | 5.972 | 1,805 | 208 | 9.37 |
| A8 | 149 | 150 | 148 | 6.106 | 1,846 | 236 | 10.56 |
| Mean value | | | | | 1,840 | | 11.52 |
| Conditional standard deviation | | | | | 38.1 | | 2.25 |

A characteristic value of compressive strength was calculated from the measured data according to ISO 2394. General principles on reliability for structures using k- coefficient for eight specimens k = 2. The characteristic strength of recycled aggregate fibre reinforced composite is $f_k$ = 6.99 MPa.

IOP Conf. Series: Materials Science and Engineering **379** (2018) 012017 doi:10.1088/1757-899X/379/1/012017

Mean value of tensile splitting strength is 0.7 MPa with standard deviation 0.12.

Non-standard prism specimens 150/150/600 mm and 150/150/1,200 mm were loaded in the direction of their axis in compression machine. Figures 2 to 4 show load-deformation diagrams of both specimens and the state of specimens after finishing of tests. Cracking and development of visible cracks occurred after reaching of ultimate load for both specimens.

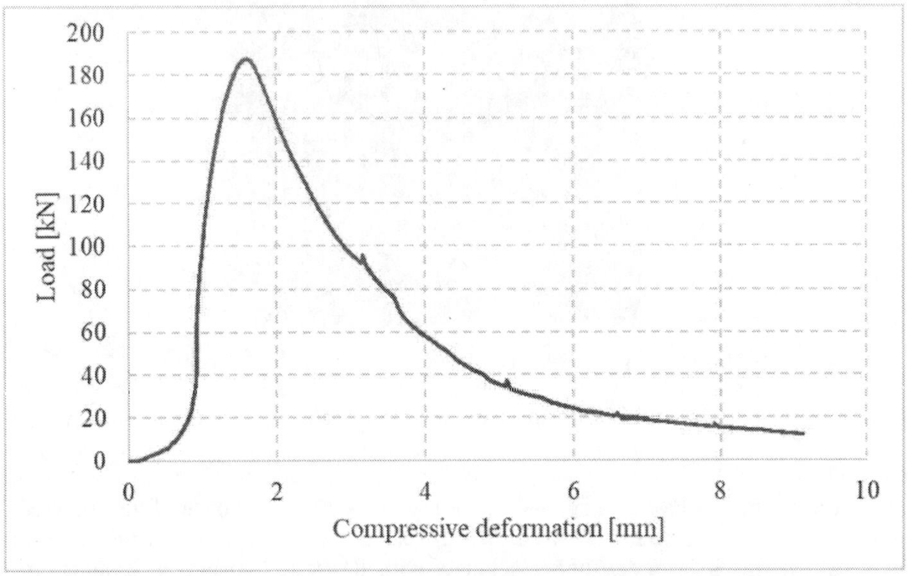


**Figure 2.** Load – deformation curve of specimen 150/150/600 mm.

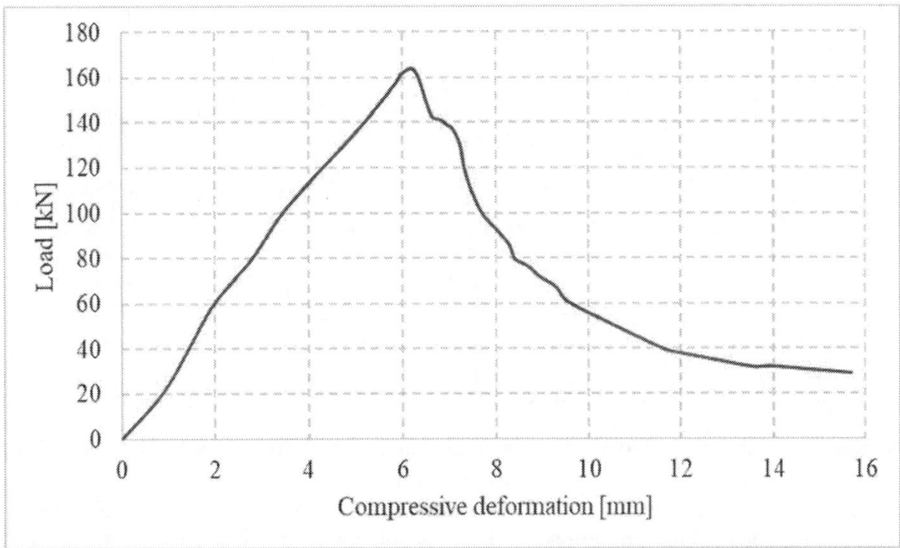


**Figure 3.** Load – deformation curve of specimen 150/150/1,200 mm.

**Figure 4.** Specimens after laboratory tests 150/150/1,200 mm.

The high porosity of the composite led to assumption of efficacious thermal properties. Volumetric thermal capacity and thermal conductivity were measured with a device Isomet on cubes 100/100/100 mm.

The measurement was based on an analysis of the temperature response of the analysed composite to heat flow impulses. The heat flow was excited by electrical heating of a resistor heater inserted into the probe which was in a direct thermal contact with the tested specimen. Evaluation of thermal conductivity and volumetric heat capacity is based on periodically sampled temperature records as a function of time, provided that heat propagation occurs in unlimited medium. Results of measurements are in Table 2.

**Table 2.** Thermal parameters.

| Specimen | Thermal conductivity $\lambda$ (W/mK) | volumetric *thermal* capacity $c$ ($10^6$ J/m$^3$K) |
|---|---|---|
| 1 | 0.6741 | 1.4244 |
| 2 | 0.7505 | 1.5288 |
| 3 | 0.5892 | 1.3925 |
| 4 | 0.7689 | 1.5213 |
| 5 | 0.6005 | 1.4068 |
| 6 | 0.5926 | 1.4083 |
| Mean value | 0.6630 | 1.4470 |
| Standard deviation | 0.0820 | 0.0610 |

Thermal transmittance (U-value) was calculated from the measured thermal parameters. The calculation assumed wall of thickness 44 cm with 15 mm lime plaster on the interior side of wall and perlite plaster of thickness 15 mm on the exterior side of wall. The calculations were performed according to EC ISO 6946. Properties of the recycled aggregate composite wall are compared to a masonry wall of the same thickness.

U-value of the aggregate composite wall was U = 1.0 W/mK; the referential masonry wall had U-value U = 1.2 W/mK.

Material with properties described above is applicable in constructions for structures with moderate loads and non-loadbearing structures. These considerations resulted in an idea of using the recycled aggregate composite as a substitution to masonry in loadbearing walls. The thermal technical parameters of the composite are better compared to solid masonry. Nevertheless, without additional insulation, the composite wall does not meet standard requirements. Considering the thermal characteristics and required resistance capacity, the composite can be used for loadbearing walls in low-storey building.

## 5. Conclusions

Designing of materials for structural applications is restricted by many specifications. E.g. the dosage of recycled aggregate is limited and blends of natural and recycled aggregate are used. The investigations focused on recycled aggregate concrete made entirely with recycled aggregate. The intention of the approach is maximizing of the exploitation of secondary aggregates in concrete for structural applications and changing the perspective and offering the alternative view of designing cement composites from recycled aggregate concrete.

A new porous fibre reinforced composite was proposed, and its material characteristics were tested.

Several potential applications from the composite were suggested and their benefits assessed. The composite used for strengthening of earthen body may result in decrease of farmland occupation and reduce excavations. Strengthening layers contribute to stability of the sloping earth body and increase the resistance. In earthen flood dams the stabilizing fibre reinforced composite layers increase the bearing life in case of overflow in floods. The composite can be applied for loadbearing walls in low-storey buildings.

Certainly the limitations of the proposed composite must be considered, too. The input material – recycled concrete aggregate possesses heterogeneous properties, what affects the scatter of resulting composite characteristics. The composite is a porous low-strength material with higher dispersion of material parameters, what must be considered in the choice of appropriate application.

The limitations are balanced by favourable environmental aspects. Increased utilization of recycled concrete contributes to saving of raw materials, elimination of the need for disposal of construction waste and low carbon production. High economic and ecological effects of the composite is provided by maximising of recycled aggregate use and minimising of cement consumption. Application of the fibre reinforced composite with recycled concrete aggregate preserves primary resources of natural aggregate, decreases depletion of the environment and reduction of construction waste disposal in landfills.

## Acknowledgements

This work was supported by project TH02030649 Environmentally Efficient Construction and Demolition Waste for Structures.

## References

[1] Mas B, Cladera A, del Olmo T and Pitarch F 2012 Influence of the amount of mixed recycled aggregates on the properties of concrete for non-structural use *Constr. Build. Mater.* **27** (1) pp 612–622
[2] Ajdukiewicz A and Kliszczewicz A 2002 Influence of recycled aggregates on mechanical properties of HS/HPC *Cem. Concr. Compos.* **24** pp 269–279
[3] Leiva C, Solís-Guzmán J, Marrero M and Arenas CG 2013 Recycled blocks with improved sound and fire insulation containing construction and demolition waste *Waste Management* **33** pp 663–671
[4] Surya M, Kanta Rao VVL and Lakshmy P 2013 Recycled aggregate concrete for transportation infrastructure *Procedia – Social and Behavioural Sciences* **104** pp 1158–1167
[5] Hansen TC 1992 Recycled aggregates and recycled aggregate concrete, recycling of demolished concrete and masonry *RILEM Report No. 6* (London: RILEM)
[6] Kou SC, Poon CS and Chan D 2007 Influence of fly ash as cement replacement on the properties

of recycled aggregate concrete *J. Mater. Civ. Eng.* **19** (9) pp709–717

[7] Puthussery JV, Kumar R and Garg A 2017 Evaluation of recycled concrete aggregates for their suitability in construction activities: An experimental study *Waste Management* **60** pp 270-276

[8] Tavakoli M and Soroushian P 1996 Strengths of aggregate concrete made using field demolished concrete as aggregate *ACI Mater. J.* **93** (2) pp 182–190

[9] McNeil K and Kang TH-K 2013 Recycled concrete aggregates: a review *Int. J. Concr. Struct. Mater.* **7** (1) pp 61–69

[10] Exteberria M, Vásquez E and Marí AR 2007 Influence of amount of recycled coarse aggregates and production process on properties of recycled aggregate concrete *Cem. Concr. Res.* **37** pp 735–742

# Fracture parameters of fine-grained composites based on the alkali-activated slag

**V Bílek Jr[1], B Kucharczyková[2], H Šimonová[2], D Kocáb[2] and Z Keršner[2]**

[1] Brno University of Technology, Faculty of Chemistry, Purkyňova 464/118, 602 00 Brno, Czech Republic
[2] Brno University of Technology, Faculty of Civil Engineering, Veveří 331/95, 602 00 Brno, Czech Republic

Email: simonova.h@vutbr.cz

**Abstract**. The paper deals with the experimental determination of fracture parameters of the composites based on the alkali-activated slag (AAS) at different ages. The main aim of the experimental investigation was to verify the effect of the addition of shrinkage reducing admixture (SRA) on the overall progress of length changes and fracture properties during AAS composites ageing. Two AAS composites, which differed only in the presence of SRA, were prepared for the purpose of experiments. The mechanical fracture parameters were determined using evaluation of fracture tests carried out on 40 × 40 × 160 mm beam specimens with an initial central edge notch. The monitored parameters were modulus of elasticity, effective fracture toughness and specific fracture energy. These parameters were determined at the age of 3, 28 and 90 days. The obtained results are supplemented by the results of relative length changes measurements performed on both studied composites.

## 1. Introduction

Although cement based composites are the most commonly used quasi-brittle structural materials in the building industry in the world, the environmental aspects related with their production and using of cement and concrete are increasingly being discussed. Their manufacturing process significantly contributes to the global emissions of $CO_2$ [1]. Therefore, there has been an increasing effort to develop and use some alternative binders instead of ordinary Portland cement (OPC) during the recent decades. The alkali-activated aluminosilicate materials are one representative of such type of binders. These binders are created through the mixing of some non- or poorly-crystalline aluminosilicate based material, such as blast furnace slag (BFS) or fly ash (FA), with an alkaline activator (hydroxides, carbonates or the most preferably silicates) and water [2, 3]. Type and dosage of the activator as well as the way of curing have a significant effect on the hydration process and resulting mechanical properties [4]. The major disadvantage of composites based on alkali-activated matrix (AAM) is an increased shrinkage during hardening period, caused by both autogenous and drying shrinkage, which finally results in volume contraction, microcracking and deterioration of tensile and bending properties [5]. For the purpose of reduction of the negative effect of the high shrinkage on the development of physical and mechanical characteristics of alkali-activated composites, shrinkage reducing admixtures (SRA) or fibres are the mostly used [6, 7].

IOP Conf. Series: Materials Science and Engineering **379** (2018) 012018 doi:10.1088/1757-899X/379/1/012018

Because there is not a sufficient amount of available information about fracture behaviour of composites based on AAM, the attention of this paper is besides others also paid on mechanical fracture parameters determined based on three-point bending fracture test evaluation.

## 2. Material and specimens

In the experiment described in this paper, two fine-grained composites based on alkali-activated slag (AAS) were tested. The difference between mixtures was only in the presence of SRA. The experimental and numerical analysis was aimed at the investigation of the effect of SRA on the overall progress of length changes and fracture properties during AAS composites ageing. Ground granulated blast furnace slag with Blaine fineness around 400 $m^2/kg$ was activated by water-glass with silicate modulus of 2.0 to produce alkali-activated binder. The amount of water-glass was adjusted to provide 10% of $Na_2O$ by slag weight and the water-to-slag ratio was 0.42 including extra added water and water present in water-glass. To prepare fine-grained composites, the silica sand with maximum grain size of 2 mm was mixed with slag and activating solution in the dose three times higher than the slag weight. Commercially produced SRA, originally developed for Portland cement based concretes (Stachement AC 600), was chosen, its amount in second mixture was 2% by weight of slag. The sets of specimens were labelled as follows: AAS and AAS_SRA, specimens without and with SRA, respectively.

The AAS composites were mixed by a laboratory mixer with a mixing speed of 30 revolutions per minute and then were cast into polypropylene moulds to prepare prismatic specimens with dimensions of 40 × 40 × 160 mm intended for the fracture test. After 72 hours, specimens were demoulded and stored until the start of experiments. Curing conditions (ambient temperature of 21 ± 2°C and relative humidity of 60 ± 10%) were the same before and after demoulding. Such severe curing conditions were selected to intensify shrinkage and to evaluate the effect of the used SRA. In total, 18 specimens were made from each mixture; six specimens were tested at each investigated age of specimens. The fracture parameters were determined at the age of 3, 28 and 90 days.

Furthermore, length changes of both investigated mixtures on the specimens with dimensions of 100 × 60 × 1000 mm were continuously determined using apparatus with movable front walls. Free movement of specimens in the moulds was ensured by polyethylene foam mat (MIRELON). Curing conditions were the same as were described above for specimens intended for fracture tests. Length changes measurement continued also after demoulding of these specimens (72 hours) up to approximately 90 days. Then the length changes measurement was finished and additional specimens with nominal dimensions 40 × 40 × 160 mm were cut from these bigger specimens and also subjected to the fracture tests. The specimens obtained in this manner are designated as "cut".

## 3. Fracture tests

All above mentioned test specimens were provided by an initial central edge notch with depth about one third of specimen height and subsequently loaded in three-point bending fracture test configuration. The tests were performed using the mechanical testing machine Heckert FP 10/1 which allows loading with controllable speed of displacement increment. For the purpose of this experiment, loading speed was set to value of 0.02 mm/min. The displacement increment loading of the specimen allows recording the load vs. displacement (deflection in the middle of span length) diagrams using an inductive sensor connected to HBM SPIDER 8 data logger during the testing. The measured *L–d* diagrams are introduced in Figure 1. One diagram for each mixture and investigated age of specimens is shown for illustration.

The first (almost linear) parts of the recorded *L–d* diagrams were used for the estimation of the elasticity modulus value. Furthermore, the effective fracture toughness values were determined using the Effective Crack Model [8], which combines the linear elastic fracture mechanics and crack length approaches. The work of fracture and afterwards specific fracture energy values were obtained from the *L–d* diagrams according to the RILEM method [9].

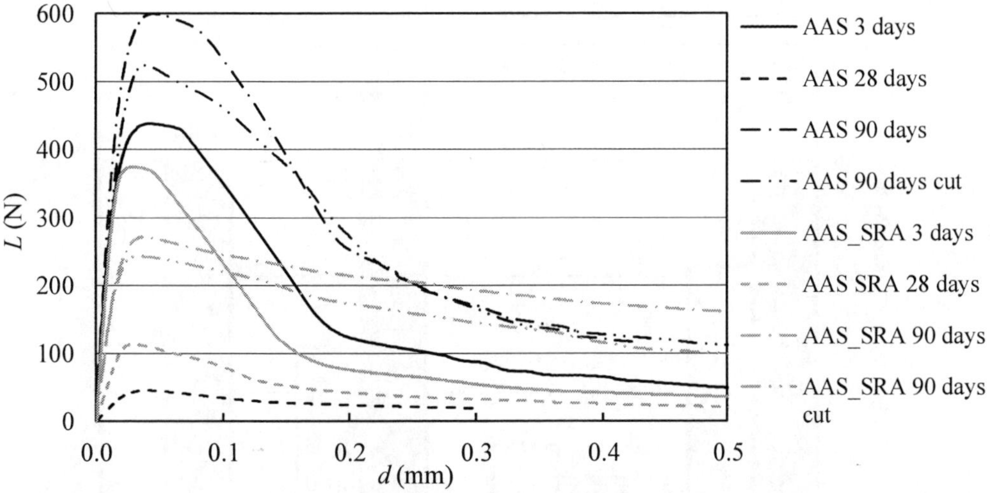


**Figure 1.** The selected *L–d* diagrams.

## 4. Results

The mean values and standard deviations (usually determined from 6 independent measurements) of selected mechanical fracture parameters obtained from the recorded *L–d* diagrams using above described approach [8, 9], i.e. modulus of elasticity, effective fracture toughness and specific fracture energy, are summarized in Figure 2, 3 and 4. The obtained values are introduced for specimens at the age of 3, 28 and 90 days.

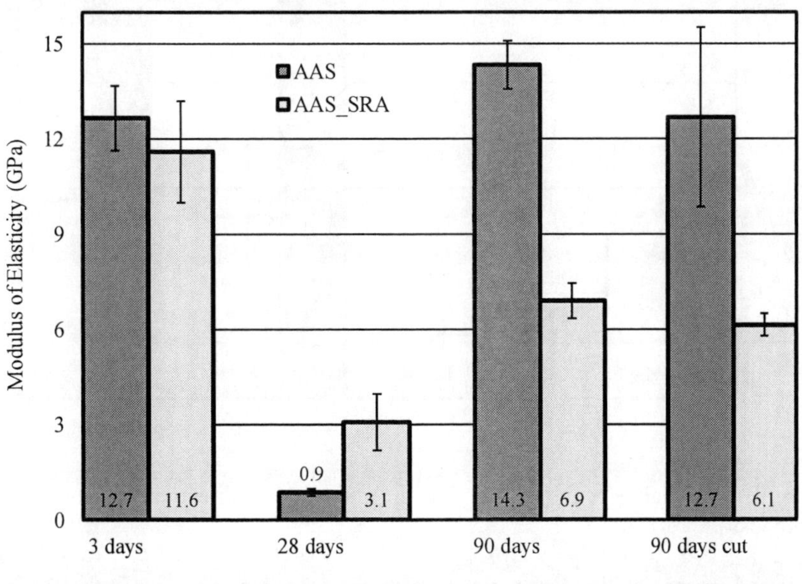


**Figure 2.** The values of modulus of elasticity of alkali-activated fine-grained composites.

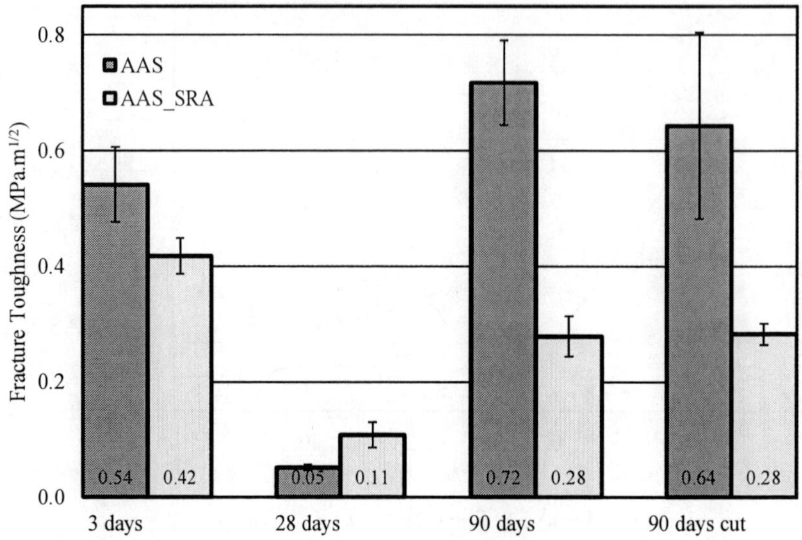


**Figure 3.** The values of the effective fracture toughness of alkali-activated fine-grained composites.

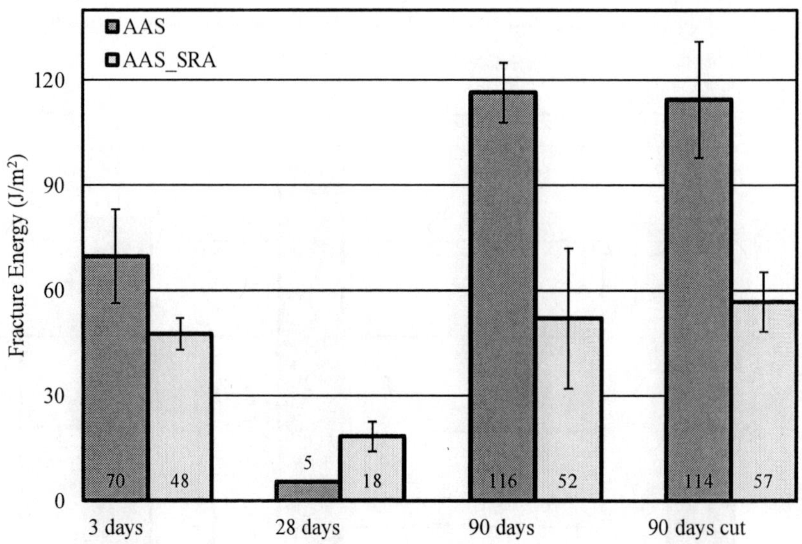


**Figure 4.** The values of the specific fracture energy of alkali-activated fine-grained composites.

All investigated mechanical fracture parameters show unexpected decrease in their values at the age of 28 days. In the case of AAS specimens, the values decreased to less than 1/10 in comparison with the values at age of 3 days. In the case of AAS_SRA specimens, the decrease was not so significant – about 75% for modulus of elasticity and fracture toughness and about 60% for fracture energy. Contrary to these parameters, compressive strength values of both AAS and AAS_SRA specimens noticeably increased between 3$^{rd}$ and 28$^{th}$ day. Such evolution of mechanical properties can be explained by microcracking since fracture parameters are more sensitive to any cracks compared to compressive strength. This is also supported by the results from length changes measurements where beneficial effect of SRA on AAS shrinkage was observed and the decrease in fracture parameters of SRA containing mortar between 3rd and 28th day is lower compared to the reference.

At the age of 90 days, the values of mechanical fracture parameters increased. In the case of AAS specimens, the increase was about 13, 33 and 67% for modulus of elasticity, fracture toughness and fracture energy, respectively, in comparison with values at the age of 3 days. In the case of AAS_SRA specimens, the values were 2–3 times higher than the values at the age of 28 days. In comparison with the values at the age of 3 days, only the fracture energy value increased by about 10%, the modulus of elasticity and fracture toughness values decreased by about 40 and 33%, respectively. Figure 5 shows the details of crack propagation from the initial notch for specimen after fracture tests at the age of 28 days (left) and at the age of 90 days (right). Obviously, the surface of the specimen is disturbed with the cracks, due to a high susceptibility of AAS composites to shrinkage. These cracks are visible at the age of 28 days, while no shrinkage-cracks are visible at the age of 90 days. Supposedly, self-healing occurred in the AAS composites, its origin will be a topic for further study. Note that all the test specimens were stored in the climate control chamber with the temperature of 21 ± 2°C and relative humidity of 60 ± 10% until their testing and were not protected from drying during whole time of ageing.

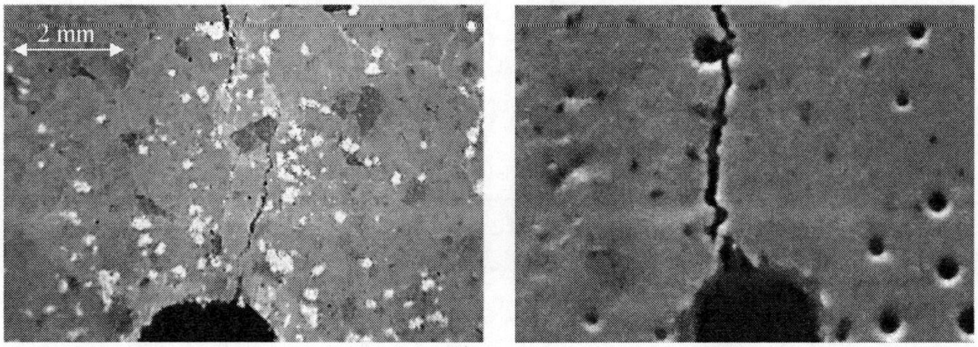


**Figure 5.** Details of crack propagation from initial notch, specimens at the age of 28 (left) and 90 days.

The influence of the specimens manufacturing process was assessed at the age of 90 days. The charts show that there are not significant differences between the results obtained from testing of the specimens casted into the moulds and specimens which were prepared by cutting from bigger specimens primarily used for length changes measurements. In most cases, the difference was about 10%.

The informative compressive strength values were determined on halves of specimens remaining after the fracture tests were performed. The compressive strength values gradually increased with age of specimens for both studied composites (see Figure 6), which is not in correspondence with obtained unexpected decrease of all investigated mechanical fracture parameters values at the age of 28 days.

The obtained mechanical fracture parameters are supplemented by the results of relative length changes measurements performed on both investigated composites. Figure 7 shows the values of relative length changes at the same ages when the fracture tests were performed. Note that in the case of AAS composite, the results are only for one specimen, because another two specimens cracked during the first days in the mould and it was not possible to continue the length changes measurements correctly. In the case of AAS_SRA composite, the mean values and standard deviations obtained from three independent measurements are introduced. The results show that the addition of SRA reduced the relative length changes by about 25% for all investigated ages of specimens. Details regarding the development of the length changes during the whole measurement can be found in [10].

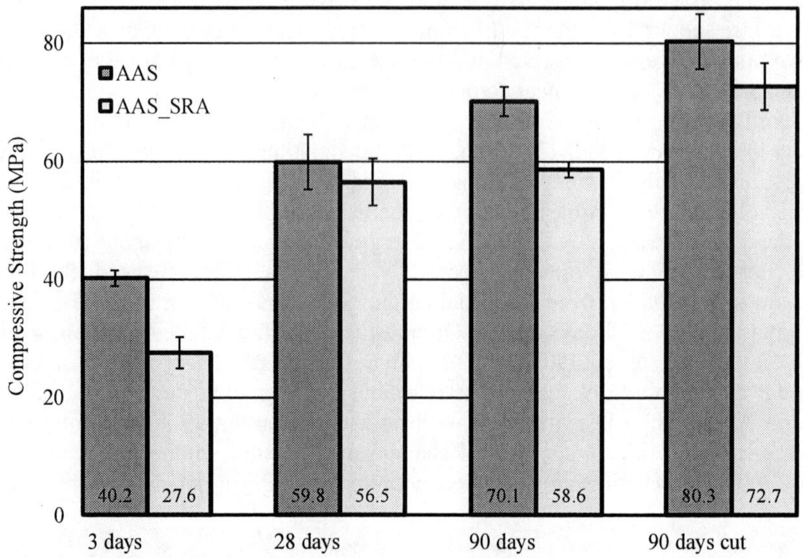


**Figure 6.** The values of informative compressive strength of alkali-activated fine-grained composites.

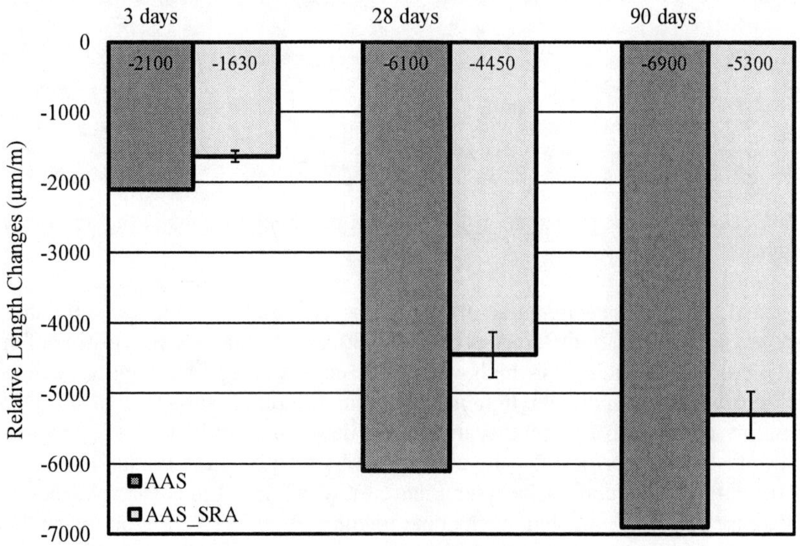


**Figure 7.** The values of length changes of alkali-activated fine-grained composites.

## 5. Conclusions

The paper deals with the investigation of the values of mechanical fracture parameters of AAS composites. In particular, the effect of the shrinkage-reducing admixture (SRA) was studied. It was found that the addition of SRA reduced the relative length changes by about 25% at the age of 90 days. On the contrary, the addition of SRA had a negative impact on the values of the mechanical fracture parameters that is reflected in the decrease of values of about 50–60%. An exception was the results obtained at the age of 28 days, when the monitored mechanical fracture parameters were higher in the case of AAS composite with SRA than in the case of AAS composite without SRA.

If the composites are evaluated separately and the trend of development of the fracture parameters is assessed, an unexpected decrease of all investigated parameters was observed for both tested AAS composites at the age of 28 days. It is suggested that more measurements have to be performed to confirm presented results and to clarify the reason for the decrease of monitored parameters values at the age of 28 days. Therefore, similar experiment will be designed and performed.

**Acknowledgements**

This outcome has been achieved with the financial support of the Czech Science Foundation under project No 18-12289Y.

**References**

[1] Aïtcin P C and Mindess S 2011 *Sustainability of concrete* (New York: Spon Press)
[2] Provis J L and van Deventer J S (eds) 2014 *Alkali activated materials: state-of-the-art report, RILEM TC 224-AAM* (Dordrecht: Springer)
[3] Shi C, Krivenko P V and Roy D 2006 *Alkali-activated cements and concretes* (London: Taylor & Francis)
[4] Shi C and Day R L 1996 Some factors affecting early hydration characteristics of alkali-slag cements *Cement Concrete Res* **26** (3) pp 439–447
[5] Cincotto M A, Melo A A and Repette W L 2003 Effect of different activators type and dosages and relation with autogenous shrinkage of activated blast furnace slag cement *Proc. of the 11th International Congress on the Chemistry of Cement* eds G Grieve and G Owens (Durban: Cement Concrete Institute of South Africa) pp 1878–1888
[6] Vilaplana J L, Baeza F J, Gala O, Alcocel, E G, Zornoza E and Garcés P 2016 Mechanical properties of alkali activated blast furnace slag pastes reinforced with carbon fibers *Constr Build Mater* **16** pp 63–71
[7] Rovnaník P, Šimonová H, Topolář L, Bayer P, Schmid P and Keršner Z 2016 Carbon nanotube reinforced alkali-activated slag mortars *Constr Build Mater* **119** pp 223–229
[8] Karihaloo B L 1995 *Fracture Mechanics and Structural Concrete* (New York: Longman Scientific & Technical)
[9] RILEM TC-50 FMC Recommendation 1985 Determination of the fracture energy of mortar and concrete by means of three-point bend test on notched beams *Materials & Structures* **18** pp 285–290
[10] Kucharczyková B, Bílek V Jr., Kocáb D and Karel O 2018 Shrinkage of Fine-grained Composites Based on Alkali-activated Slag *Key Engineering Materials: Non-Traditional Cement and Concrete* eds Š Nenadálová et al (Switzerland: Trans Tech Publications) **761** pp 7–10

# Effect of particle size of waste brick powder on the properties of alkaline activated materials

**J Brînduş-Simuţ[1], M Vyšvařil[1], P Bayer[1], M Keppert[2] and P Rovnaníková[1]**

[1] Faculty of Civil Engineering, Brno University of Technology, Veveří 331/95, Brno 602 00, Czech Republic
[2] Faculty of Civil Engineering, Czech Technical University, Thákurova 7/2077, Praha 166 29, Czech Republic

E-mail: rovnanikova.p@fce.vutbr.cz

**Abstract.** The properties of alkaline activated brick dust mixed with water glass with a silicate modulus (Ms) of 1.0 were studied in fresh mixtures as well as hardened products in connection with the particle size of the dust. The brick dust (ground waste brick) was separated into 5 fractions according to particle size. The influence of particle size on the rheological properties of the fresh mixtures was studied via measurements of the yield stress and viscosity. The flexural and compressive strengths of the hardened products were determined after storage under laboratory conditions for 7, 14 and 90 days; the influence of 7 days of curing at 65°C on strength was evaluated as well. The microstructure of the hardened products was studied by means of porosimetry and SEM.

## 1. Introduction

Bricks are produced by the burning of clay minerals. During this process, a certain amount of amorphous phases of aluminosilicates are formed. Amorphous phases are necessary for pozzolanic activity, i.e. the reaction of brick powder with an alkaline solution. Brick powder is waste material from the production of calibrated brick elements; it can be also obtained from the processing of demolition waste. The effects of different conditions for the alkali-activation of brick powder have been studied. Reig et al. [1] activated red clay brick waste using sodium hydroxide and sodium silicate solutions. They studied the effects of several variables, mainly concentration of NaOH, $SiO_2/Na_2O$ ratio and water/binder ratio on the porosity, composition and morphology of reaction products. With the optimal composition and a curing temperature of 65°C, a compressive strength of 50 MPa was achieved. Robayo et al. [2] reported similar results. Komnitsas et al. [3] focused on the effects of the particle size of waste brick powder and the temperature of curing on the properties of hardened pastes. They studied the aging period, the temperature of curing and the effect of freeze-thaw cycles on compressive strengths. When studying the composition of the product, they found a new mineral phase, pirrsonite. Sassoni et al. [4] designed a mixture of alkali-activated brick powder for use as a repair material for brick masonry. Molar ratios of $SiO_2/Al_2O_3$ ranging from 0.4 to 1.4 and curing temperatures of 20 and 50°C were investigated. Pastes with $SiO_2/Al_2O_3$ ratios of 0.8 and 0.9 exhibited open porosity and water vapour permeability fairly similar to those of historic lime-based mortars.

The surface area of the waste brick powder is an important factor in the course and outcome of the reaction. Different granulometric classes of calcined clay (125 μm, 100 μm, 80 μm and 63 μm) were used as an aluminosilicate precursor for the fabrication of geopolymers. The influence of particle size on the kinetics of geopolymerization reactions and on the mechanical properties of geopolymers was studied [5]. As the particle size of the waste brick powder decreased, the rate of reaction increased.

The rheological properties of alkali-activated materials are different from those of Portland cement-based pastes [6–8]. The mechanisms of the formation of an aluminosilicate gel at an early age are strongly dependent on the chemical composition of the mixture, i.e. on the type of precursor and the composition of the activator. The presence of Ca in solid precursors, for example in slags or fly ashes, could dramatically change the chemistry of the system, and this will change the rheological properties of fresh pastes [9].

A precursor is a waste material that can vary in particle size. Therefore, the aim of the study was to monitor the effect of particle size on the properties of fresh mixtures and hardened products.

## 2. Materials and methodology

### *2.1 Characterization of raw materials*

The investigated brick powder originates from the grinding of calibrated brick elements from the HELUZ Brickworks factory, v.o.s., located in Hevlín (Czech Republic). The chemical composition of the brick powder was obtained via XRF analysis (PANalytical) and is presented in Table 1; the mineralogical composition was obtained via XRD analysis (SAXS - PANalytical Empyrean) and is specified in Table 2. The proportions of the components were calculated via Rietveld analysis. Sodium water glass (Vodní sklo, a. s., Czech Republic) contains 26.43% $SiO_2$, 16.61% $Na_2O$, and 56.96% $H_2O$. The ratios of the elements contained in samples of alkali-activated brick powder are Si/Al = 0.44, Ca/Si = 0.23, (Na+K)/Si = 0.27, (Na+K)/Al = 0.12, (Na+K+Ca+Mg)/(Si+Al) = 0.17.

**Table 1.** Chemical composition of brick powder.

| | $SiO_2$ | $Al_2O_3$ | $Fe_2O_3$ | CaO | MgO | $K_2O$ | $Na_2O$ | $P_2O_5$ | $TiO_2$ | $SO_3$ | Sum |
|---|---|---|---|---|---|---|---|---|---|---|---|
| Mass % | 55.5 | 17.0 | 5.8 | 10.5 | 2.4 | 2.8 | 0.7 | 0.2 | 0.8 | 1.9 | 97.6 |

**Table 2**. Mineralogical composition of brick powder.

| Mineral | Formula | [%] |
|---|---|---|
| Quartz | $SiO_2$ | 26.2 |
| Hematite | $Fe_2O_3$ | 2.3 |
| Albite | $NaAlSi_3O_8$ | 13.0 |
| Microcline | $KAlSi_3O_8$ | 3.6 |
| Orthoclase | $KAlSi_3O_8$ | 3.5 |
| Muscovite | $KAl_2(AlSi_3O_{10})(OH)_2$ | 12.5 |
| Illite | $K_{0.65}Al_2(Al_{0.65}Si_{3.35}O_{10})(OH)_2$ | 3.8 |
| Diopside | $CaMgSi_2O_6$ | 4.4 |
| Akermanite | $Ca_2MgSi_2O_7$ | 2.8 |
| Amorphous | - | 27.8 |

The pozzolanic activity of the brick powder was determined using the modified Chapelle test. A mixture of freshly annealed lime and water in a tightly closed container is stirred with a magnetic stirrer at a temperature of 93°C for 24 h inside a thermobox. The residual calcium hydroxide is determined via the saccharide method. The result represents the amount of calcium hydroxide consumed in the reaction of 1 g of brick powder per 24 hours. The pozzolanic activity of the used brick powder is 356 mg $Ca(OH)_2$ per 1 gram of brick powder. Pozzolanic activity is an important parameter for the alkaline activation of brick powder.

The particle size distribution of the brick powder with grains of less than 1 mm in diameter is shown in Figure 1, where $d_{10}$ = 2.094 μm, $d_{50}$ = 18.750 μm, and $d_{90}$ = 302.256 μm. The brick powder was sorted into 5 fractions: 0–90 μm, 0–125 μm, 0–250 μm, 0–500 μm and 0–1000 μm.

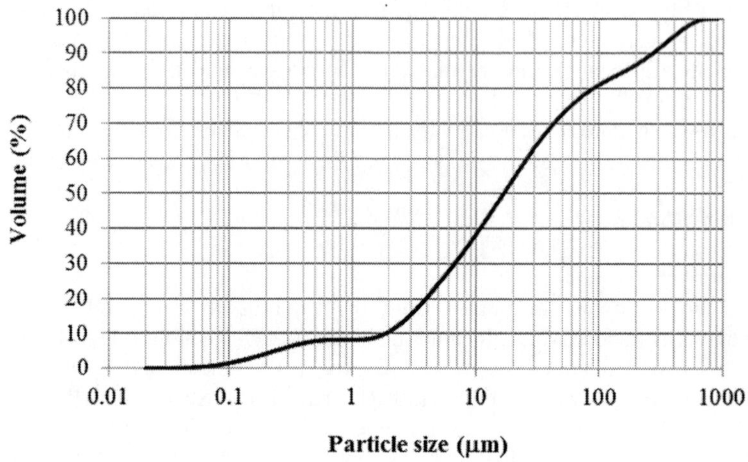


**Figure 1**. Particle size distribution of the brick powder.

*2.2. Composition of tested mixtures*

The mixtures containing alkali-activated brick powder were prepared from 200 g of brick powder of an appropriate particle size, and an alkaline solution which contains 44 g of sodium water glass with a silicate modulus ($M_s$) of 1.6, 6.04 g of sodium hydroxide, and 56 g of water. The total mass of the raw materials was kept constant for all samples, and the silicate modulus of the alkaline solution was $M_s$ = 1.0 for all mixtures. Five mixtures with different brick powder particle sizes (Table 3) were prepared for subsequent testing. First, the alkaline activator was prepared by mixing sodium hydroxide solution with sodium water glass; second, the brick powder and alkaline activator were blended together.

**Table 3.** Particle size of brick powder in mixtures.

| | G1 | G2 | G3 | G4 | G5 |
|---|---|---|---|---|---|
| Grain size [mm] | 0 – 0.09 | 0 – 0.125 | 0 – 0.250 | 0 – 0.500 | 0 – 1.000 |

The fresh mixtures were poured into moulds; they were then covered with PE foil and stored for 1 day in the laboratory at 20 ± 1°C. After 1 day the test specimens were demoulded and wrapped in the PE foil again. Some of the specimens were stored at a laboratory temperature of 20 ± 1°C for 7, 28 and 90 days, while the remainder were stored at a temperature of 65°C for 7 days.

## 3. Experimental procedure

The flow properties and viscoelastic characteristics of the pastes were evaluated through rheological measurements with the aid of a Discovery HR-1 (TA Instruments) hybrid rheometer and TRIOS 4.0.2.30774 software. The measurements were performed in a Peltier Concentric Cylinder system with a DIN rotor at 25°C. The paste was introduced into the measurement system at the end of the mixing cycle. After a 60s period of rest, the rheological measurements were started. The testing routine comprised a shear rate increase (from 0.1 to 100 $s^{-1}$) applied in 30 steps with 15 s of measuring time at each shear rate. This was followed by a gradual reduction in the shear rate performed in the same manner. The results were expressed as shear rate vs. shear stress (flow curves) and the Herschel-Bulkley model (1) was applied to the descending branches of the flow curves in order to fit the experimental data. It was also used to describe the rheological behaviour of the pastes [10]:

$$\tau = \tau_0 + k\dot{\gamma}^n \quad (1)$$

where $\tau$ is the shear stress, $\dot{\gamma}$ is the shear rate, $\tau_0$ corresponds to the yield stress, $k$ is the consistency coefficient and $n$ is the fluidity index which characterizes the shear-thinning ($n < 1$) or shear-thickening ($n > 1$) behaviour of the material. The time evolutions of the viscoelastic properties of the pastes were evaluated by small amplitude (0.01%) oscillation tests at 1 Hz frequency. These tests measured the dynamic moduli $G'$ and $G''$, which represent the elastic and viscous behaviour of the material. If the elastic behaviour is dominant ($G' > G''$), the sample exhibits the characteristics of a gel or solid state. Conversely, if $G'' > G'$, the sample shows the properties of a liquid.

The bending and compressive strengths of 20×20×100 mm test specimens were determined. Prior to the testing of their mechanical properties, the specimens were weighed and their dimensions measured. Bulk densities were calculated from the measured values. The porosity of the alkali-activated brick powder was assessed using a PoreSizer 9310 mercury porosimeter (Micromeritics). Pore diameter, cumulative intruded volume, cumulative pore surface and differential intruded volume were determined. Detailed microstructure images were taken via a MIRA3 XMU environmental scanning electron microscope (Tescan) [11, 12]. Three samples, designated G1, G2 and G3, were monitored.

## 4. Results and discussion

### *4.1. Rheological properties*

The shape of the descending branches of the flow curves (Figure 2) indicates the dilatant character of the brick powder pastes. The yield stress of the pastes varied with the particle distribution in the brick powder (Table 4). With increasing particle sizes up to 0.25 mm the yield stress decreased, which was expected, but grew again with larger particles. The yield stresses of the pastes were very high and considerably higher than the yield stresses of metakaolin geopolymers [13]. The consistency coefficients of the pastes (Table 4) were on the same value level as for metakaolin geopolymer pastes [14], and they decreased with increasing brick powder particle size. This is due to the decrease in amorphous phase content as brick powder particle size increases. Mixtures with high crystalline content are very reluctant to flow. The fluidity indexes of pastes were higher than 1, which proved these mixes were shear-thickening. The shear-thickening character of the pastes became more significant with increasing brick powder particle size.

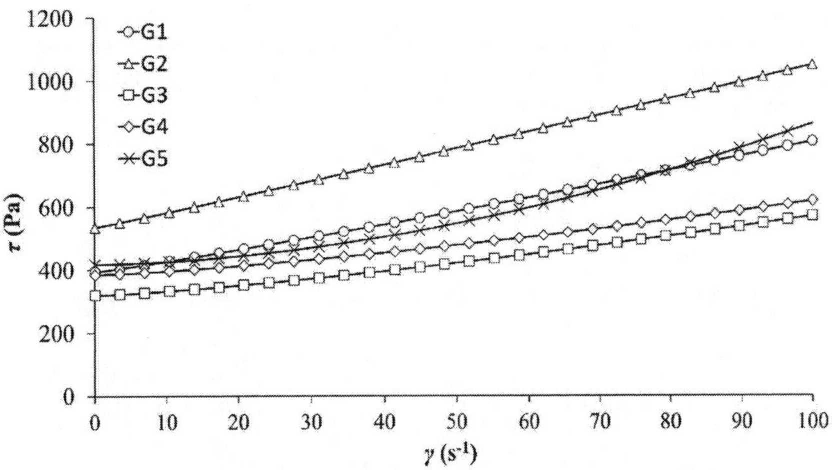


**Figure 2.** Descending branches of flow curves of brick powder mixtures.

Figure 3 shows the time evolution of the viscoelastic parameters $G'$ and $G''$ during a small amplitude (0.01%) oscillation test. It is evident that $G'$ decreased and $G''$ increased with increasing brick paste particle size. This means that with increasing particle size, the viscous component grows and the paste structure becomes less stiff, which is also displayed via the diminishing viscosity (consistency coefficient) of the pastes.

**Table 4**. Flow parameters of brick powder mixtures obtained from the Herschel-Bulkley model.

| | $\tau_0$ (Pa) | $k$ (Pa s) | $n$ (-) |
|---|---|---|---|
| G1 | 534.80 | 4.19 | 1.04 |
| G2 | 393.70 | 2.54 | 1.10 |
| G3 | 320.01 | 0.59 | 1.31 |
| G4 | 384.65 | 0.50 | 1.33 |
| G5 | 417.67 | 0.12 | 1.79 |

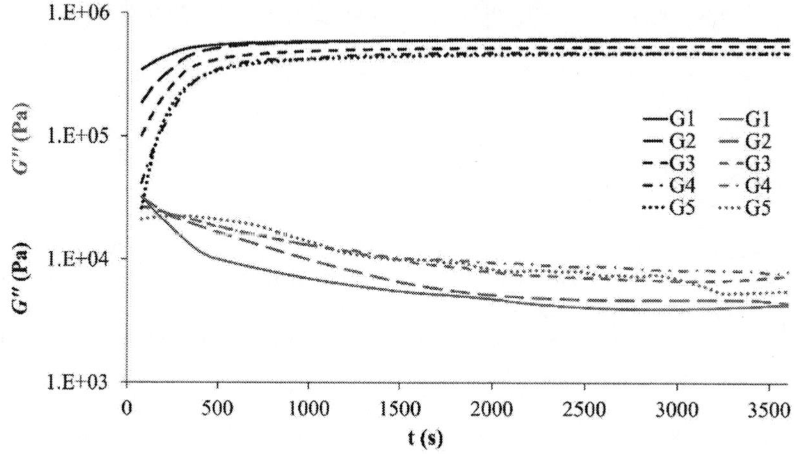


**Figure 3.** Time evolution of the storage modulus ($G'$) and loss modulus ($G''$) of brick powder mixtures.

*4.2. Mechanical properties*

The bulk densities were independent of the time and temperature of curing. The specimens cured at 20°C had higher bulk densities (1860 to 1910 kg·m$^{-3}$) compared to the specimens cured at a temperature of 65°C for 7 days (1720 to 1820 kg·m$^{-3}$).

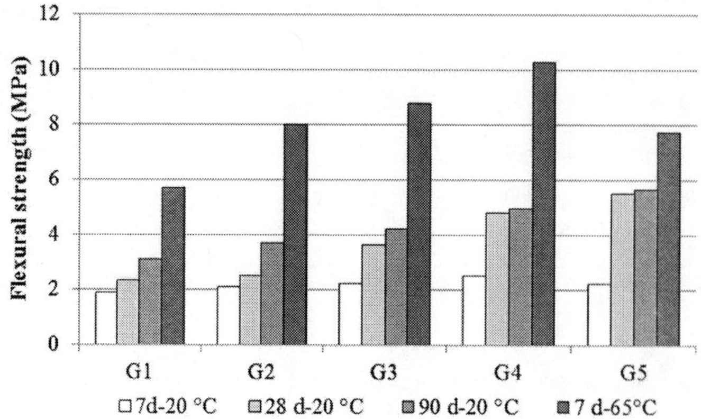


**Figure 4**. Flexural strength over time.

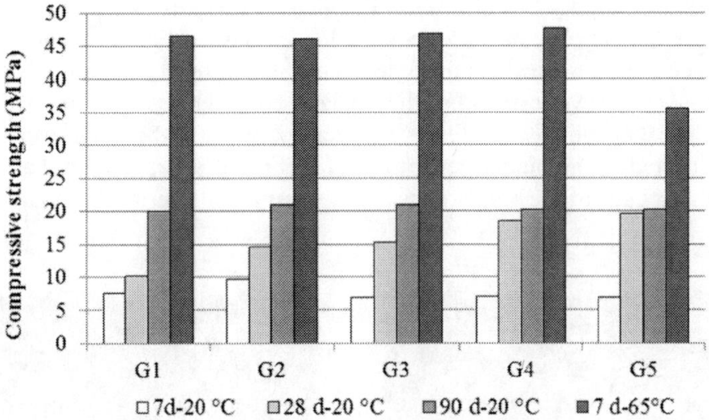


**Figure 5**. Compressive strength over time.

The flexural strengths (Figure 4) of pastes increased with curing time at 20°C, and with larger brick grain size. Significantly higher strengths reached specimens cured at 65°C. Compressive strength (Figure 5) also increased with time of curing up to 28 days, but in contrast to flexural strength all samples exhibit approximately similar compressive strengths at 90 days. The compressive strength of specimens cured at 65°C was markedly higher.

*4.3. Porosimetry*

Cumulative pore volume included both pores and cracks; it was determined on samples matured for 28 days at 20°C. The total pore and crack volumes corresponded to the 28 day strengths, in particular to the flexural strength. The G1 mixture with the lowest flexural strength of 2.32 MPa had the largest total pore and cracks volume (0.169 $cm^3 \cdot g^{-1}$).

On the other hand, the G5 mixture with a 5.28 MPa flexural strength had a total pore and crack volume of 0.135 $cm^3 \cdot g^{-1}$. The curves showing cumulative pore and cracks volume display specific pore diameter and crack intervals in which most of the total number of pores and cracks in all specimens are present. The G1 mixture had a range of pore diameters from 0.25 to 0.03 μm, G3 had diameters of 0.15 to 0.03 μm, and the G5 mixture had 1 to 0.1 μm. Sample G5 contained larger pores and cracks, but their cumulative volume was lower than in the case of samples G1 and G3 (Figure 6).

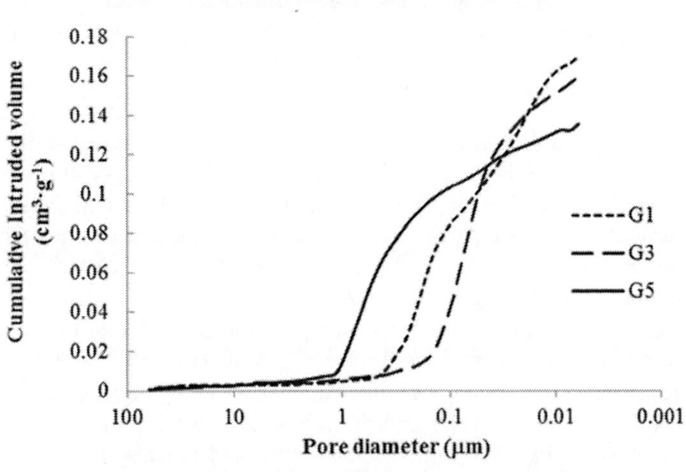


**Figure 6.** Cumulative volume of pores.

### *4. 4. Microstructure*

Micrographs of samples G1, G3 and G5 matured for 28 days at 20°C are present in Figures 7–9. Sample G1 contained grains below 90 μm in size and this led to the formation of a compact binder. The pores in the binder were very small but the binder's compactness was broken by the formation of shrinkage cracks. The G3 sample had fewer shrinkage cracks, and the G5 sample had the fewest. The larger grains located in the binder perform the function of an aggregate that transmits the stress resulting from the shrinkage of the binder phase. The micrographs verify the results of the porosimetry measurements.

**Figure 7.** Sample G1.

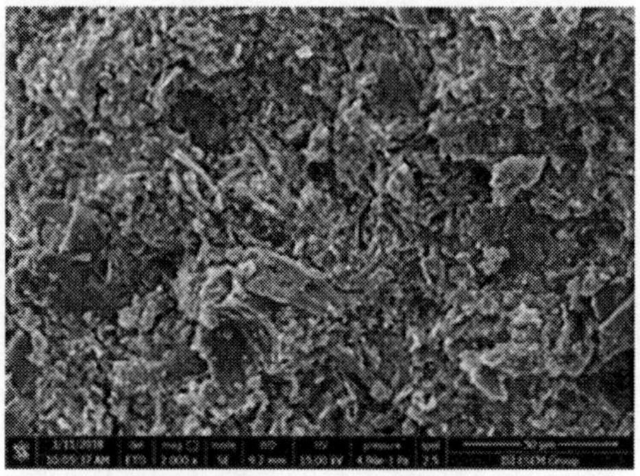

**Figure 8.** Sample G3.

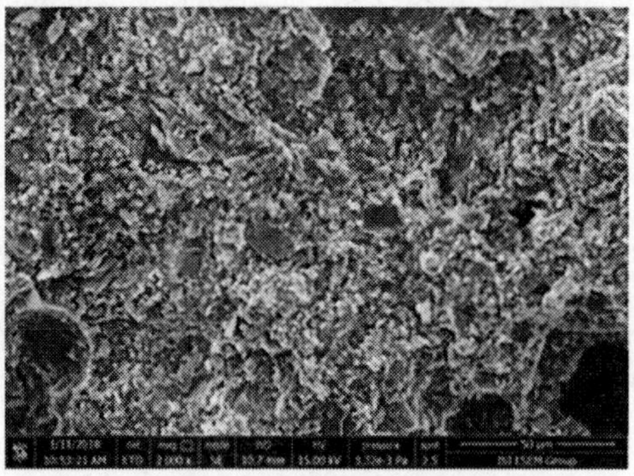

**Figure 9.** Sample G5.

## 5. Conclusions

This paper has focused on the properties of fresh and hardened alkali-activated brick powder pastes in connection with the grain size of the powder. The main conclusions are as follows:

- The pastes were shear-thickening suspensions with very high yield stress and low viscosity; the paste structure became less stiff with increasing particle size of brick powder.
- The flexural strength of samples cured at a temperature of 20°C increased over time and with increasing brick dust grain size.
- The compressive strength increased depending on the grain size of the brick powder up to 28 days; the strengths of all samples were similar after 90 days.
- Both of the strengths of the samples treated at 65°C were significantly higher.
- The mixtures prepared with the smallest grain size of brick powder showed the largest total pore and cracks volume. The high value of pore and crack volume is due to formation of shrinkage cracks.

IOP Conf. Series: Materials Science and Engineering **379** (2018) 012019 doi:10.1088/1757-899X/379/1/012019

- The compactness of the binder and significant shrinkage cracks decreasing with grain size of brick powder.
- The porosimetry measurements, tensile strengths and microstructure images were mutually consistent.

**Acknowledgments**
This research has been supported by the Czech Science Foundation under project No 16-02862S.

**References**

[1] Reig L, Soriano L, Borrachero M V, Monzó J and Payá J 2013 *Constr. Build. Mater.* **43** pp 98–106
[2] Robayo R A, Mulford A, Munera J and Mejía de Gutiérrez M 2016 *Constr. Build. Mater.* **128** pp 163–9
[3] Komnitsas K, Zaharaki D, Vlachou A, Bartzas G and Galetakis M 2015 *Adv. Pow. Tech.* **26** pp 168–186
[4] Sassoni S, Pahlavan P, Franzoni E and Bignozzi M C 2016 *Ceram. Int.* **42** pp 14685–14694
[5] Louati S, Baklouti S and Samet B 2016 *Appl. Clay Sci.* **132-133** pp 571–8
[6] Puertas F, Varga C and Alonso M M 2014 *Cem. Concr. Comp.* **53** pp 279–288
[7] Singh B, Ishwarya G, Gupta M and Bhattacharyya S K 2015 *Constr. Build. Mater.* **85** pp 78–90
[8] Kashani A, Provis J L, Qiaoa G G and van Deventer J S J 2014 *Constr. Build. Mater.* **65** pp 583–591
[9] Favier A, Habert G and de Lacaillerie J B E 2013 *Cem. Concr. Res.* **48** pp 9–16
[10] Mezger T G 2014 The Rheology Handbook 4th ed. (Hannover: Vincentz Network)
[11] Navrátilová E and Neděla V 2016 *Microsc. Microanal.* **22** (S3) pp 1862–3
[12] Navrátilová E and Neděla V 2017 *Microsc. Microanal. 23* (S1) pp 2186–7
[13] Poulesquen A, Frizon F and Lambertin D 2011 *J. Non-Cryst. Solids.* **357** pp 3565–3571
[14] Aboulayt A, Riahi M, Anis S, Ouazzani T M and Moussa R 2014 *Int. J. Innov. Appl. Stud.* **7** pp 1170–7

IOP Conf. Series: Materials Science and Engineering **379** (2018) 012020 doi:10.1088/1757-899X/379/1/012020

# Utilization of by-products for vacuum insulation panels production

**J Zach[1], J Peterková[1] and V Novák[1]**

[1] Brno University of Technology, Faculty of Civil Engineering, Centre AdMaS, Purkyňova 139, 612 00 Brno, Czech Republic

E-mail: zach.j@fce.vutbr.cz

**Abstract.** The paper deals with the possibilities of utilization of alternative raw materials from the textile industry for the production of vacuum insulation panels with extremely low thermal conductivity value. The paper presents current research findings on the behavior of fibrous structures on an organic basis at extremely low pressure. In addition, the paper presents the practical findings and results of experimental work in the production area and the behavior of vacuum insulation panels with an alternative core insulator based on organic fibers.

## 1. Introduction

The development in the field of thermally insulating materials in the civil engineering is very closely related to the current situation in the area of thermal protection of buildings and the demands on building structures from the point of thermal protection. Due to the recent implementation of the European Parliament and committee directive from 2010/31/EU [1] on the development of progressive insulating materials, there's an effort to achieve better thermal insulation properties in their development.

At the same time, however, it is an attempt to use as many renewable materials and by-products in the production of building materials as possible.

From the progressive technologies of thermal properties point of view, the vacuum insulation panels (VIP) are at present the representatives of materials with the lowest equivalent value of the thermal conductivity coefficient. These products reach extremely low values of the thermal conductivity coefficient, which in practice is at the level of 3 mW/(m.K) for the glass wool based core insulators and 5 mW/(m.K) for $SiO_2$– based insulators (in the form of aerogels or pyrogenic $SiO_2$). The current problem with the use of VIP is their relatively high price and also their limited life, which is especially problematic when using the materials in civil engineering [2]. In recent years, alternative core insulation, for example on the basis of melamine-formaldehyde fibers [3] are being used in the field of VIP as well as insulants based on by-products.

At the Brno University of Technology, Faculty of Civil Engineering has been researching for several years the possibility of using alternative (especially) organic raw materials in the production of VIP. The aim is primarily to design optimal insulators using natural fibers and fibers based on by-products. A problem with the design and manufacture of these alternative insulators is the selection of suitable fibers with the desired thickness and high purity so that there is no outgassing in the resulting VIP after vacuum or the subsequent degradation of its thermal insulation properties. Also important is the proper orientation of the fibers during the production and design of the insulator at optimal bulk density, so as to avoid rapid degradation of the thermal insulating properties when the pressure inside the panel is increased [4, 5].

IOP Conf. Series: Materials Science and Engineering **379** (2018) 012020 doi:10.1088/1757-899X/379/1/012020

## 2. Design and production of VIP

A selection of suitable feedstock was carried out in the framework of the research. The following raw materials were chosen: hemp fibers, flax fibers, cotton fibers, sheep wool fibers, polyester fibers (primary and secondary).

These samples were studied by their properties and microstructure (see Figures 1 and 2). The thickness and length of the fibers were determined and their microscopic analysis was performed. Bast fiber (hemp and flax) samples have been found to exhibit relatively inappropriate properties for theproduction of vacuum insulations. Classical treatment of plants by mechanical defibering does not cause defibering on primary fibers, but on technical fibers, their thickness is on average more than 90 microns (in terms of fiber thickness, the difference between hemp and flax fibers is not essential).

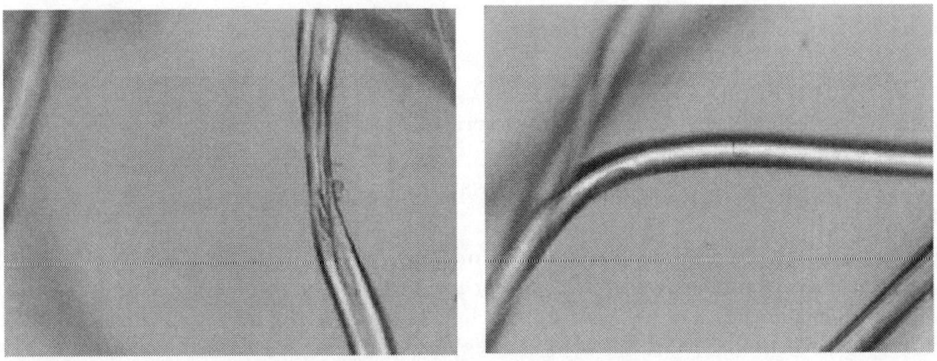

**Figure 1.** Photo of cotton fiber (left) and PES fiber (right) at 40x magnification.

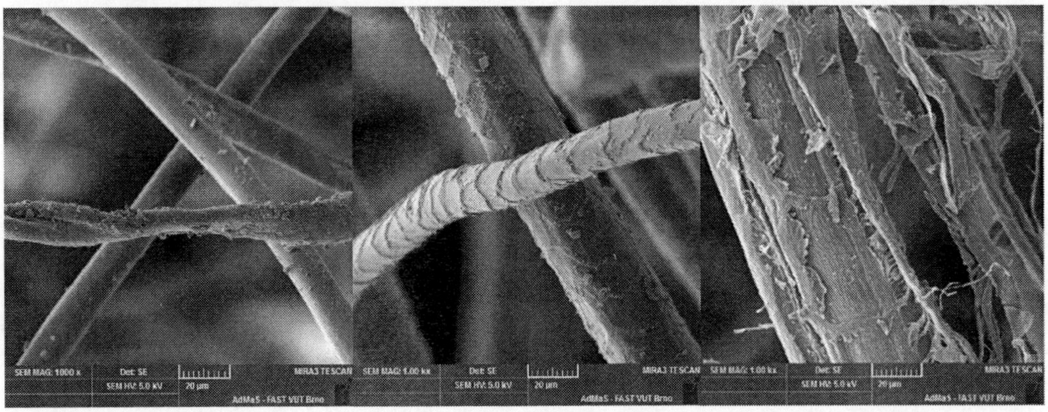

**Figure 2.** REM picture – detail of structure of insulation based on textile/cotton fibers (left), sheep wool (in the middle) and flax (right).

Further on, it has been discovered that the thickness of the bast/natural fibers can be further adjusted by intensive mechanical separation, whereby it can thus reach up to 10 microns of fiber thickness. In these cases, however, the fiber length is reduced rapidly. The dependence between the length and thickness of the fibers was noted, as seen in Figure 3.

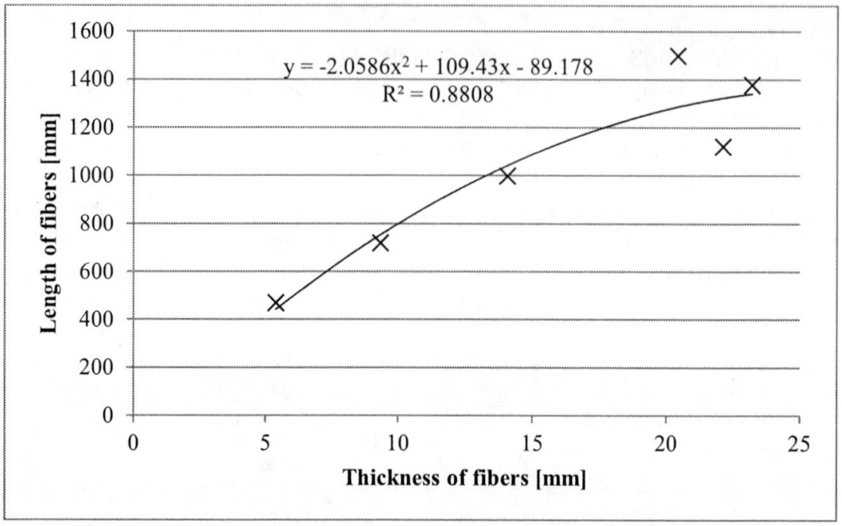


**Figure 3.** Dependence of length of flax fibers on thickness of flax fibers.

Reducing the thickness of flax fiber to 5.4 microns resulted in a fiber length of 469 microns (0.47 mm), whereby the fiber is inappropriate for making the core insulation using the bicomponent technology of binding by the air-lay method. High organic contamination was discovered during a microscopic analysis of sheep wool, which makes these fibers unsuitable for vacuum.

**Table 1**. Overview of thicknesses and lengths of fibers.

| Type of Fiber | Thickness [μm] | Length [mm] |
|---|---|---|
| recycled PES | 12.41 | 49.9 |
| primary PES | 12.26 | 35.5 |
| cotton | 10.80 | 26.1 |
| hemp | 90.51 | 41.2 |
| flax | 91.24 | 38.5 |
| sheep wool | 24.12 | 28.3 |

The primary and secondary polyester fibers, cotton fibers and flax fibers were selected to produce vacuum insulation (according to results – see in table 1 and availability of fibers). Core insulators were created by thermal bonding method with the addition of 15% connective bi-component PES fibers.

## 3. Results and discussion

Test samples (size: 200 x 200 mm) were created from the manufactured insulations for laboratory experimental measurements. Determination of the following key properties was carried out: bulk density, thickness, linear dimensions, stress at 10% deformation [6–10]. The bulk density of samples ranged between 81–150 kg/m$^3$. The lowest value was found for cotton samples and the highest value for hemp samples. The samples showed a higher compressibility of over 10%, which is the standard for determining the mechanical properties according to EN 826. Therefore, compression after vacuuming of panels in all cases exceeds the 10% standard of compression. Samples of insulators were produced for a nominal thickness of 10 mm.

On the test samples, determination of thermal conductivity (according to ISO 8301 [11]) was also carried out in dependence on pressure and under vacuum in dependence on temperature too. Vacuum pressure was determined at 5 Pa/0.05 mBar (99.9% vacuum), which is the typical initial pressure at VIP production. Thermal conductivities for pressures of 1, 10 and 100 mBar were also determined. Subsequently, thermal conductivity was determined at normal atmospheric pressure. The results are shown in the following chart (Figure 4).

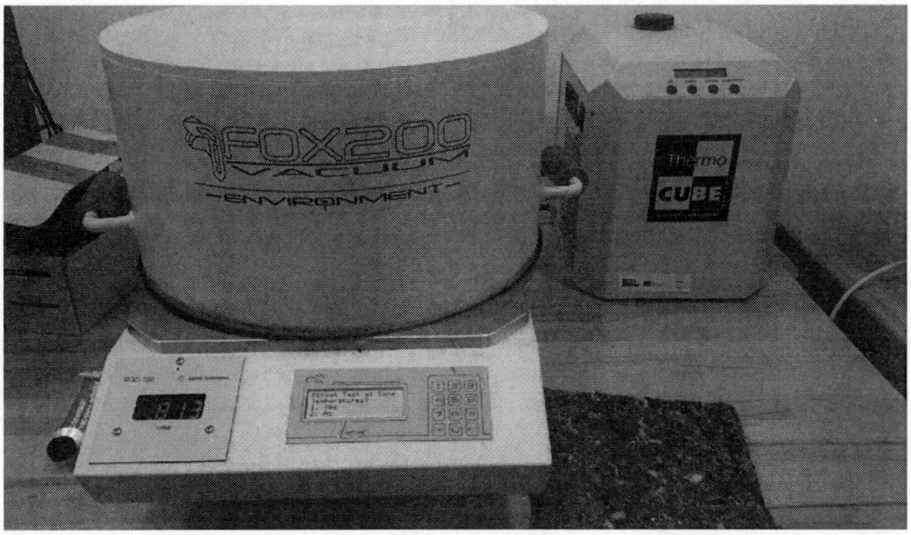


**Figure 4.** Photo of the test apparatus FOX 200 Vacuum.

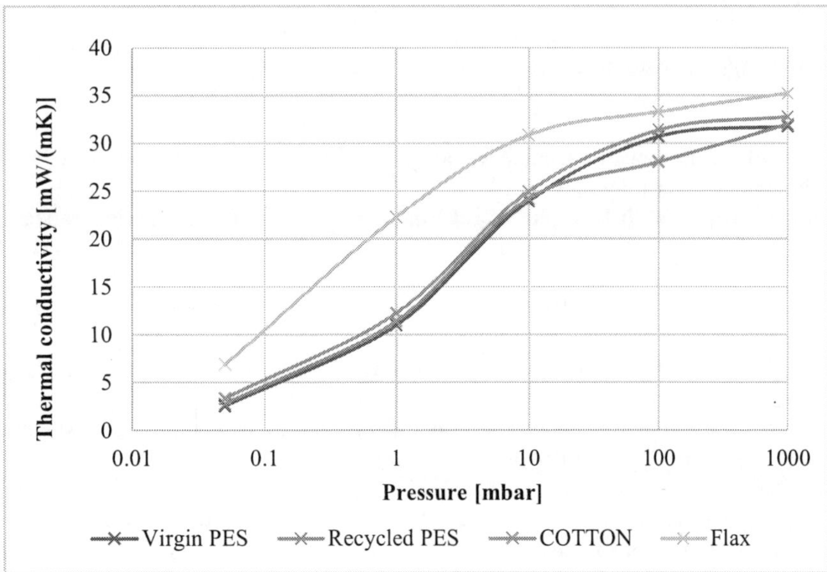


**Figure 5.** Dependence of thermal conductivity of core materials on pressure.

The results show that for all test specimens the increase in pressure led to a rapid degradation of the thermal insulation properties. The best results were achieved in PES samples, where, for both core insulators, a thermal conductivity of under 3.0 mW/(m.K) was achieved with a pressure of 5 Pa, specifically for the Virgin PES sample it was 2.5 mW/(m.K) and for the recycled PES sample it was 2.8 mW/(m.K). Those result are comparable with the best result with VIP based on very fine glass fibers [2]. In the case of cotton, the value was 3.3 mW/(m.K) with a lower thickness of fibers. This is due to worse defibering and especially the curvature of the cotton fibers, which leads to poor orientation of fibers in the production of mats and to the creation of more contact surfaces in the direction of the heat flow. The flax-based sample showed the worst heat-insulating properties due to the high fiber thickness in both the vacuum (a value of 6.8 mW/(m.K) was reached), and the subsequent pressure increase, when even the pressure of 1 mBar caused increase in thermal conductivity to 22 mW/(m.K). In case of flax fibers is possible to see negative aspect of higher thickness of fibers – in accordance with results of other scientific work [12].

## 4. Conclusion

As can be seen from the obtained results, primary and recycled PES samples can be used to produce VIP with very good thermal insulation properties, especially as an alternative to a glass wool based core VIP. An advantage of the recycled PES is the use of the by-product in the production of advanced thermal insulation, as well as a lower price compared to very fine fibers based on glass wool.

As is obvious from the dependence of thermal conductivity on pressure, all insulators have a rapid degradation of thermal insulation properties in the area of higher pressure. The degree of degradation of the thermal insulation properties in dependence on pressure is practically identical for the primary PES, recycled PES, and cotton based insulations too. VIPs made from these insulators, therefore have a lower durability and are particularly suitable for use in refrigeration and for example in packaging technology (here the advantage is, in particular, the consequent recyclability of these insulators and possible reuse for next VIP production).

In addition, even better results can be achieved by optimizing the inner structure of the insulators (targeted fiber orientation) and by optimizing the bulk density of the insulators. This is next direction of research currently underway, and it is highly probable that the value of the thermal conductivity for fiber-based insulators on an organic basis will be effectively reduced to at least about 2.0 mW/(m.K).

## Acknowledgements

This paper was elaborated with the financial support of the project GA 17-00243S "Study of the behavior of insulating materials under extremely low pressure".

## References

[1] Directive of the European Parliament and of the Council 2010/31/EU on energy efficiency of buildings Another reference

[2] Brunner S, GhaziWakili K, Stahl T and Binder B 2014 Vacuuminsulation panels for building applications — Continuous challenges and developments *Energy and Buildings* **85** pp 592–596

[3] Nemanič V and Žumer M 2015 New organic fiber-based core material for vacuum thermal insulation *Energy end Buildings* **90** p 137–141

[4] Caps R, Beyrichen H, Kraus D and Weismann S 2008 Quality control of vacuum insulation panels: Methods of measuring gas pressure *Vacuum* **82** pp 691–699

[5] Boafo F E, Zhaofeng Ch, Chengdong L and Tengzhou X 2014 Structure of vacuum insulation panel in building system *Energy and Buildings* **85**

[6] EN 822 Thermal insulating products for building applications. Determination of length and width

[7] EN 12085 Thermal insulating products for building applications. Determination of linear dimensions of test specimens

[8] EN 823 Thermal insulating products for building applications. Determination of thickness

[9] EN 1602 Thermal insulating products for building applications. Determination of the apparent density

[10] EN 826 Thermal insulating products for building applications. Determination of compression behaviour

[11] ISO 8301 Thermal insulation - Determination of steady-state thermal resistance and related properties -- Heat flow meter apparatus

[12] Reichenauer G, Heinemann U and Ebert H-P 2007 Relationship between pore size and the gas pressure dependence of the gaseous thermal conductivity *Colloids and Surfaces A: Physicochemical and Engineering Aspects* **300** pp 204–210

IOP Conf. Series: Materials Science and Engineering **379** (2018) 012021 doi:10.1088/1757-899X/379/1/012021

# Effects of the impact of 7.62 mm small-arms projectiles on various building materials

**J Kratochvíl[1], O Koutný[1] and M Drdlová[2]**

[1] Brno University of Technology, Faculty of Chemistry, Purkyňova 118, 612 00 Brno, Czech Republic
[2] Senior Researcher, Research Institute for Building Materials, Hněvkovského 65, 617 00 Brno, Czech Republic

E-mail: xckratochvilji@fch.vut.cz

**Abstract.** The presented study deals with the experimental assessment of the effects (i.e. penetration depths) of 7.62 mm x 39 and 7.62 mm x 54 R multipurpose projectiles on various structural and shielding/cover materials. The materials selected for testing were hardwood, softwood, brick and sand. The primary focus of the testing was the effect of projectiles on samples during perpendicular impact at different impact velocities and therefore various impact energies. The secondary aim was to determine and observe projectile properties after target perforation. The observed parameters include the residual energy/velocity and structural properties of the projectiles (e.g. fragmentation).

## 1. Introduction

The security situation in Europe is rapidly worsening. There have been several severe terrorist attacks in European countries and political relations between NATO and Russia are deteriorating. The higher risk of terrorism in particular means the protection of critical infrastructure has become a key issue. The majority of critical infrastructure consists of civilian structures/buildings which therefore usually lack any form of ballistic and blast protection. The aim of this paper is to evaluate some common building materials such as bricks and wood from the point of view of their behaviour under ballistic load. Also, one type of natural material was tested, namely siliceous sand. Testing was performed using assault rifle calibre and sniper rifle/universal machine gun calibre ammunition. The calibre of the selected assault rifle ammunition was 7.62 mm x 39, while 7.62 mm x 54R was the calibre of the chosen universal machine gun/sniper rifle ammunition. The official designation of these cartridges often varies. The 7.62 mm x 39 cartridges are marked 7.62–43, M43 or simply Kalashnikov, while the 7.62 mm x 54R cartridges are labelled 7.62–59, 7.62 Mosin – Nagant or Dragunov. Both selected projectiles are designated as Multipurpose ball-type ammunition [1]. This slightly contradicts established ballistic testing standards. For example, STANAG 4569 usually requires armour piercing (AP) or armour piercing incendiary (API) ammunition. On the other hand, testing mode was selected to simulate an attack performed by an individual or a group, both of which tend to be usually far less well-equipped when compared to military troops and don´t have access to specialized ammunition. Ball ammunition is widespread: during the Cold War billions of such cartridges were manufactured, and they are often available for civilian use.

The mechanical response of solids during high speed dynamic loads can be described using formulas commonly used for liquids. Rankine-Hugoniot formulas are especially useful for this purpose. These formulas describe change of momentum, change of mass and internal energy [2],[3],[5].

Mass conservation:

$$\rho_0 U_S = \rho(U_s - u_p) \tag{1}$$

Momentum conservation:

$$p = \rho_0 U_S u_p \tag{2}$$

Energy conservation:

$$\Delta e = \frac{p}{2}\left(\frac{1}{\rho_0} - \frac{1}{\rho}\right) \tag{3}$$

where ρ, p and $\Delta e$ are density, pressure and change in internal energy of the shocked material. Subscript stands for ambient conditions. $U_S$ is shock velocity and $u_p$ is particle velocity.

The basic empirically based approach to the calculation of the penetration or perforation of a selected projectile into a given material is based on energy equilibrium. The usual "starting" point for such calculations is the following formula:

$$E = kd^m t^n \tag{4}$$

The loss of kinetic energy of the projectile after sample perforation can be calculated using the following empirical formula:

$$E_R = E_I - sk = 0.5mv_I^2 - sk = 0.5mv_R^2 \tag{5}$$

where $E_R$ is the residual kinetic energy of the projectile (J),
$E_I$ is the impact energy of the projectile (J),
s is the thickness of the sample (if the sample was perforated) or depth of penetration (if the sample was penetrated) (mm),
k is the material coefficient ($J \cdot mm^{-1}$),
m is projectile weight (kg),
$v_I$ is impact velocity ($m \cdot s^{-1}$),
$v_R$ is residual velocity ($m \cdot s^{-1}$).

If the projectile does not perforate the sample, residual velocity is zero ($v_R = 0$) and depth of penetration can be calculated from the following formula:

$$s = 0.5mv_I^2 k^{-1}\ (mm) \tag{6}$$

The material coefficient depends on two main factors. The first factor is the type of projectile – its geometry, velocity and weight, while the second factor is the impact energy of the projectile. This means that material coefficients can be gained for same type of projectile with different impact energies.

A similar formula was determined for softwood and hardwood:

Softwood:

$$k = 1.8037 + 2.501 \cdot 10^{-4} E_I - 1.51273 \cdot 10^{-7} E_I^2 \tag{7}$$

Hardwood:

$$k = 2.8756 + 3.95775 \cdot 10^{-4} E_I + 9.07583 \cdot 10^{-8} E_I^2 \tag{8}$$

The determination of the material coefficient for bricks (and generally for other hard and brittle materials) is much more difficult due to two factors:

1) The projectile often breaks and fragments upon impact – this leads to an increase in the diameter of the damaged zone, but also a much lower depth of penetration.

2) The sample often disintegrates, preventing measurement of the depth of penetration of the projectile Therefore, the formula for depth of penetration was not calculated using the material coefficient *k*; instead, impact energy was used (note that this formula is available for 7.62 x 54R):

$$s = -0.0269603 + 0.0152704 E_I - 3.10331 \cdot 10^{-4} E_I^2 \tag{9}$$

The main purpose of this paper is to determine the terminal ballistic properties of various structural materials.

## 2. Experimental programme

The presented study deals with the determination of the response of various structural materials to ballistic load expressed as kinetic energy loss of a projectile after perforation of a panel. The tested wooden samples were fabricated from softwood and hardwood and of a defined thickness. The loss of projectile kinetic energy was determined using the known mass of the projectile and the measured impact velocity and residual velocity of the projectile after target penetration. The correlation between impact velocity ($v_I$) and kinetic energy ($E_K$) loss was monitored. The Depth of Penetration (DoP) test was performed for the same materials. The DoP test was also performed for the brick and sand sample as they represent other common types of structural and shielding materials. The DoP value corresponds to the total kinetic energy of the projectile absorbed during the process of penetration into the tested material. The ratio of the total kinetic energy of the projectile to the DoP value represents the material's capability to resist the penetration of the selected ammunition.

## 3. Materials and methods

Common structural materials were selected for ballistic testing, namely hardwood (oak), softwood (spruce) and brick. Siliceous sand was chosen as a representative of shielding materials. Wooden beams were glued together in order to achieve thicknesses of 100, 200 and 300 mm. The same approach to sample preparation was chosen for the DoP test, for which wood samples with a thickness of 800 mm were prepared. CP-20 bricks were used for the tests, with the dimensions 240 x 140 x 60 mm. Assuming that brick is an isotropic material, ten bricks were placed in a row one behind another to reach a total length of 600 mm and screwed together into a steel frame to prevent any disintegration of the whole assembly. One shot was fired for each such brick assembly. The sand was dried to a constant weight and sieved through a 5 mm sieve before being placed in wooden box with the dimensions 600 x 600 x 2000 mm.

The aim of this paper is to describe the ballistic testing of materials using normal impact. The testing was performed using ballistic test barrels placed in a UZ-2002 universal breech mounted on a firing rest. The barrels used for testing were a 7.62 mm Soviet Model 1943 for 7.62 mm x 39 testing and a 7.62 mm Mosin – Nagant for the 7.62 mm x 54R. Figure 1 shows the kinetic energy of the observed projectiles based on their impact velocity.

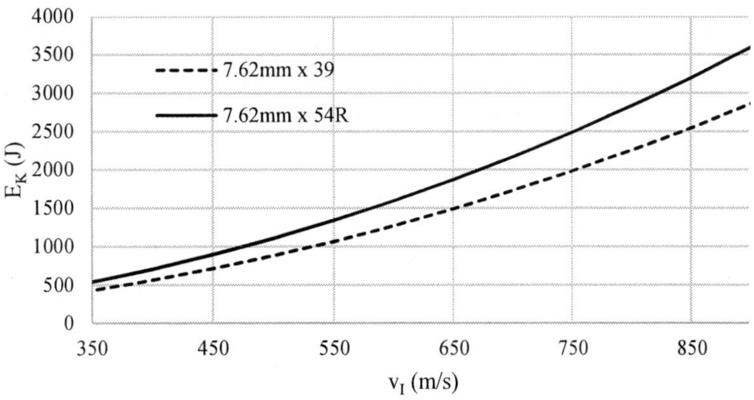


**Figure 1**. Kinetic energy of 7.62 mm x 39 and 7.62 mm x 54R projectiles based on their impact velocity.

The tests were performed on an indoor range. The air temperature was between 18–22°C, with the test munitions being tempered to the air temperature. Humidity was between 60 and 80%. Impact velocity was measured using LS-06 optical light gates. Distance from muzzle to target was 10 m; the velocity of the projectile was measured at 8 m from the muzzle. Residual velocity was determined from the high-speed camera record. The standard velocity at 8 m from the muzzle is 855 m/s for the 7.62 mm x 54R and 695 m/s for the 7.62 mm x 39 projectile. These were also set as the higher limit of impact velocity for each projectile. These velocities are assumed to be nominal, i.e. the most probable impact velocity which can be reached in a real situation according to STANAG 4569.

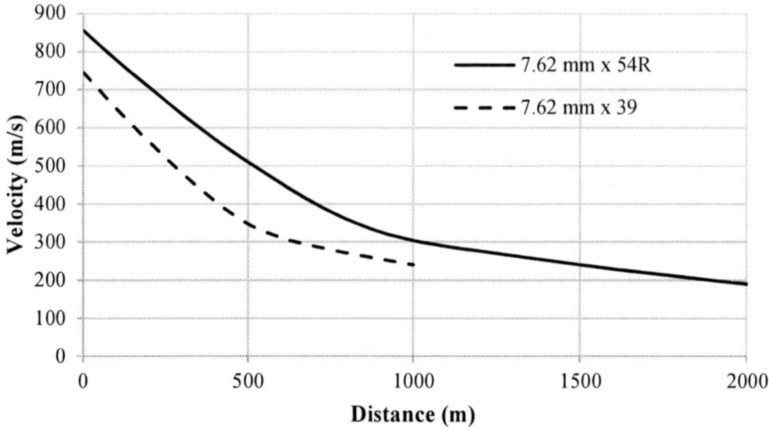


**Figure 2.** Projectile velocity at specific distances.

In order to achieve lower $v_I/E_I$, the powder charge was reduced according to ballistic tables. Lower $E_I$ simulates shots being fired at longer ranges. The lower limit of projectile impact velocity was set as low as possible while keeping it high enough to provide perforation of the sample.

## 4. Results and discussion

The residual velocity of each projectile after perforation of the target was monitored. The kinetic energy loss was determined in comparison to the impact velocity. This value represents the absorbed energy of the projectile during the perforation process through a target of defined material and thickness. Figure 3, Figure 4 and Figure 5 show the dependence of absorbed kinetic energy on the impact velocity of chosen projectiles for softwood panels with thicknesses of 100 mm, 200 mm and 300 mm. It is necessary to note that both projectiles have the same cross-section geometry, i.e. they are both 7.62 mm in diameter.

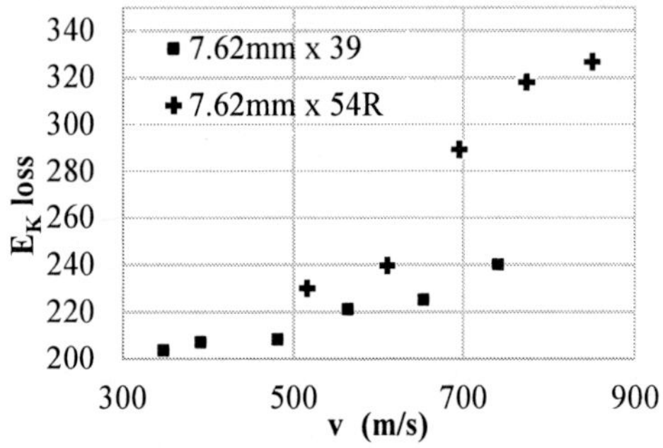


**Figure 3.** Kinetic energy loss – 100 mm softwood panel.

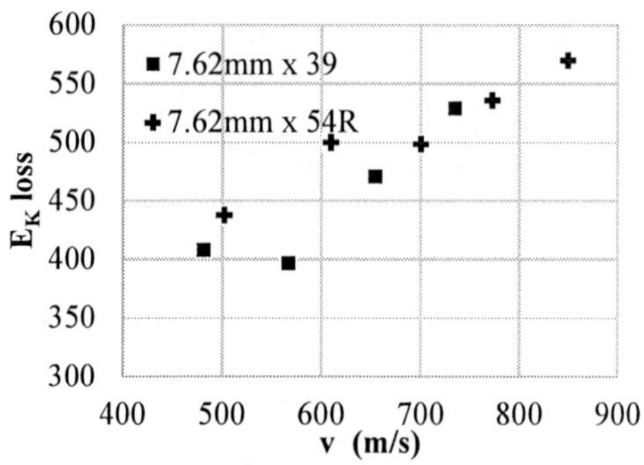


**Figure 4.** Kinetic energy loss – 200 mm softwood panel.

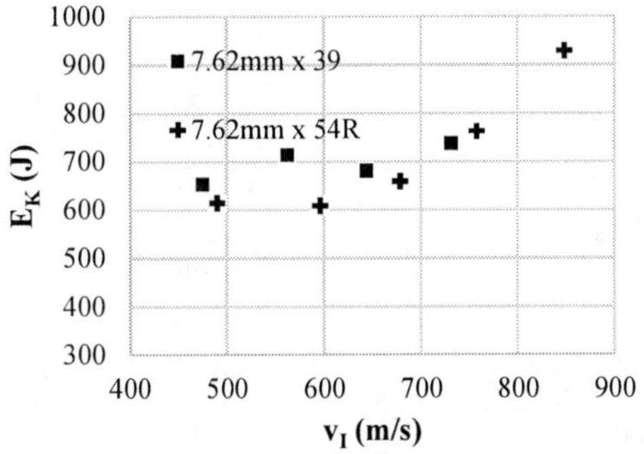


**Figure 5.** Kinetic energy loss – 300 mm softwood panel.

Generally, it can be concluded that with increasing impact velocity the kinetic energy loss increases. With softwood the kinetic energy loss rises in proportion to increasing impact velocity. This effect could be related to the fact that nearly no deformation of the projectile occurs, and neither is there almost any destabilization of its trajectory. Moreover, no hardening mechanism can be assumed in the softwood material [6]. This behaviour is fully compliant for both projectiles up to an impact velocity of approx. 700 m/s. For the higher impact velocity which was used for the 7.62 mm x 54R projectile there is a substantial deviation from the linear trend of the plotted dependence. This deviation is more evident in the case of the samples with higher thickness. This behaviour can be attributed to the secondary effects occurring during the penetration process; namely the destabilization of the projectile and the separation of the projectile jacket. The first of those effects makes a major contribution when the value of kinetic energy loss is higher than that which is expected according to the linear trend. Due to the destabilization of the projectile, the cross-section perpendicular to the flight trajectory is higher and therefore more energy is absorbed during penetration. The second effect occurs when the kinetic energy loss is lower than the expected value given by linear approximation. Due to jacket separation, the cross-section of the projectile is reduced, which results in higher pressure at the tip of the penetration tunnel and consequently lower dissipation of kinetic energy. Comparing the resistance of the tested material against the penetration of projectiles for different thicknesses it can be concluded that the kinetic energy of a 7.62 mm x 39 projectile with a nominal impact velocity of 695 m/s can be reduced by 14% by a 100 mm softwood panel. The reduction in kinetic energy seems to be proportional to panel thickness according to the test results: a 29% reduction for the 200 mm panel and a 42% reduction for the 300 mm panel. Similar behaviour can be observed in the case of the 7.62 mm x 54R projectile with a nominal impact velocity of 855 m/s. For the panel thicknesses of 100 mm, 200 mm and 300 mm, the reduction of kinetic energy was determined to be 10%, 18% and 29%, respectively.

Figure 6, Figure 7 and Figure 8 display the results from the ballistic testing of hardwood panels with thicknesses of 100 mm, 200 mm and 300 mm.

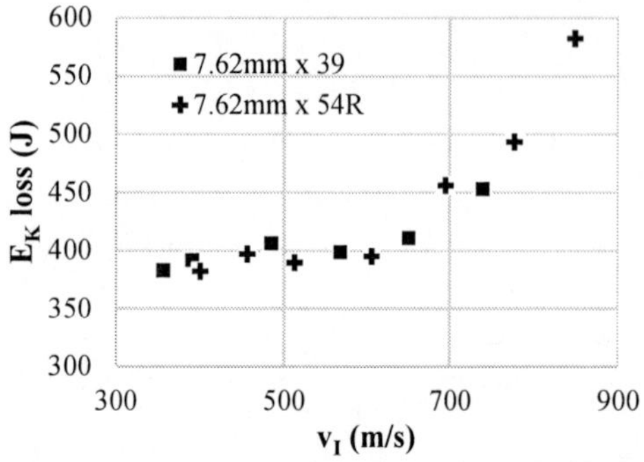


**Figure 6.** Kinetic energy loss – 100 mm hardwood panel.

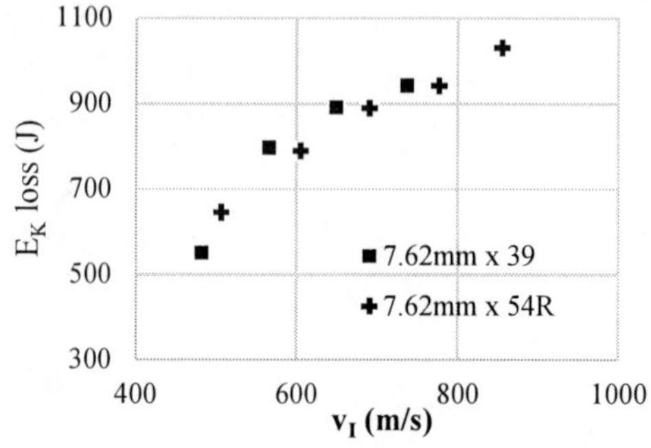


**Figure 7.** Kinetic energy loss – 200 mm hardwood panel.

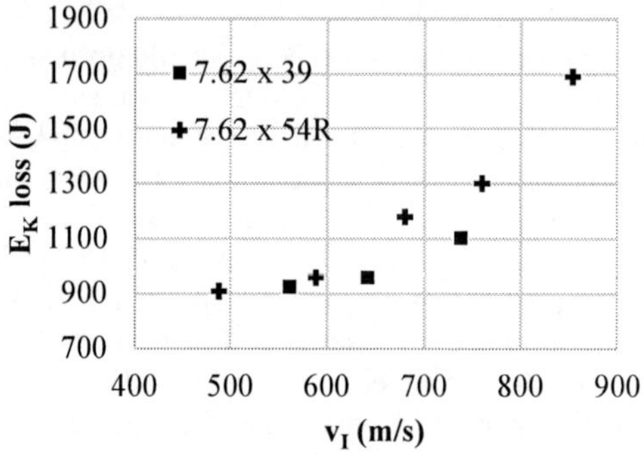


**Figure 8.** Kinetic energy loss – 300 mm hardwood panel.

The kinetic energy of projectiles absorbed by hardwood is approximately two times larger compared to softwood panels of 100 mm thickness. This ratio proportionally decreases with increasing panel thickness to a factor of 1.55 for 300 mm thickness. It is evident that there is a similar trend to that of the previous experiment with regard to rising kinetic energy loss with increasing impact velocity for both projectiles. However, the trend cannot be designated as proportional i.e. linear to the increasing impact velocity. For the 100 mm panel it is seems to be the case that the difference in kinetic energy loss is constant in the range of impact velocity up to 600 m/s for both projectiles. It is necessary to note that a projectile with an impact velocity lower than 350 m/s is not able to perforate the sample. The kinetic energy loss of projectiles with an impact velocity larger than 600 m/s proportionally increases with impact velocity. The results for hardwood panels with thicknesses of 200 mm and 300 mm correspond to the results for softwood, i.e. a similar effect which is responsible for positive or negative deviation from the linear trend occurs. The negative deviation of the linear trend is evident from the larger impact velocities for the 200 mm hardwood and softwood panels. This could be also attributed to the destabilization of the projectile and therefore the increase in cross-section perpendicular to the flight trajectory. For the 300 mm thick hardwood panel a positive deviation in the linear trend occurs at larger impact velocities. This effect could be attributed to the partial separation of the jacket and the subsequent decrease in projectile cross-section, resulting in better penetration capability, as in the case of the 300 mm softwood panel. Comparing the material's resistance against the penetration of projectiles for different thickness it can be concluded the kinetic energy of a 7.62 mm x 39 projectile with a nominal impact velocity of 695 m/s can be reduced by a 100 mm hardwood panel by 25%. The reduction in kinetic energy is proportional to the panel thickness up to 200 mm, which reduces the projectile's kinetic energy by 52%. For larger material thicknesses, a negative deviation from the linear trend arises, which is evident from the kinetic energy reduction of 62% for the 300 mm panel.

Similar behaviour can be observed with the 7.62 mm x 54R projectile, with a nominal impact velocity of 855 m/s. For the panel thicknesses of 100 mm, 200 mm and 300 mm, the reduction in kinetic energy can be determined as 18%, 32% and 33%, respectively, and again a negative deviation from the linear trend is seen for the largest thickness.

The depth of penetration of selected projectiles was monitored in wooden samples via the DoP test. The results from the DoP testing of softwood are summarized in Figure 9, which expresses the dependence between DoP and $E_I$.

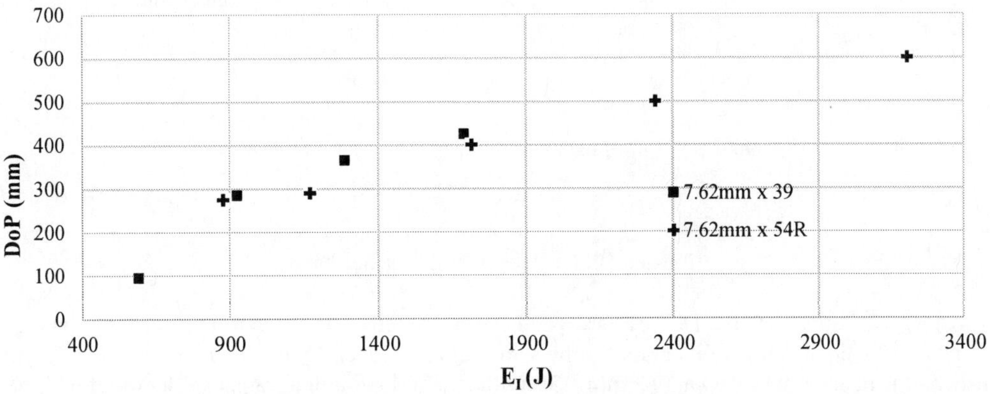


**Figure 9.** DoP test results for softwood.

As can be seen, the perforation of softwood material by selected projectiles can be related to their impact kinetic energy in the selected range of impact velocities. Generally, it is evident that over 430 mm of softwood is necessary to stop a 7.62 mm x 39 projectile at its nominal impact velocity of 695 m/s, i.e. with an impact kinetic energy of 1700 J. In the case of the 7.62 mm x 54R projectile, over 600 mm of softwood is necessary to stop it at its nominal impact velocity of 855 m/s, i.e. with an impact kinetic energy of 3200 J. The higher slope of the DoP – $E_I$ dependence is evident at low impact kinetic energy (up to approximately 1000 J). In other words, there is a negative deviation from the linear trend resulting in a lower DoP than could be expected. According to the statement that the mechanical properties of wood are not strain-sensitive, i.e. that there is no physical reason for any strain hardening or strain softening behaviour by wood-like materials at assumed strain rates ($10^4$ – $10^6$ $s^{-1}$), a different effect responsible for the decrease in DoP from linear trend is presented: The destabilization of the projectile and increase in yaw may be attributed to this effect due to the lower impact velocities involved, which are far below the nominal velocities. Lower impact velocity and therefore lower momentum causes high yaw and the projectile is more sensitive to destabilization from its flight trajectory. This results in a higher cross-section perpendicular to the flight trajectory and a lower DoP due to the faster absorption of the projectile's kinetic energy by the surrounding material.

The results from the DoP testing of hardwood are summarized in Figure 10, which expresses the dependence between DoP and $E_I$.

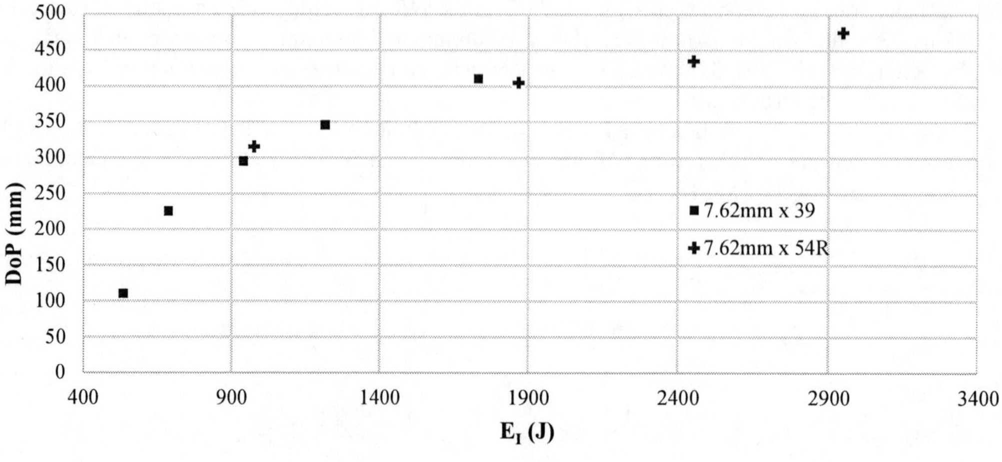


**Figure 10.** DoP test results for hardwood.

Just as with softwood, the DoP of hardwood can be directly attributed to the impact kinetic energy of a projectile independent of its mass for the same calibre. Generally it is evident that over 400 mm of hardwood is necessary to stop a 7.62 mm x 39 projectile at its nominal impact velocity of 695 m/s, i.e. impact kinetic energy is 1700 J. At the same time, over 470 mm of hardwood is necessary to stop a 7.62 mm x 54R projectile at its nominal impact velocity of 855 m/s, i.e. impact kinetic energy is 3200 J. It is evident that the same DoP was obtained for both types of wood up to an impact kinetic energy of 1000 J. From this value a linear trend is evident. The slope of the linear trend for hardwood is twice as steep as in the case of softwood. This parameter can be directly attributed to the material properties responsible for ballistic loading resistance. A similar trend occurs in the development of DoP with increasing projectile impact kinetic energy. At low impact velocities, i.e. low projectile kinetic energies of up to 1000 J, a negative deviation from the linear trend is evident. This phenomenon can be attributed to low projectile stability due to its lower impact velocity and the destabilization of the projectile's flight trajectory at the initial phase of penetration due to low momentum, just as in the case of the softwood samples. This results in higher absorption of the projectile's kinetic energy by surrounding material. This assumption can be confirmed by the same DoP up to 1000 J of impact kinetic energy for both types of wood, as was mentioned above.

The DoP test was also performed on brick, another common structural material. The results are summarized in Figure 11.

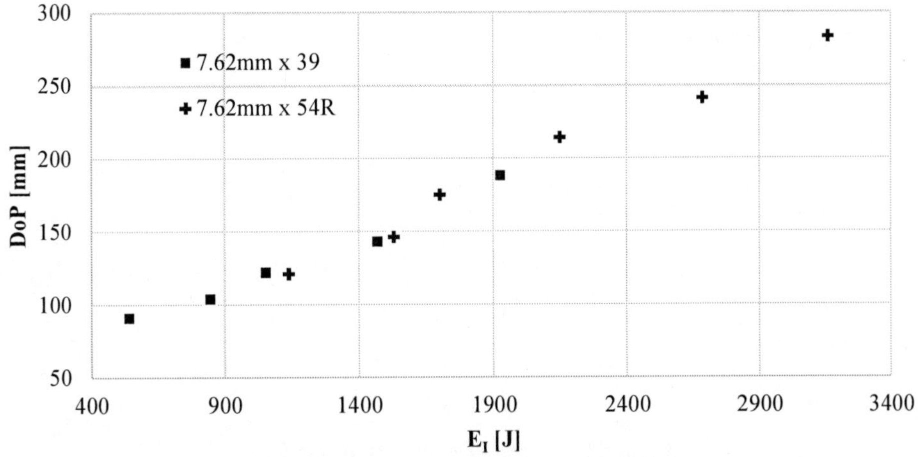


**Figure 11.** Depth of Penetration – brick.

The DoP of brick can be directly attributed to the impact kinetic energy of a projectile independent of its mass for the same calibre, which corresponds to the previous assumption applied to the results for the wooden samples. According to the DoP results it can be concluded that the minimum thickness of a standard brick wall that will stop a 7.62 mm x 39 projectile moving at its nominal velocity of 695 m/s is more than 190 mm. The minimum thickness of a brick wall that will stop a 7.62 mm x 54R projectile moving at its nominal velocity of 855 m/s is more than 280 mm. In contrast to the previous tested materials (softwood and hardwood), the linear trend of the DoP is evident throughout the whole range of monitored impact kinetic energy. In other words, no negative deviation of DoP occurs at low impact velocity (i.e. low impact kinetic energy), in contrast to the wooden samples. No severe destabilization of the projectile in the initial phase of penetration can be considered. As the brick has much higher mechanical properties than any wooden materials, the destabilization of the projectile during the initial phase of penetration is suppressed by the surrounding material. The DoP – $E_I$ dependence and its linear character show that no strain strengthening or strain softening mechanism occurs. This corresponds to the general assumption that a brittle or quasi-brittle material mechanical response will occur under high-rate dynamic loading. [7]

The last considered material from the testing session is a siliceous sand, which was selected as the material most likely to be used in temporary shielding systems. The results from the DoP test are summarized in Figure 12.

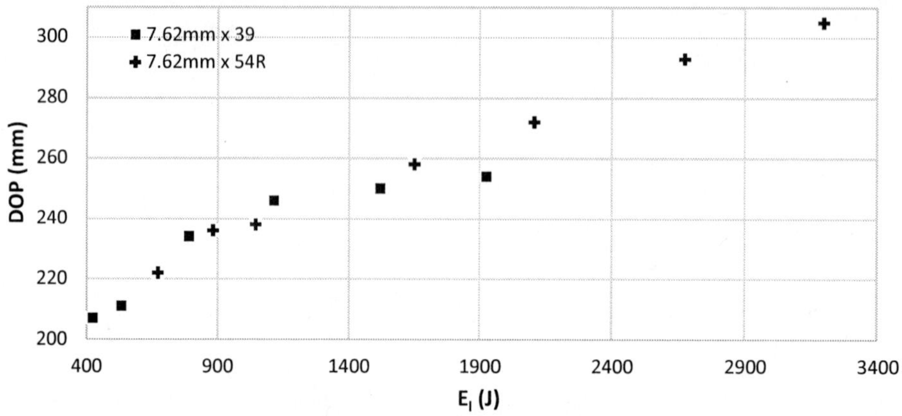


**Figure 12.** Depth of Penetration – sand.

From the DoP results it can be concluded that there is a direct relation between DoP and $E_I$ for projectiles of the same calibre. However, due to the heterogeneous character of sand there are visible deviations from the general trend in contrast to the other tested materials, which is especially valid for the 7.62 mm x 39 projectile. At high impact velocities the expected DoP for this projectile should be higher. According to the DoP results it can be concluded that the minimum sand layer thickness needed to stop a 7.62 mm x 39 projectile moving at its nominal velocity of 695 m/s is more than 250 mm. The minimal sand layer thickness necessary to stop a 7.62 mm x 54R projectile moving at its nominal velocity of 855 m/s is more than 310 mm. The negative deviation of the linear trend is visible from the low impact velocity (i.e. low impact kinetic energy), just as with the wooden samples. The same effect is assumed to occur due to the low mechanical properties and stability of the tested sample, which is a particulate material. [8]

## 5. Conclusion

In the first part the loss of kinetic energy of 7.62 mm x 39 and 7.62 mm x 54R projectiles during the penetration of softwood and hardwood panels was monitored and the following conclusions can be derived.

- It can be concluded that with rising projectile impact velocity the kinetic energy loss linearly increases. However, at impact velocities of over 700 m/s in the case of softwood and 600 m/s in the case of hardwood a deviation occurs in the linear trend. Two effects are assumed to be responsible for this behaviour. Positive deviation from the linear trend is attributed to the destabilization of the projectile during the penetration process and an increase in the projectile cross-section perpendicular to the flight trajectory. Negative deviation from the linear trend can be related to the separation of the projectile jacket during the penetration process and the subsequent decrease in the projectile cross-section perpendicular to the flight trajectory.
- The loss of kinetic energy is proportional to the thickness of the panel for softwood samples, which corresponds with the results observed for the hardwood samples up to panel thicknesses of 200 mm. For thicker hardwood panels a negative deviation from the linear trend should be expected.
- According to the data obtained for kinetic energy loss, the ability of a material to resist damage from ballistic loading by assumed projectiles is two times higher in the case of hardwood compared to softwood when panel thickness is 100 mm. This ratio proportionally decreases with increasing thickness to a factor of 1.55 for 300 mm thick panels.

In the second part the Depth of Penetration (DoP) was monitored for softwood, hardwood, brick and siliceous sand impacted by 7.62 mm x 39 and 7.62 mm x 54R projectiles. The following conclusions can be drawn.

- Generally, it can be concluded that the DoP is a linear function of impact kinetic energy for projectiles with the same calibre. However, negative deviation from the linear trend occurs at low impact kinetic energy. This effect could be attributed to the low stability of the projectile due to its low momentum and destabilization during the initial phase of penetration. The second effect is significantly reduced in the case of the brick sample because the material surrounding the projectile during the penetration process has sufficiently high physical-mechanical properties to prevent any severe destabilization.
- The DoP test provides a basic assumption regarding the minimum layer of monitored material which is needed to stop the selected projectiles. From the DoP results it can be concluded that the minimum softwood panel thicknesses needed to stop 7.62 mm x 39 and 7.62 mm x 54R projectiles are 430 mm and 600 mm, respectively. In the case of hardwood, the minimum thicknesses are reduced to 400 and 470 mm respectively for the selected projectiles. Brick has the greatest ability out of all the investigated materials to resist damage from ballistic loading. 190 mm of brick material is necessary to stop a 7.62 mm x 39 projectile, while 280 mm is needed for 7.62 mm x 54R ammunition. The DoP results also show that particulate materials such as siliceous sand of the 0–5 mm fraction provide relatively good protection against the selected projectiles: a 250 mm layer of sand is needed to stop a 7.62 mm x 39 projectile and a 310 mm layer is needed for 7.62 mm x 54R.
- The results of the DoP are fully in accord with the general formulation of high-strain rate dynamic loading described by the Rankine-Hugoniot relation, where the density or volume weight of materials has a direct effect on the fracturing of the body of a material. Brick (i.e. the tested material with the highest density) shows the best performance under ballistic loading of all the materials investigated in this study.

**Acknowledgements**

As doctoral students of the Materials Research Centre at FCH BUT- Sustainability and Development, REG LO1211, with financial support from National Programme for Sustainability I (Ministry of Education, Youth and Sports), authors want to thank the foundation for its financial support. The authors also wish to express their gratitude and sincere appreciation to the authority of The Technology Agency of the Czech Republic, project No. TE02000162 - Centre of advanced materials and technologies for protection and safety enhancement, for financial support.

**References**

[1] Carlucci D E and Jacobson S S 2013 Ballistics: Theory and Design of Guns and Ammunition Second Edition 2 (CRC Press)
[2] Rosenberg Z and Dekel E 2012 Terminal ballistics (Berlin: Springer)
[3] Buchar J and Vodička J 2003 Terminální balistika (Prague: Academia)
[4] STANAG, STANAG 2280. 1. 2008 (Brussels: NATO Standardization Agency)
[5] Morison C M 2004 Dynamic response of walls and slabs by single-degree-of-freedom analysis—a critical review and revision (DOI: 10.1016/j.ijimpeng.2004.11.008)
[6] Johnson W Historical and present-day references concerning impact on wood (DOI: 10.1016/0734-743X(86)90003-5)
[7] Gilbert M, Hobbs B and Molyneaux T C K The performance of unreinforced masonry walls subjected to low-velocity impacts: experiments (DOI: 10.1016/S0734-743X(01)00049-5)
[8] Chian S C, Tan B C V and Sarma A Projectile penetration into sand: Relative density of sand and projectile nose shape and mass (DOI: 10.1016/j.ijimpeng.2017.01.002)

IOP Conf. Series: Materials Science and Engineering **379** (2018) 012022 doi:10.1088/1757-899X/379/1/012022

# Destructive and non-destructive assessment of the frost resistance of concrete with different aggregate

**D Kocáb[1], B Kucharczyková[1], P Daněk[1], T Vymazal[1], P Hanuš[1] and R Halamová[1]**

[1] Brno University of Technology, Faculty of Civil Engineering, Veveří 331/95, Brno, Czech Republic

Email: dalibor.kocab@vutbr.cz

**Abstract**. The paper focuses on determining the frost resistance of concrete using selected destructive and non-destructive test methods. The experiment was performed with four specially made concretes, which differed only in the origin of sand and coarse aggregate. The primary outcome of the experiment are the differences in frost resistance as measured by different methods. In addition, it briefly describes the influence that the choice of aggregate has on the frost resistance of concrete.

## 1. Introduction

In recent years, new trends have been introduced into the design of reinforced concrete structures. Current concepts of structures and especially materials from which the structures are constructed place emphasis on properties related to environmental impact, specifically, durability, reliability and sustainability [1]. Durability is broadly used as a general term meaning the resistance of concrete to harsh environment. The character, intensity and mechanism of each influence may, however, vary considerably, making the concept of concrete durability somewhat vague. Moreover, no available method can be used to determine the durability of concrete in general terms [2]. The degradation of concrete in a structure is influenced by various factors such as carbonation, mechanical wear, cracks formed by (for example) inappropriate curing, etc.

Frost resistance is thus another important factor that needs to be considered during the design of concrete structures. The effects of frost are one of the most frequently observed forms of damage to the surface layer of concrete as well as its internal structure. When concrete is dry during exposure to frost, the structure does not usually sustain significant damage. However, if the structure or its part is in contact with water (road panels, railway sleepers, water reservoirs, etc.), the concrete absorbs it into its pore structure. If the water freezes, there is a risk of cracking and other defects caused by the expansion of ice in the pore structure. Water located in the capillary pores of concrete begins to freeze at approximately -0.5°C. The volume of the ice thus produced is larger by approximately 9% compared to the original water. This generates high internal strain in the concrete and, as a result, causes damage to its internal structure, which irreversibly leads to the reduction of basic material properties [3].

Frost resistance of concrete is influenced by many factors, most of which are related to its composition. The most important factor is the distribution and size of air pores in the concrete structure [4]. Air-entraining agents are the most reliable means of increasing the resistance of concrete against frost, because they form a system of closed spherical air voids inside the concrete. These provide space for the ice to expand, thus reducing the risk of damage to the internal structure of the concrete [3].

Another important aspect of the frost resistance of concrete is the amount, quality, type and grain size of the aggregate used [5].

Although standards do not prescribe a method for determining the frost resistance of concrete in structures, this property can be evaluated in several ways in laboratories using test specimens. These testing procedures are, for example, those described in ASTM C 666-03 [6], GB / T 50082-2009 [7], BS 5075-2 [8], ČSN 73 1322 [9] or ČSN 73 1380 [10]. The principle is always to monitor the decrease in the value of a certain property of concrete after a freeze-thaw cycle (F-T cycle) in the presence of water. According to [6], the most appropriate means of determining the frost resistance of concrete is to observe the changes in its dynamic modulus of elasticity after F-T cycles, as this method can detect initial defects in the internal structure of concrete much earlier than during the observation of the decrease in other properties, particularly strength. In principle, this involves determining the percentage of the value of the modulus determined after a given number of F-T cycles and the original value of the modulus of elasticity determined before the cyclic freezing and thawing. The result is the value of the relative dynamic modulus of elasticity (RDM).

## 2. Experiment

The experiment aimed primarily to compare the suitability of each of the available methods for measuring the frost resistance of concrete. A secondary aim was to determine the influence of aggregate on the concrete's frost resistance.

### *2.1. Material and test specimens*

Four types of concrete with designations D, E, F and G were produced for the purpose of the experiment. These concretes (their composition is shown in Table 1) differed mainly in the type and origin of aggregate (sand as well as coarse aggregate). The other components were always the same.

**Table 1.** Composition of the concrete used in the experiment.

| | kg / 1 m$^3$ of fresh concrete | | | |
|---|---|---|---|---|
| Component | Concrete D | Concrete E | Concrete F | Concrete G |
| CEM II/B-S 32.5 R (Mokrá) | 400 | 400 | 400 | 400 |
| Water | 185 | 185 | 185 | 185 |
| Sand 0-4 mm (Bratčice) | 760 | - | 760 | - |
| Sand 0-4 mm (Tovačov) | - | 745 | - | 745 |
| Aggregate 4-8 mm (Luleč) | 135 | 135 | - | - |
| Aggregate 4-8 mm (Tovačov) | - | - | 135 | 135 |
| Aggregate 8-16 mm (Luleč) | 460 | 460 | - | - |
| Aggregate 8-16 mm (Tovačov) | - | - | 450 | 450 |
| Aggregate 11-22 mm (Luleč) | 345 | 345 | - | - |
| Aggregate 11-22 mm (Tovačov) | - | - | 335 | 335 |
| Plasticiser (Sika ViscoCrete-4035) | 2.00 | 1.80 | 1.80 | 1.60 |
| Air-entraining agent (Sika LPSA 94) | 0.25 | 0.25 | 0.25 | 0.25 |

Two types of sand and two types of coarse aggregate were used. Both sands 0/4 mm consisted of naturally mined aggregate with one from the Bratčice site and the other from Tovačov. One type of the coarse aggregate (consisting of grading 4/8, 8/16 and 11/22) was a naturally crushed aggregate from Luleč quarry (petrographic name of the raw material: greywacke) and the second type of the coarse aggregate (of the same grain sizes) was a naturally mined aggregate from Tovačov (gravel). The composition of the concrete was designed to combine both sands with both coarse aggregates. In order to achieve similar consistency of each concrete, plasticiser was added with minor variations in amount. The slump test values ranged within (200 ± 20) mm for all the concretes, see Table 2.

**Table 2.** Properties of each concrete in the fresh (FC) and hardened (HC) state.

| | Concrete | | | |
|---|---|---|---|---|
| Property | D | E | F | G |
| Slump test (FC) [mm] | 200 | 220 | 180 | 220 |
| Air content (FC) [%] | 5.5 | 5.5 | 6.5 | 6.0 |
| Density (FC) [kg/m$^3$] | 2230 | 2250 | 2190 | 2220 |
| Compressive strength after 28 days (cube, HC) [N/mm$^2$] | 51.4 | 54.6 | 44.9 | 49.9 |
| Density (HC) [kg/m$^3$] | 2270 | 2270 | 2230 | 2270 |
| Flexural strength before F-T cycles (prism, HC) [N/mm$^2$] | 5.1 | 6.1 | 4.2 | 4.5 |
| Tensile splitting strength before F-T cycles (fragment, HC) [N/mm$^2$] | 4.60 | 5.25 | 4.05 | 4.20 |

Six test specimens were made from each concrete with the nominal dimensions of 100 × 100 × 400 mm. The test specimens aged for 24 hours in special plastic (Hakorit) moulds covered with a PE foil. They were then removed from the moulds and stored in an environment with the air temperature of 20 ± 2°C and humidity of at least 95%. At the age of 25 days, the specimens were submerged for three days in water with the temperature of 20 ± 2°C.

### *2.2. Testing methods*

At the age of 28 days, the test prisms were removed from the water bath and subjected to measurements of the dynamic modulus of elasticity using the ultrasonic (US) pulse velocity test and the resonance method. The ultrasonic pulse transit time was the first to be measured. The measurements were always carried out along three longitudinal lines evenly spaced over the height of the test specimen using a Pundit PL-200 ultrasonic device. The transit times thus recorded were used to calculate first the propagation of the US wave through the concrete and then the dynamic modulus of elasticity $E_{cu}$ according to ČSN 73 1371 [11]. Using a Handyscope HS4 oscilloscope (with an acoustic emission sensor and software, which works on the basis of a fast Fourier transform), the natural frequencies of longitudinal, flexural, and torsional vibrations were then determined. These values were then used to calculate the dynamic modulus of elasticity $E_{crL}$ (from longitudinal vibrations) and $E_{crf}$ (from flexural vibrations) in accordance with the standard ČSN 73 1372 [12]. More information on the pulse velocity test and the resonance method can be found for example in [13].

Based on the results of these measurements, the test specimens of each concrete type were divided into two groups of three, so that the average value of the dynamic modulus of elasticity was approximately the same for both groups. One group of specimens was used for reference. These specimens were first used for the determination of the flexural strength $f_{cf}$ according to EN 12390-5 [14] and the tensile splitting strength $f_{ct}$ measured on their fragments according to the standard EN 12390-6 [15]. The second group of test specimens was subjected to cyclic freezing in accordance with the standard [9]. One hundred and fifty freeze-thaw cycles were conducted with the test specimens and the value of the dynamic modulus of elasticity was determined after each 25$^{th}$ cycle. Based on the changes in its average value, the relative dynamic modulus of elasticity (RDM) was determined. It represents the ratio between the value of the modulus of elasticity after F-T cycles and the value of the modulus of elasticity before the test of frost resistance. RDM (U) stands for the relative modulus of elasticity determined using the US pulse velocity test, RDM (FL) was measured using the resonance method of the natural frequency of the longitudinal vibrations and RDM (FF) was calculated from the natural frequency of flexural vibrations. After 150 freeze-thaw cycles, the test specimens were subjected to the flexural strength test and the fragments to the tensile splitting strength test. The strength parameters thus determined were then used to calculate the ratio of their values before and after the freeze-thaw cycles ($f_{ct}$ ratio and $f_{cf}$ ratio).

## 3. Results and discussion

Figure 1 presents the RDM (U) results for all four types of concrete (D, E, F, and G) in relation to the number of F-T cycles performed. It is in fact a record of the decrease of the dynamic modulus of elasticity $E_{cu}$ during freezing and thawing. Figures 2 and 3 show the relative dynamic moduli of all the concrete types determined using the resonance method, again in relation to the number of freeze-thaw cycles.

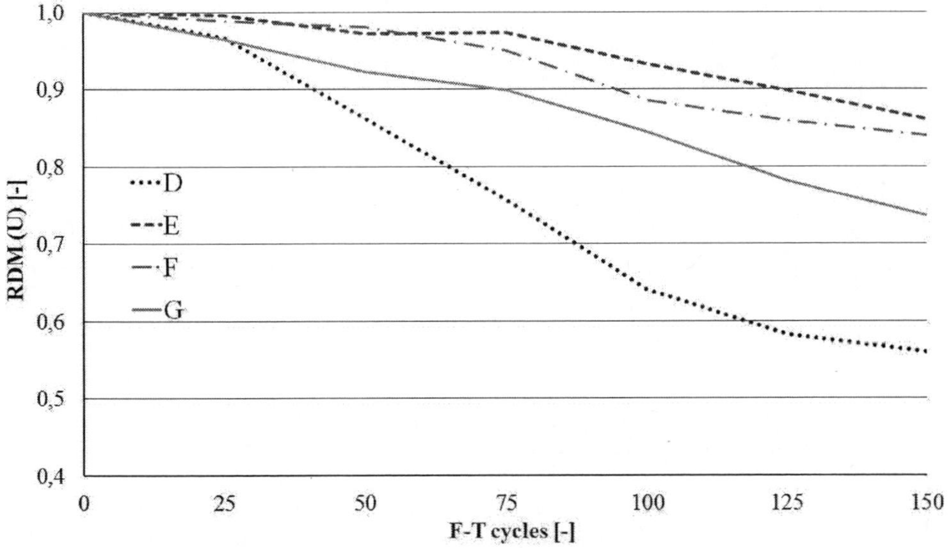


**Figure 1.** Progress of RDM (U) for concrete D, E, F and G in relation to the number of F-T cycles.

Table 3 lists the ratios of all the moduli of elasticity and of both strength parameters determined before and after 150 freeze-thaw cycles. The higher the value of this ratio, the higher the resistance of the concrete.

The measurement results show that the choice of aggregate has a major effect on the frost resistance of concrete. However, it is not clear which of the aggregates is the best in this respect. Concrete E (made with Tovačov sand and coarse aggregate from Luleč) reached the highest RDM values, while the highest strength was achieved by concrete F (with Bratčice sand and Tovačov coarse aggregate). The worst results, both in destructive and non-destructive tests, were determined for concrete D (with Bratčice sand and Luleč coarse aggregate).

What is more interesting and more important is the comparison of the different testing procedures used for the determination of frost resistance of the concrete. The harshest criterion is the decrease in flexural strength of concrete (the ratio between values before and after F-T cycles), which applies to all four concretes. On the other hand, the ultrasonic pulse velocity test seems to be the least sensitive method for assessing the degree of internal damage to concrete after F-T cycles. It can also be stated that all three observed RDM exhibit a similar trend, while measurement by the resonance method using the natural longitudinal and transverse vibration frequency is almost identical. Thus, when using the resonance method, it could be sufficient to assess frost resistance only on the basis of $E_{crL}$ (the measurement of the natural frequency of longitudinal vibration is not dependent on the dimensional accuracy of the specimen's cross-section). Yet, it is still advisable to measure the natural frequencies of all types of vibration and calculate the dynamic Poisson's ratio with them. This can be useful e.g. for calculating the dynamic modulus of elasticity $E_{cu}$ determined by the ultrasonic pulse velocity test.

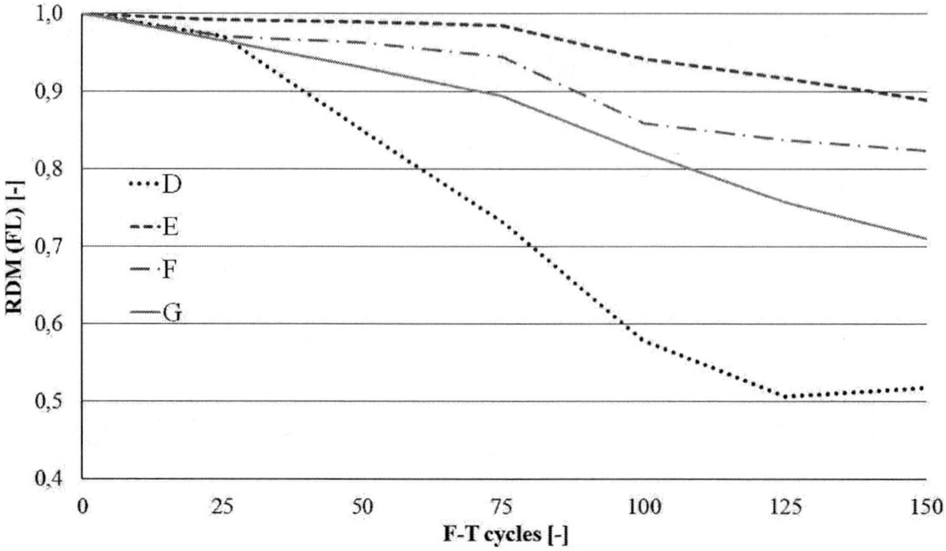


**Figure 2.** Progress of RDM (FL) for concrete D, E, F and G in relation to the number of F-T cycles.

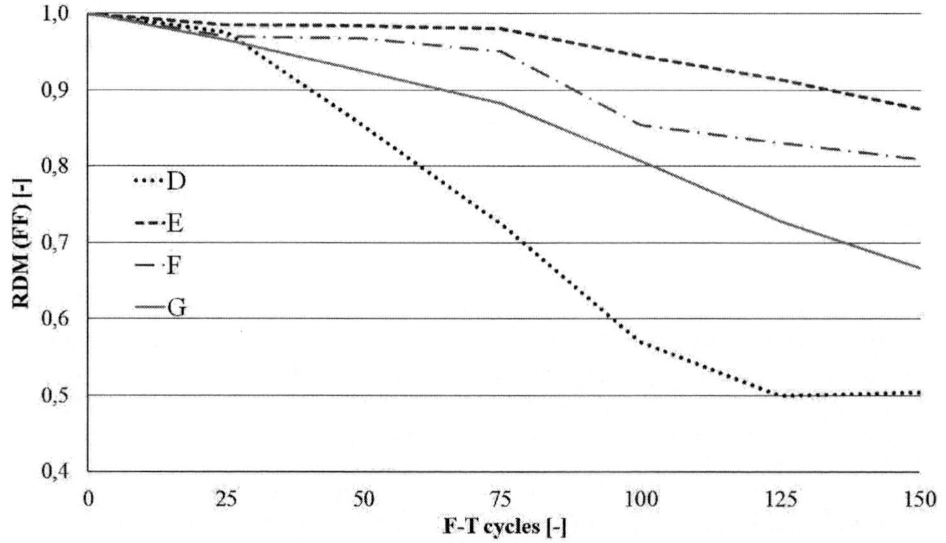


**Figure 3.** Progress of RDM (FF) for concrete D, E, F and G in relation to the number of F-T cycles.

**Table 3.** Results of RDM and ratios $f_{ct}$ and $f_{cf}$ after 150 F-T cycles (in percent).

| | Concrete | | | |
|---|---|---|---|---|
| Parameter | D | E | F | G |
| RDM (U) | 56.0 | 86.1 | 84.1 | 73.7 |
| RDM (FL) | 51.8 | 88.9 | 82.4 | 71.0 |
| RDM (FF) | 50.4 | 87.6 | 80.9 | 66.7 |
| $f_{ct}$ ratio | 55.8 | 75.2 | 78.9 | 75.8 |
| $f_{cf}$ ratio | 42.6 | 66.5 | 72.5 | 59.9 |

## 4. Conclusion

The experiment results show that RDM values measured non-destructively during frost resistance testing correspond more to a decrease in tensile splitting strength than to loss of flexural strength. A crucial advantage of non-destructive test methods in terms of assessing frost resistance is the possibility to observe the progress of internal damage sustained by the specimens throughout the duration of the test. Destructive tests may give information about this damage only when using a much higher number of specimens, and even so, the results will be affected by the fact that every testing session will be performed with a different (new) set of specimens. The loss of flexural strength compared to RDM or tensile splitting strength is more severe and in terms of assessing concrete for frost resistance also sets a much stricter standard for failure.

**Acknowledgments**
This work was supported by the Czech Science Foundation [grant No. GA17-14302S].

**References**

[1] fib Bulletin No. 28 2004 *Environmental Design* (Lausanne: fib) 80
[2] Aïtcin P-C 1998 *High-Performance Concrete* Stern B (New York: E&FN Spon) 624
[3] Collepardi M 2006 *The New Concrete* (Treviso: Grafiche Tintoretto) 421
[4] Neville A 2011 *Properties of concrete* (New York: Pearson) 846
[5] Zarauskas L, Skripkiūnas G and Girskas G 2017 Influence of Aggregate Granulometry on Air Content in Concrete Mixture and Freezing - Thawing Resistance of Concrete, *Procedia Engineering* **172** pp 1278-1285
[6] ASTM C666 / C666M – 15 Standard Test Method for Resistance of Concrete to Rapid Freezing and Thawing (West Conshohocken, PA: ASTM International)
[7] GB/T 50082-2009 Standard for test methods of long-term performance and durability of ordinary concrete (English Version) (Beijing: Codeofchina Inc.)
[8] BS 5075-2:1982 Concrete admixtures. Specification for air-entraining admixtures (London: BSI)
[9] ČSN 73 1322 Determination of frost resistance of concrete (Prague: UHM)
[10] ČSN 73 1380 Testing the freeze-thaw resistance of concrete - Internal structural damage (Prague: CNI)
[11] ČSN 73 1371 Non-destructive testing of concrete – Method of ultrasonic pulse testing of concrete (Prague: UNMZ)
[12] ČSN 73 1372 Non-destructive testing of concrete – Testing of concrete by resonance method (Prague: UNMZ)
[13] Kocáb D, Králíková M, Cikrle P, Misák P and Kucharczyková B 2017 Experimental analysis of the influence of concrete curing on the development of its elastic modulus over time *Mater. Technol.* **51** pp 657-665
[14] EN 12390-5 Testing hardened concrete – Part 5: Flexural strength of test specimens (Brussels: CEN)
[15] EN 12390-6 Testing hardened concrete – Part 6: Tensile splitting strength of test specimens (Brussels: CEN)

# The mechanical properties of flax and hemp fibres reinforced geopolymer composites

**K Korniejenko[1], M Łach[1], M Hebdowska-Krupa[1] and J Mikuła[1]**

[1] Institute of Materials Engineering, Faculty of Mechanical Engineering, Cracow University of Technology, Jana Pawła II 37, 31-864 Cracow, Poland.

E-mail: kinga.korniejenko@mech.pk.edu.pl

**Abstract.** The article aim is to analyse the influence of addition of different kinds of flax and hemp fibres on the mechanical properties of the geopolymer composites. The paper describes the mechanical properties of fly-ash based geopolymer reinforced with short natural fibres made from the plants that have grown in Poland. The samples were prepared using sodium promoter and fibres (1% by mass of the composite). The empirical part of the research is based on the compressive strength tests, flexural strength tests and detailed microstructure examination. The research involved the samples reinforced by different kind of fibres - the flax and hemp fibres were prepared by different kind of treatment before addition to composites. The results show the influence of previous treatment of the fibres on the mechanical properties of the composites.

## 1. Introduction

The research on the composites based on geopolymers reinforced with the natural fibres is rather new area of investigation connected with the development of sustainability policy [1, 2]. Nowadays, there is the tendency to develop new and eco-friendly composites to replace the technologies based on non-renewable resources and technologies that are energy consuming such as Portland cement technology [3, 4]. The most promising alternative solutions for construction industry are technologies based on alkali-activated and/or geopolymer composites reinforced with fibres [5, 6]. The addition of artificial fibres usually makes mechanical properties better than the addition of natural ones [2], but from the environmental point of view, the addition of renewable fibres is more beneficial. The replacement of the artificial fibres with their natural counterparts decreases significantly $CO_2$ emission calculated on the product – ecological footprint [7, 8].

The most important research on natural fibres as a reinforcement for geopolymers matrix was made with following fibres [1, 2]: cotton fibres [2, 9], coir [2, 10], sorghum fibres [11], wool fibres [12, 13], raffia [1], jute [14, 15], fique / sisal fibres [2, 16], corn husk [17], abaca [18], rice stem [19], pineapple leaf fibres [6] and bamboo [21]. The research results show that natural fibres can be a valuable additive for geopolymer composites and there is a high chance that they find the application as materials for construction, energy and other industries in the nearest future [15]. In comparison to materials reinforced by synthetic fibres they have numerous of advantages such as [2, 3]:

- Lower density than synthetic fibres.
- They are no toxic.
- They are easy for processing.
- The cost of production is lower than synthetic fibres.

- They have high strength-to-mass ratio in early stage.
- They have good tensile strength.
- They are environmentally friendly, especially renewable and recyclable.

Some research has been made on flax and hemp fibres [4, 7]. The authors present that this kind of fibres can be used for the geopolymer composites, because they are relatively resistant to the alkaline environment of the matrix [7, 21]. The mechanical properties of such composites are better than unreinforced geopolymer matrix [7, 21]. Additionally, the reinforced composites show graceful failure, unlike the brittle failure of the matrix. In case of flax, the research shows that the composites are also thermally resistant up to 400°C (the flax fibres are protected by geopolymer matrix) [4, 7].

The flax and hemp fibres were selected to be investigated as they are easily available/reachable in Poland, due to the suitable climate and environmental conditions. The mechanical characteristics of mentioned fibres are suitable to be used in building materials. [22, 23]. Moreover, the ecological footprint reduction is possible only when locally available raw materials are used. Thus, the transport of raw materials can be significantly reduced, with positive impact on environment. It has also justification on the economic site – the elimination of the cost of transportation [24, 25].

## 2. Experimental procedure

Specimens with 1% by mass of natural fibres were investigated. The variables include different types of fibres: flax and hemp.

### *2.1. Materials*

*2.1.1. Geopolymer matrix.* The geopolymer matrix was prepared from fly ash from Skawina, Poland CHP plant and sand in ratio 1:1. This kind of fly ash is suitable for manufacturing geopolymer because of its physical properties and chemical composition. It has been confirmed by previous research, including SEM observations, EDS analysis, analysis of chemical composition (elements as well as oxides) and physical properties such as density [1, 2]. This fly ash form CHP plant in Skawina (Poland) is rich in oxides such as $SiO_2$ (47,81%), $Al_2O_3$ (22,80%), which is advantageous for production of geopolymers. The fly ash particles morphology was typical for such by-products of coal combustion and suitable for the process of alkali-activation [1, 2].

*2.1.2. Flex fibres.* As an additive, three kinds of flax fibres were applied: two kinds of flax tow and flax hurds. Tow is a coarse, broken fibre, removed during flax processing. They are usually shorter than 30 cm and are semi-products for textile fibres production. The hurds, called sometimes as hards, are the coarse parts of flax that adhere to the fibre after it is separated. They are usually residual from fibre production process. The following flax fibres were applied to compositions:

- Tow flax fibres after dew retting process (Figure 1) – the fibre is rather thin, but with some small hard part of stiffness elements (similar to flax hurds).
- Flax hurds (Figure 2) – the dimension is usually between 3-7 mm. The parts are stiff and 'hard'.
- Green tow flax fibres (Figure 3) – the fibre is similar to tow flax fibre after dew retting process but is more stiff.

The fibres have been purchased from the Institute of Natural Fibres and Medicinal Plants in Poland.

IOP Conf. Series: Materials Science and Engineering **379** (2018) 012023 doi:10.1088/1757-899X/379/1/012023

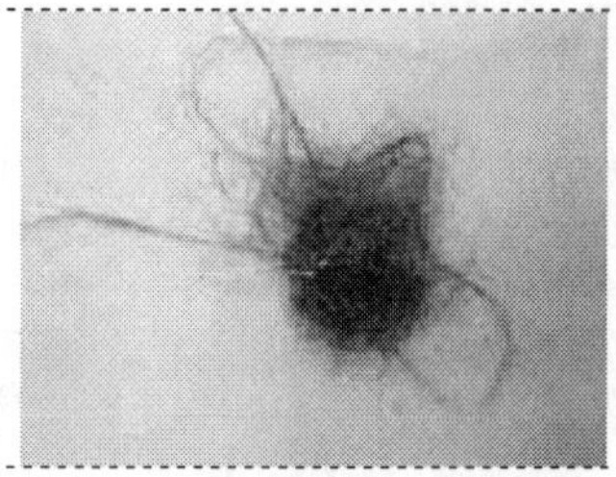

**Figure 1.** Flax tow after dew retting process.

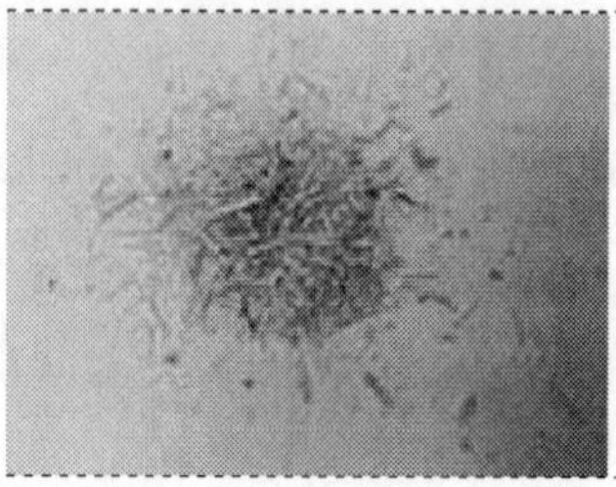

**Figure 2.** Flax hurds.

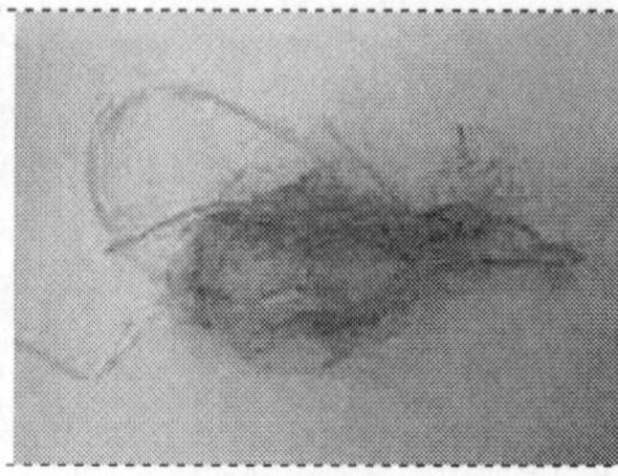

**Figure 3.** Green tow flax fibre.

*2.1.3. Hemp fibres.* As an additive, three kinds of hemp fibres were applied: they were in different forms and had different level of processing. The following hemp fibres were applied to compositions:

- Hemp hurds (Figure 4) – the dimension is usually between 5–10 mm (a little bit larger than flax hurds). The parts are more stiff than other investigated fibres.
- Decorticated hemp fibres (Figure 5) – thin fibre dedicated for decorticated purpose, without residues.
- Hemp microfibres (Figure 6) – very thin fibre with small number of thicker residues.

**Figure 4.** Hemp hurds.

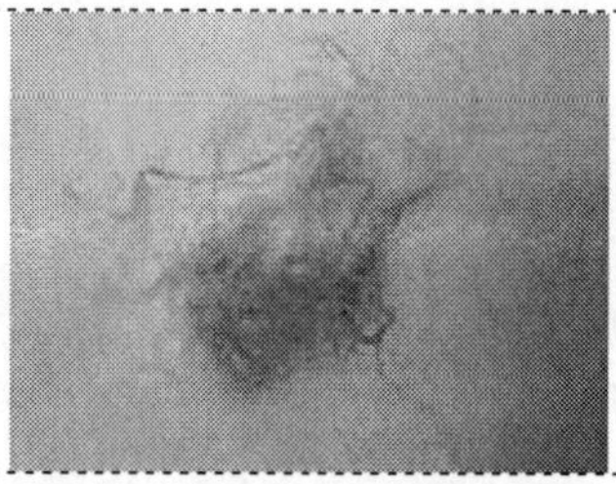

**Figure 5.** Decorticated hemp fibre.

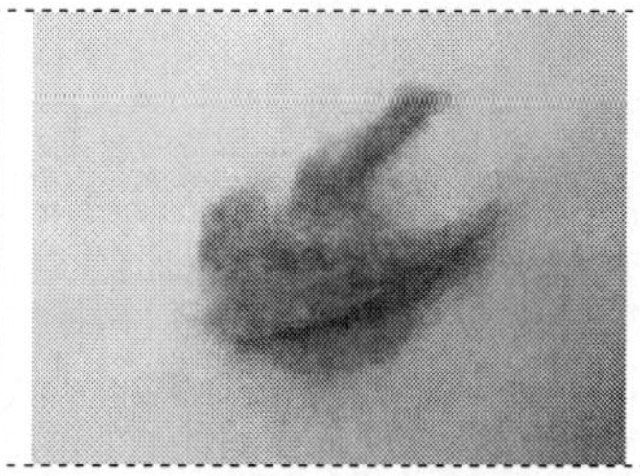

**Figure 6.** Hemp microfibres.

The fibres have been purchased from the Institute of Natural Fibres and Medicinal Plants in Poland.

*2.2. Specimens*

Samples were prepared using sodium promoter, fly ash, sand and various flax and hemp fibres (1%, by mass of the composite). The process of activation has been made by 8M sodium hydroxide solution combined with the sodium silicate solution (liquid glass at a ratio of 1:2.5). To produce geopolymers, flakes of technical sodium hydroxide were used and an aqueous solution of sodium silicate (R-145) which molar module was 2.5 and density was about 1.45 g/cm$^3$. The tap water was used instead of the distilled one. The alkaline solution was prepared by means of pouring the aqueous solution of sodium silicate over the solid sodium hydroxide. The solution was mixed and left until its temperature became stable and the concentrations equalized, which took about 2 hours. The fly ash, sand, alkaline solution and fibres were mixed about 20 minutes by using low speed mixing machine (to receive the homogeneous paste). Next, it was poured into two sets of plastic moulds. The first set consisted of the moulds dedicated to undergo compressive strength tests and the second set consisted of the moulds dedicated to undergo flexural strength tests. The samples were hand-formed and then subjected to vibratory removal of air bubbles. Tightly closed moulds were heated in the laboratory dryer for 24h at 75°C. Then, the samples were unmoulded. They were investigated after 28 days.

## 3. Analytical procedure

Scanning electron microscope (SEM) type JEOL JSM 820 with EDS has been used for microstructure research. The research has been made for fibres and the samples previously broken during compressive or flexural strength tests. The samples were covered with a thin layer of gold with JEOL JEE-4X vacuum sputter. The investigations were made at various magnifications (between 20–100x).

Compressive strength tests were carried out according to the methodology described in the standard EN 12390-3. ('Testing hardened concrete. Compressive strength of test specimens'), because of the lack of separate standards for geopolymer materials. The tests involved at least 6 samples. Tests were performed using the universal testing machine - single-point load (Instron type 4465). Samples used to the compressive strength test had cubic shape and dimensions (approx.): 50 mm × 50 mm × 50 mm.

Flexural strength tests were carried out according to the methodology described in the standard EN 12390-5 ('Testing hardened concrete. Flexural strength of test specimens'). The calculations were based on following equation:

$$f_{cf} = \frac{Fl}{d_1 d_2^2} \tag{1}$$

where:

$f_{cf}$ – compressive strength, MPa

F – maximal load, N

l – space between supporting points, mm (for conducted tests:140 mm)

$d_1$, $d_2$ – sample dimensions, mm

Samples used to the flexural strength test had prismatic shape and dimensions (approx.): 200 mm × 50 mm × 50 mm. Tests were performed using the universal testing machine - single-point load (Instron type 4465).

## 4. Experimental results

### *4.1. Microstructure research*

The SEM observations were made for fibres as well as for composition reinforced by fibres. The images were made at various magnifications - between 20–100x, depending on the particular fibre. The observation of microstructure of composites gave a preliminary information about the coherency of fibres (filler) with the geopolymer matrix (Figure 7). The fibre distribution was also evaluated.

In the Figure 8, there is presented the morphology of the flax tow after dew retting process. This kind of fibre is built with some smaller fibres that can be separated during the composition manufacturing. Flax hurds (Figure 9) have other morphology. They are rather compact and do not split themselves during the mixing process and composite production.

**Figure 7.** Geopolymer with flax.fibre.

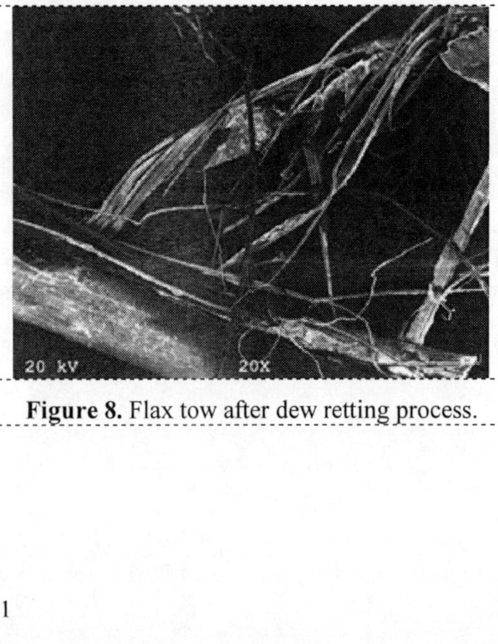


**Figure 8.** Flax tow after dew retting process.

The green tow flax fibre (Figure 10) has very interesting structure. On the one hand it is quite compact, on the other the fibre structure is rough. The rough structure should allow the good coherence with composite matrix.

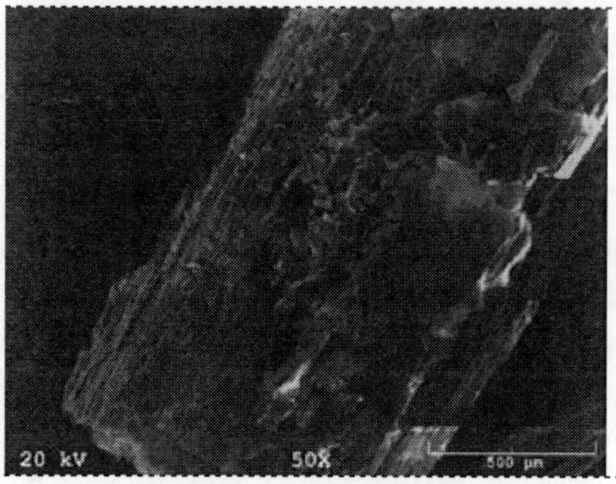


**Figure 9.** Flax hurds.

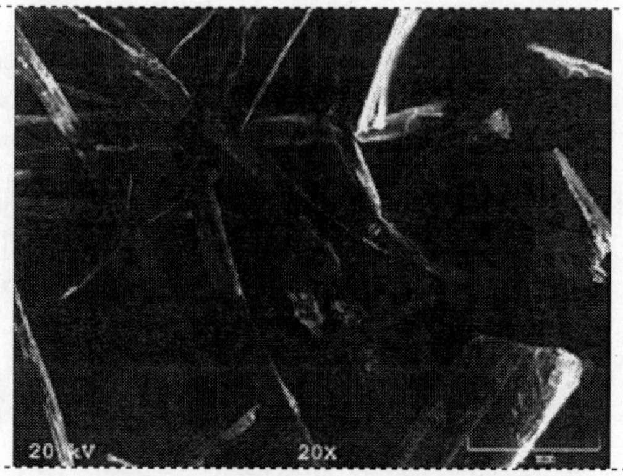


**Figure 10.** Green tow flax fibre.

The decorticated hemp fibres (Figure 11) and the hemp microfibres (Figure 12) have thin fibres in their structure. It can be advantageous for composite, but it also requires carefully mixing process to ensure proper fibres distribution in the matrix. Thin fibres have tendency to create the agglomerates.

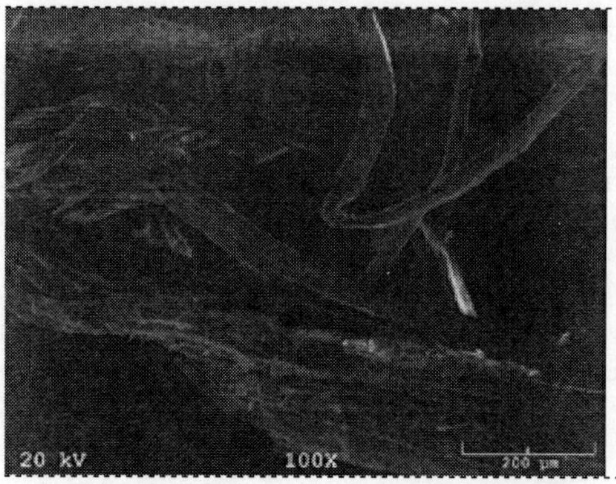


**Figure 11.** Decorticated hemp fibre.

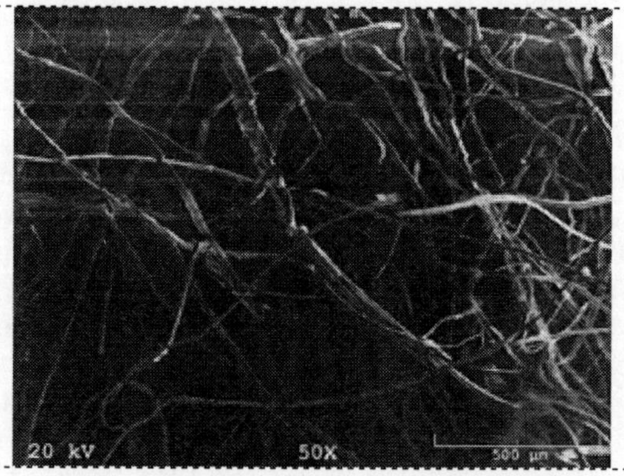


**Figure 12.** Hemp microfibres.

The fibre distribution in the matrix heavily influences the properties of the specific composite, as the fibres aggregation can decrease its mechanical properties. In case of investigated composites, the fibres distribution was regular. The microstructural observation also allows to notice that the structure is coherent (good adhesion the fibres to the matrix, Figure 7).

### *4.2. Compressive strength test*

The results of the compressive strength test are shown in Table 1. The results of both fibre reinforced and plain (without fibres) specimens are presented.

**Table 1.** Compressive strength test after 28 days.

| Sample | MPa | standard deviation |
|---|---|---|
| **Geopolymer without fibre** | 42.70 | 3.38 |
| **Geopolymer with tow flax fibres after dew retting process (1%)** | 31.92 | 1.80 |
| **Geopolymer with flax hurds (1%)** | 30.72 | 3.36 |
| **Geopolymer with green tow flax fibres (1%)** | 33.46 | 3.12 |
| **Geopolymer with hemp hurds (1%)** | 35.90 | 2.00 |
| **Geopolymer with decorticated hemp fibres (1%)** | 42.65 | 4.17 |
| **Geopolymer with hemp microfibres (1%)** | 35.20 | 7.03 |

Better results were attained through reinforcement by hemp fibres than flax ones. The best results have been achieved for decorticated hemp fibres. It is more than 42 MPa and it is comparable with geopolymer without reinforcement. Despite the decreasing of compressive strength, this result seems to be promising for future applications. The results for different flax fillers are very similar. The differences are statistically not important.

### *4.3. Flexural strength test*

The results from the flexural strength tests are shown in Table 2. The results of both fibre reinforced and plain (without fibres) specimens are presented.

**Table 2.** Flexural strength test after 28 days.

| Sample | MPa | standard deviation |
|---|---|---|
| **Geopolymer without fibre** | 5.22 | 0.25 |
| **Geopolymer with tow flax fibres after dew retting process (1%)** | 4.67 | 0.06 |
| **Geopolymer with flax hurds (1%)** | 5.83 | 0.80 |
| **Geopolymer with green tow flax fibres (1%)** | 5.12 | 0.37 |
| **Geopolymer with hemp hurds (1%)** | 5.28 | 0.27 |
| **Geopolymer with decorticated hemp fibres (1%)** | 6.09 | 0.13 |
| **Geopolymer with hemp microfibres (1%)** | 6.01 | 0.63 |

Similarly to the compressive strength tests, better results were attained through reinforcement by hemp fibres than flax ones. Moreover, the results with hemp fibres are slightly better than for geopolymer matrix without reinforcement. The best results give thin fibres such as decorticated hemp fibres and hemp microfibres than thick fillers. The flexural strength is important parameter for application in building industry, especially for more advanced construction. The achieved results show that new composites are comparable with contemporary used materials such as cementitious materials.

## 5. Discussion

There is lack of literature about research on influence of different kind of hemp and flax fibres on mechanical properties of the composites. Some research on the short flax fibres admixtures has been made based on geopolymers made from kaolinite clay with different amount of flax fibres: 4, 7 and 10 wt.% [7]. The research results pointed out improving mechanical properties of geopolymers by fibre additives, especially flexural strength. However, some of them seem to include incoherence like increasing of the flexural strength from 5.8 MPa to 70.2 MPa in connection with 10% flex fibre addition [7]. Those values of flexural strength were much higher compared to the results presented in our manuscript, which could be due to the different research methodology. 70 MPa was probably achieved at total failure of the sample and not at the "first crack". Such test probably shows a flexural strength of fibres not of the whole composite. The other research is focused on flax fabric reinforcement [4].

There is also lack of complex research on hemp fibres. These fibres have been investigated in composites based for complex matrix that included: fly-ash, waste glass, ordinary Portland cement (OPC) and recycled aggregates [25]. The geopolymer mortar was reinforced by short hemp fibres (ca. 20–30 mm) up to 8% by mass of the composite Investigation shows decrease of mechanical properties caused by fibre additives (about 50% in the compressive strength due to addition of 8% hemp fibres). The compressive strength for the matrix was 45 MPa, after 8% fibre addition it was only 12 MPa. The similar situation was also in the case of flexural strength - it was 3,45 MPa for geopolymer matrix and 2.13 MPa for composite. Despite the decreasing mechanical properties for the composites, the authors stress the advantageous for the new material, especially environmental benefits – reduce the carbon footprint, connected with using the natural fibres (the use of at least 8% hemp fibres leads to carbon negative emissions -19.7 $kgCO_2eq/m^3$) [25]. The decrease of mechanical properties was attributed to the high content of the fibres. Limitation to 1 or 2% by mass of the composite will be more advantageous for mechanical properties.

## 6. Further research

The results of the research and literature review show that the geopolymer composites can be low-cost and effective construction materials. Further tests should be performed in order to optimize the mechanical properties of the composites as well as to investigate other properties such as absorptivity and resistance for different environments. The possible application for this kind of composites is construction industry, so the planned research should involve frost resistance as well as heat resistance and also durability issues. In the long perspective, it is also recommended to design some prototype elements, exemplary bricks and make the LCA analysis for the chosen products.

## 7. Conclusions

Geopolymer composites reinforced with various flex and hemp fibres have been produced and characterized. The samples were prepared using sodium promoter and six various types of fibre additives (1% by mass of the composite). The matrix was geopolymer cement (fly ash and sand in ratio 1:1).

The observed differences between the fibres influence the mechanical properties of the whole composite. The composites with hemp fibres outperformed the flex reinforced ones in terms of mechanical properties - higher values of both compressive and flexural strength were observed when hemp was involved. The best results were achieved in case of fine fibres such as decorticated hemp fibres and microfibres. This study shows that it is possible to produce the composites of reasonable properties from the industrial wastes (fly ash) and renewable resources – natural fibres.

Even the fibre addition does not improve the final strength of the composite significantly, the overall mechanical parameters are positively influenced; the presence of the fibres changes the character of failure mechanism from brittle to ductile. Additionally, using the natural fibres instead of industrial ones has positive impact on environment. Thus, the designed composites have potential to be used as new eco-friendly material e.g. in construction industry.

## Acknowledgments

This work was supported by the ERANet-LAC 2nd Joint Call (http://www.eranet-lac.eu) and funded by National Centre for Research and Development, Poland, under grant: "Development of eco-friendly composite materials based on geopolymer matrix and reinforced with waste fibres".

## References

[1] Korniejenko K, Łach M and Mikuła J 2018 Mechanical Properties of Raffia Fibres Reinforced Geopolymer Composite *Advances in Natural Fibre Composites* eds. R. Fangueiro S Rana (Springer, Cham) pp 135-144

[2] Korniejenko K, Frączek E, Pytlak E and Adamski M 2016 *Procedia Engineering, 2016* **151** 388

[3] Alomayri T, Shaikh FUA and Low IM 2013 *J Mater Sci.* **48** 6746
[4] Assaedi H, Shaikh FUA and Low IM 2017 *Journal of Asian Ceramic Societies* **5** 62
[5] Sankar K and Kriven WM 2015 *Cer Engr Sci Proc* **38(10)** 39
[6] Yan L, Kasal B and Huang L 2016 *Compos. Part B-Eng* **92** 94
[7] Alzeer M and MacKenzie K 2013 *Applied Clay Science* **75–76** 148
[8] Naidu AL, Jagadeesh V and Bahubalendruni MVAR *2017 International Journal of Advanced Research in Engineering and Technology* **8(1)** 56
[9] Alomayri T, Shaikh FUA and Low IM. 2013 *Compos. Part B-Eng* **50** 1
[10] Patel R and Joshi R. 2016 *International Advanced Research Journal in Science* **3(12)** 171
[11] Chen R, Ahmari S and Zhang L. 2014 *J Mater Sci* **49** 2548
[12] Yliniemi J, Nugteren H, Illikainen M, Tiainen M, Weststrate R and Niinimäki J 2016 *Int. J. Miner. Process.* **149** 42
[13] Alzeer M and MacKenzie K 2012 *J Mater Sci.* **47** 6958
[14] Trindade ACC, Arêas IOM., Almeida DCT., Alcamand HA, Borges PHR and Silva FA 2018 Mechanical Behavior of Geopolymeric Composites Reinforced with Natural Fibres *Strain-Hardening Cement-Based Composites* eds. V Mechtcherine, V Slowik, P Kabele P. (Springer, Dordrecht) pp 383–39
[15] Trindade ACC, Alcamand HA, Borges PHR and Silva FA 2017 *J. Ceram. Sci. Technol.* 389
[16] Sankar K and Kriven WM 2015 *Cer Engr Sci Proc* **38(10)** 61
[17] Musil SS, Keane PF and Kriven WM 2014 *Cer Engr Sci Proc* **34(10)** 123
[18] Malenab RAJ, Ngo JPS and Promentilla MAB 2017 *Materials* **10(6)** 579
[19] Heo UH, Sankar K, Kriven WM and Musil SS 2015 *Cer Engr Sci Proc* **38(10)** 87
[20] Ribero RAS, Ribero MGS, Sankar K and Kriven WM 2016 *Constr Build Mat* **123** 501
[21] Galzerano B, Formisano A, Durante M, Iucolano F, Caputo D and Liguori B 2017 *Journal of Composite Materials* 1
[22] Murri AN, Medri V and Landi E 2017 *Journal of the American Ceramic Society* **100(7)** 2822
[23] Bribián IZ, Capilla AV, Usón AA 2011 *Building and Environment* **46** 1133
[24] Pacheco-Torgal F 2014 *Construction and Building Materials* **51** 151
[25] Mastali M, Abdollahnejad Z and Pacheco-Torgal F 2018 *Construction and Building Materials* **160** 48

# Blast resistance of slurry infiltrated fibre concrete with hybrid fibre reinforcement

**M Drdlová[1], P Bibora[1] and R Čechmánek[2]**

[1] Senior Researcher, Research Institute for Building Materials, Brno, Czech Republic
[2] Head of department of Applied Research and Development, Research Institute for Building Materials, Brno, Czech Republic

E-mail: drdlova@vustah.cz

**Abstract.** This paper investigates experimentally the blast resistance of slurry infiltrated fibre concrete (SIFCON) with hybrid fibre reinforcement realized by various combinations of polypropylene (PP), aramid, carbon, glass, wollastonite and steel fibers. The binary reinforcement systems were designed using the combinations of steel/PP, steel/aramid, steel/glass, steel/wollastonite and steel/carbon fibres with various fibre geometry. Six different batches (including one reference specimen) were examined with aim to determine the influence of the parameters of hybrid reinforcement on blast resistance of the concrete. Mechanical parameters at quasistatic load (flexural strength, compressive strength) were determined and load-deflection curves were captured. Real blast tests were performed on the slab specimens (500 x 500 x 40 mm) according to the modified methodology M-T0-VTU0 10/09. The change of the ultrasonic wave transit time before and after the blast load in certain measurement points, damage of the slab, the weight of fragments and their damage potential were evaluated and compared. The results indicate, that hybrid fibre reinforcement, when properly designed, can highly enhance overall blast resistance of SIFCON and can significantly reduce the spalling. Steel-aramid combination brought the best blast resistance.

## 1. Introduction

High performance concrete itself as a quasi-brittle material does not conform to the blast resistance requirements. Fiber reinforcement is often used to provide enhanced toughness and ductility to brittle cementitious matrix. By the addition of steel fibres to conventional concrete, the impact strength value can be increased 2–15 times depending on the amount of the fibres [1]. Research on the use of fibres to increase both the blast and impact resistance of concrete has typically been focused to steel fibres and, less, on polypropylene fibres. [2–4] Some studies have been found dealing with glass, nylon, ceramic and carbon fibres reinforced concrete related to blast or impact resistance of the material [5–9]. These studies confirm better performance of fibre reinforced concrete (FRC) under impact load compared to plain concrete without reinforcement. Ref. [10] is comparative study of concrete with high spectra of different kind of fibres (PP, glass, basalt, steel) and fibre blends in the same matrix and their performance while subjected to the blast load. According to this study, the blast resistance of the concrete is connected with the tensile strength of the incorporated fibre. Fibres with a low tensile strength are not able to resist the high shear and flexural strains generated by blast loads. [10]

The reinforcement of concrete with a single type of fiber may improve the desired properties to a limited level, whereas combination of two or more types of fibres, if used in optimal way in concrete, achieves better engineering properties due to positive synergetic effect. [11] Generally, inclusion of fibers in concrete in hybrid forms improves many of its engineering properties such as toughness, ductility, energy absorption capacity and durability performance in comparison with mono fiber reinforcement. [12] The different methods of hybridization include combining different lengths, diameters, modulus and tensile strengths of fibers [13, 14]. Maalej et al. [15] found that hybrid fibre-reinforced concrete offered increased shatter resistance with reduced scabbing, spalling and fragmentation, whilst exhibiting better energy absorption.

Special type of high performance concrete with high steel fibre reinforcement is slurry infiltrated fibre concrete (SIFCON). This composite is prepared by infiltrating of the preplaced fibres by the mortar slurry. This production technology enables to incorporate large amounts of steel fibres (typically 8–10%, but in general up to about 15%) in cement-based composites [16]. Compared to ordinary high performance fiber-reinforced concrete, SIFCON exhibits enhanced ductility, energy absorption capacity and overall integrity under dynamic load [17], and thus has potential for applications in protective structures. Excellent blast resistance of SIFCON in terms of overall integrity and internal damage was concluded in [18], but some fragmentation of the matrix was admitted. This can be prevented by using small amount of micro-fiber reinforcement in the mortar, but more information in this field must be provided, as the studies comparing the dynamic properties of hybrid fibre reinforced SIFCON composites are very limited [19] and focus only on low strain rates.

The presented research is part of the development process of cement-based materials with enhanced blast resistance, which will be used in the elements for critical infrastructure protection, such as barriers and sandwich structures. The main objective of the work presented herein is thus to provide more information about the effects of various hybrid-fibre reinforcement on mechanical properties and blast resistance of SIFCON.

## 2. Experimental program

### *2.1. Materials and specimen preparation*

Fine $SiO_2$ sand (670 kg/m$^3$) with grain size of 0–1 mm, cement CEM 52.5R (990 kg/m$^3$) and water (320 kg/m$^3$) were the main components of the slurry mixture. Superplasticizer Glenium 422 (10 kg/m$^3$), produced by BASF, was added to achieve good workability. Silica fume Elkem 940U (78 kg/m$^3$), produced by BASF, with a typical particle size 100–500 nm, was used to achieve an optimized particle packing density and for pozzolanic properties. As the main reinforcement, the steel fibres KrampeHarex 50/0.8DE fibres (length 50 mm, diameter 0.8 mm) with hooked ends were selected, the dosage in each specimen was 9 vol. %. As secondary reinforcement, microfibres were used; their properties and volume fraction are summarized in Table 1. The particular micro-fibres were selected to cover the high spectra of properties. The amount of the fibres varied slightly to achieve similar workability of the slurry.

**Table 1.** Properties of incorporated fibres and their dosage.

| Designation | Fibre content (Vol. %) | Fibre material | Density of the fibre (kg.m$^{-3}$) | Length/diameter of the fibre (mm/μm) |
|---|---|---|---|---|
| ST50AR0.5 | 0.5 | Aramid | 1,400 | 1.0/12 |
| ST50GL0.7 | 0.7 | AR glass | 2,680 | 1.3/12 |
| ST50PP0.5 | 0.5 | Polypropylene | 910 | 2.2/15 |
| ST50CP0.5 | 0.5 | Carbon PAN | 1,760 | 2.0/12 |
| ST50WO2.0 | 2.0 | Wollastonite | 2,800 | 0.3/9 |

Steel moulds were used to cast the specimens. Moulds were treated with releasing agent and sealed at the edges by using silicone to prevent any leakage of cement–sand slurry from the mould. The mixing procedure of the slurry was as follows: 2/3 of the required water amount with whole dosage of superplasticiser was poured into the Power-sprays GRC 125 mixer; sand, cement and microsilica were added and stirred together for 60 seconds, to achieve homogenous paste without any bulks. As the next step, the micro-fibres were added and mixed for another 60 seconds. At the end of the mixing process, the rest of the water was added and mixed for another 30 seconds. Prepared mortar was poured over the pre-placed steel fibres. Prepared slurry was workable enough and the infiltration was possible for all tested mixes, but the length of PP fibre (2.2 mm) was determined as limit value to keep good infiltration level. The test specimens were demoulded after 24 hours and were cured for 28 days in curing water ponds. For quasistatic mechanical tests, prism specimens of dimensions 100 × 100 × 400 mm were cast, for blast tests, slabs of dimensions 500 × 500 × 40 mm were prepared.

*2.2. Quasistatic mechanical tests and bulk density*

The mechanical parameters were obtained using universal strength testing machine TIRAtest 2710, R58/02. The compressive and flexural load (four-point flexural test) was applied at quasi-static conditions (speed of 5 mm/min). The stress-deflection curves were captured to compare the post peak behaviour of the materials.

*2.3. Blast tests*

Prepared slabs were subjected to the test procedure according to the certified methodology M-T0-B VTÚO 10/09 modified to cover two tests. The methodology was developed by the Military Research Institute of the Czech Republic and is used for assessing the blast resistance of several materials. The concrete specimen is fixed in the steel frame of the stand and the stand is placed on a solid foundation. During the first test, the spheres of Semtex 10 plastic high explosive weighing 150.0 g are used as testing charges. The weight and distance (100 mm from the test specimen) of the testing charge was adjusted to be strong enough to cause the significant visible damage of the specimens, including fragmentation. Under the stand, the check panel consisting of hardboard-polystyrene-aluminium plate sandwich is placed. Aluminium plate creates the front face adjacent to the test specimen and its thickness is 0.5 mm. Observed and evaluated parameters are the weight of created secondary fragments and their destructive power expressed as the degree of damage of the check panel (without damage/indentation/perforation). Second test covers the loading of the specimen fixed in the stand with the charge 100 g of Semtex 10. The weight for this test was adjusted to cause the low-level damage (small cracks, the slab should remain integral). The change of the ultrasonic wave transit time at four measuring points before and after test was evaluated. The rate of the change of this parameter is directly connected to the internal damage of material. The scheme of the test rig is depicted in Figure 1. The distance of the charge was same as in the previous test – 100 mm. The data obtained from both tests corresponds well and give enough information to create good overview about how the particular material performs under blast load; and materials can be compared easily.

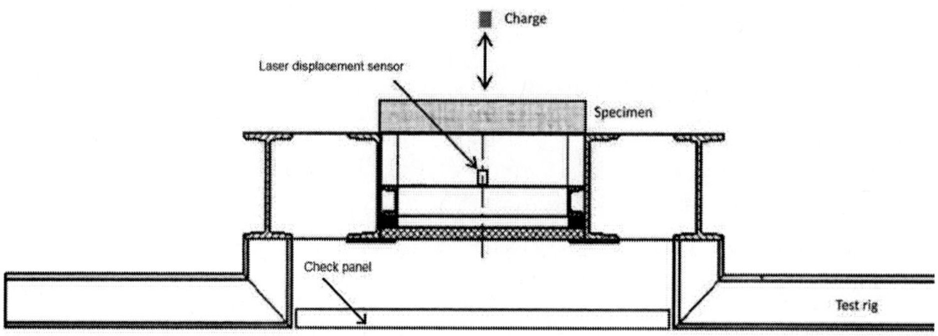


**Figure 1.** Scheme of the test rig.

## 3. Results and discussion

### *3.1. Physico-mechanical properties*

The results of determined physico-mechanical properties are summarized in Table 2. The compressive strength was positively affected by addition of secondary micro-reinforcement. On contrary, the flexural peak strength was almost not affected by secondary fibre incorporation, the values were similar - between 32–35 MPa, except for specimen with aramid fibres (42 MPa). Stress-deflection curves (see Figure 2, 3 and 4) showed different post peak behaviour for specimens with secondary reinforcement. The ascending part of the curve is similar for all specimen. All specimens with secondary reinforcement, except for specimen with wollastonite, showed no sharp flexural peak stress with plateau region and less steep shape of the decreasing part of the curve (= the post peak stress was higher at a given strain), which indicates higher load bearing capacity of the specimens. The highest post-peak load bearing capacity showed the specimen with aramid fibre reinforcement.

**Table 2.** Physico-mechanical properties of specimens with various hybrid reinforcement.

| Designation | Bulk density (kg·m$^{-3}$) | Compressive strength (MPa) | Flexural strength (MPa) |
|---|---|---|---|
| ST50AR0.5 | 2,679 | 136.0 | 42.1 |
| ST50GL0.7 | 2,680 | 132.0 | 33.3 |
| ST50PP0.5 | 2,686 | 143.1 | 32.1 |
| ST50CP0.5 | 2,697 | 150.2 | 35.1 |
| ST50WO2.0 | 2,693 | 128.1 | 33.2 |
| ST50REF | 2,681 | 122.4 | 34.1 |

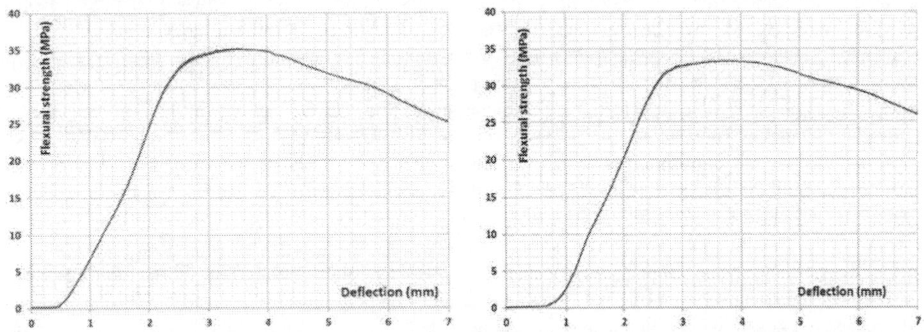


**Figure 2.** Stress-deflection curves for specimen ST50CP0.5 (left) and ST50PP0.5 (right).

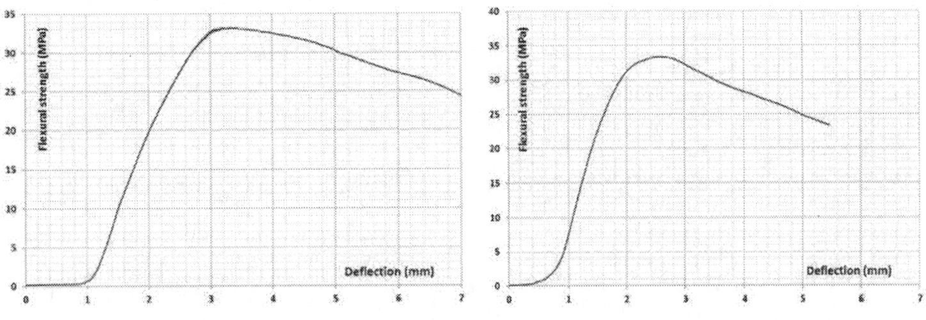


**Figure 3.** Stress-deflection curves for specimen ST50GL0.7 (left) and ST50WO2.0 (right).

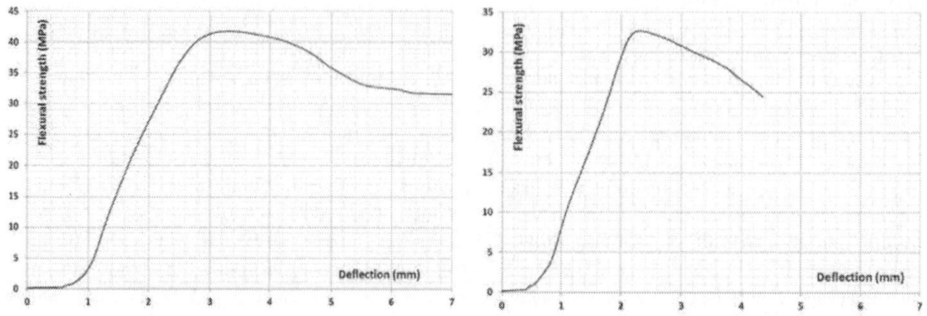


**Figure 4.** Stress-deflection curves for specimen ST50AR0.5 (left) and ST50REF (right).

### *3.2. Blast test results*

The experiments focused on the effect of different hybrid fibre mixes on the blast performance of SIFCON were carried out. Performance was determined according to the methodology described above, and the results are presented in Table 3. Two slabs were prepared with each mix. When evaluating the behaviour of the concrete under blast load, the special attention must be paid to secondary debris creation, as the bystanders can be injured by them during a blast. The destructive power of the debris was expressed as the degree of damage of the check panel.

None of the specimens exhibited a complete loss of integrity. The REF specimen with only primary reinforcement sustained the greatest amount of damage, see Table 3 and Figure 5b. Surface crater was created in the centre of the rear side of the slab. Using the polypropylene and wollastonite secondary reinforcement, the overall blast resistance of the specimens increased, the internal damage of the slab expressed as increase of US wave transit time in % was lower compared to the ST50REF specimen, but fragmentation of the slab still occurred, see Figure 5a and 5c, and created fragments caused perforation of the check panels. The low performance of specimen ST50PP0.5 with polypropylene fibres is probably caused by their low mechanical properties – the tensile strength of the PP fibre is 0.3 GPa, which is almost the order of magnitude lower than of the other used fibres. Another reason can be the weaker bond of the fibre in the hardened concrete. Observed low effectivity of PP fibres in enhancement of blast resistance of hybrid fibre reinforced concrete corresponds with findings published in [10]. In the case of ST50WO2.0 (wollastonite reinforced specimen), the low enhancement of the blast resistance was probably caused by unsuitable geometry of the used fibres – the length of the fibre was only 0.3 mm, which was not sufficient for effective strain transfer. The results of blast test were consistent with results of quasistatic tests, where the low reinforcing effect of wollastonite fibres was also observed – the flexural peak stress and post peak behaviour was almost the same as in case of REF specimen, without any enhancement.

**Table 3.** Blast test results.

| Designation | Increase of US wave transit time at four points (%) | Fragment weight (g)/ (%) | Check panel damage/quantity of perforation |
|---|---|---|---|
| ST50AR0.5 | 23/90/134/96 | 20/0.08 | Indentation/0 |
| ST50GL0.7 | 404/24/80/75 | 50/0.21 | Indentation/0 |
| ST50PP0.5 | 436/1200/75/97 | 166/0.69 | Perforation/2 |
| ST50CP0.5 | 90/98/114/102 | 30/0.13 | Indentation/0 |
| ST50WO2.0 | 423/968/86/88 | 126/0.49 | Perforation/5 |
| ST50REF | 230/2284/90/74 | 192/0.44 | Perforation/5 |

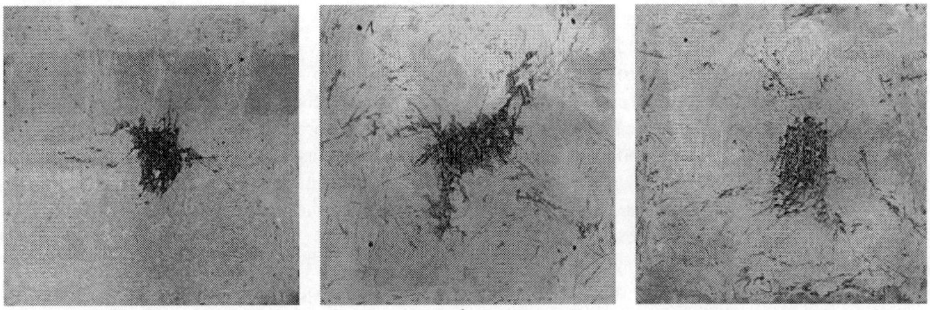

a b c

**Figure 5.** Slabs after blast test a) ST50PP0.5, b) ST50REF, c) ST50WO2.0.

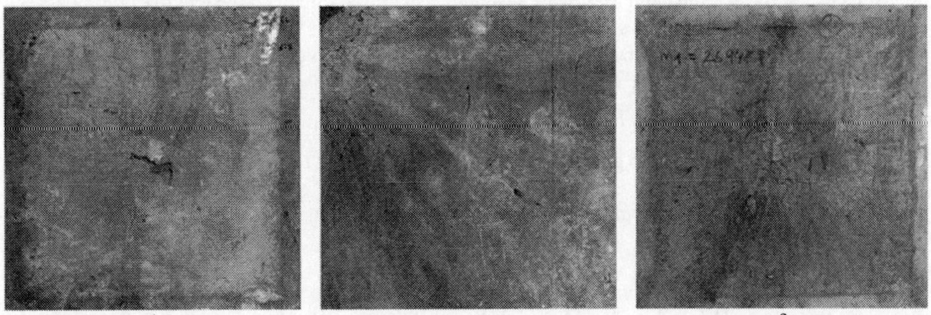

d e f

**Figure 6.** Slabs after blast test d) ST50GL0.7 e) ST50AR0.5 f) ST50CP0.5

Specimen ST50GL0.7 with glass fibre reinforcement performed well under the blast load – only very few splinters of total weight 50 g were created, without perforation of the check panel. The addition of aramid fibre as secondary reinforcement had unambiguous positive influence on the response of the slab to the blast load. Very low internal damage and almost no fragmentation (see Figure 6e) were observed in the case of specimens ST50AR0.5. The slab remained integral with only 20 g weight loss, with no visible cracks or crater creation. Similar results were achieved in the case of the specimen ST50CP0.5 with carbon fibre, just some surface splinters of total weight 30 g were created without perforation of the check panel. High enhancement of the overall blast resistance of the specimens ST50AR0.5 and ST50CP0.5 can be attributed to the mechanical properties of the fibres (high tensile strength and modulus) coupled with their shape parameters, in particular the length sufficient for good anchoring and high aspect ratio (L/d). With the same fibre length, the higher aspect ratio means higher amount of fibres included in the matrix for the same volume content of fibres, thus the matrix is more homogenously reinforced. Less areas between the fibres remain unreinforced and thus brittle, which prevents the slab from fragmentation.

## 4. Conclusions

Research on behaviour of SIFCON with various hybrid fibre reinforcement under blast load was carried out in order to obtain new knowledge necessary for designing the SIFCON type materials for protective structures. The following conclusions can be drawn:

-At quasistatic load, the addition of secondary reinforcement improves the post peak behaviour of SIFCON, except for specimen with wollastonite fibre. The hybrid fibre reinforced specimens (except for ST50WO2.0) showed plateau region around the flexural peak strength and less steep shape of

the decreasing part of the curve (= the post peak stress was higher at a given strain). Both the aforementioned statements indicate higher load bearing capacity compared to REF specimen.
-The addition of secondary micro-fibre reinforcement brought improvement of blast resistance of all specimens, as the micro-fibres reinforced and thus strengthened the matrix between the steel fibres. The contribution differed significantly depending on the type of fibre.
-The aramid fibre and PAN carbon fibre secondary reinforcement was proved to give great improvement in performance under blast load over mono-fibre reinforced specimen, in terms of reduction of mass lost, fragmentation and internal damage. This can be attributed to the high tensile strength of the fibres and suitable shape parameters (length and high L/d aspect ratio) connected with high anchoring length and high amount of fibres in the specimen.
-PP micro-fibres were proved to be less suitable for enhancement of the blast resistance, probably due to their lower mechanical parameters. Another reason can be the weaker bond of the fibre in the hardened concrete.
-Wollastonite micro-fibre inclusion was not very effective for improvement of blast resistance of the SIFCON composite, the main reason are unsuitable shape parameters – the length of the fibers is too short for effective transfer of the strain.
-The fibre length must not exceed 2.2 mm to achieve good infiltration of primary reinforcement and thus homogeneity of the specimen.

**Acknowledgments**
The authors wish to express their gratitude and sincere appreciation to the authority of The Technology Agency of the Czech Republic, project No. TE02000162 - Centre of advanced materials and technologies for protection and safety enhancement, for financial support.

**References**
[1] Schrader E K 1981 *ACI Journal* **78** p 141
[2] Wu C, Oehlers D J, Rebentrost M, Leach J and Whittaker A S 2009 *Eng. Struct.* **31** p 2060
[3] Luo X, Sun W and Chan S Y N 2000 *Cement Concrete Res.* **30** p 907
[4] Masuya H, Yamamoto M, Toyama M and Kajikawa Y 2000 *Structural Mater.* **8** p 205
[5] Tabatabaei Z S, Volz, J S, Baird J, Gliha, B P and Keener D I 2013 *Int. J. Impact Eng* **57** p 70
[6] Su H and Xu J 2013 *Constr. Build. Mater.* **45** pp 306–313
[7] Musselman E 2007 *Blast and impact resistance of carbon fibre reinforced concrete* Ph. D dissertation (Pennsylvania, USA: The Pennsylvania State University. University park)
[8] Ginter V T, Baeza F J, Ivorra S *et al* 2012 Materials and Des. **34** p 332
[9] Coughlin A M, Musselman E S, Shokkerb A J and Linzell D G 2010 *Int. J. Impact Eng* **37** p 521
[10] Drdlová M, Buchar J, Ridky R *et al* 2015 Blast resistance characteristics of concrete with different types of fibre reinforcement *Struct Concrete* **16** p 508
[11] Singh S K, Chourasia A, Dalbehera, M M *et al* 2013 Hybrid fibre reinforced concrete – A review *New Building Material & Construction World*
[12] Pakravan H R, Latifi M and Jamshidi M 2017 *Constr. Build. Mater.* **142** p 280
[13] Silva E R, Coelho J F J and Bordado J C 2013 *Constr. Build. Mater.* **40** p 473
[14] Ahmed S F U and Maalej M 2009 *Constr. Build. Mater.* **23** p 96
[15] Maalej M, Qeck S T and Zhang, J 2005 *J. Mater. Civil Eng.* **17** p 143
[16] Zhang M H, Sharif M S H and Lu G 2007 *Mag. Concr. Res.* **59** p. 199
[17] Drdlová M, Řídký R and Čechmánek R 2016 *Mater. Sci. Forum* **865** p 135
[18] Drdlová M, Sviták O and Prachař V 2017 *Mater. Sci. Forum* **908** p 72
[19] Azoom K T and Pannem R M R 2017 *Int. J. Civil. Eng. Technol.* **8** p 1123

IOP Conf. Series: Materials Science and Engineering **379** (2018) 012025 doi:10.1088/1757-899X/379/1/012025

# The behavior of cement-bonded wood-chip material under static and impact load

**M Drdlová**[1], **M Popovič**[2] **and M Šebík**[2]

[1] Applied research and development department, Research Institute for Building Materials, Brno, Czech Republic
[2] SVS FEM, Skrochova 42, Brno, Czech Republic

E-mail: drdlova@vustah.cz

**Abstract.** The response of cement-bonded wood-chip (CBWC) material under static and dynamic load was investigated both experimentally and numerically. This investigation is a part of complex development of safety road barrier with integrated noise protecting wall based on the wood-chip boards. Three types of materials with different amount of the binder were investigated in two moisture content states in order to obtain a comprehensive overview of the properties and behavior of the material at several types of loading, necessary for future design of the safety barrier. Real quasi-static (compressive, tensile and flexural strength) and dynamic experiments (IZOD test and Newton Cradle drop test) were carried out. The research revealed strong dependence of the mechanical properties on the moisture content in the material, with different trend – toughness showed ascending trend with increasing moisture content, whereas compressive, flexural and tensile strength were lower when the CBWS material was soaked. The results of the tests were used as inputs for creating and subsequently verifying the numerical material model of cement-bonded wood-chip material. The material model was used in numerical simulations of crash test of mobile noise protecting wall according to the ČSN EN 1317 standard.

## 1. Introduction

Wood chip boards are used in building industry as thermal insulation, formwork, noise and visibility protection or casing. The cement-bonded wood-chip material is a mixture of wood chips (up to 90%), Portland cement and wood mineralization agent. This mixture is pressed to the compression moulds to produce boards with various surface-structures. The final properties of the wood chip material are influenced by the mix proportion of raw materials, shape parameters of the used wood chips (their thickness and length are typically in range of 0.5–5.0 mm and 20.0–50.0 mm, respectively), wood mineralization agent and production technology.

Both short and long term mechanical properties of the cement bonded wood composites were investigated by Dinwoodie [1]. Modulus of rupture, tensile strength, density, water absorption, frost resistance and long-term durability of 7 different wood-cement composites were determined and compared. Impact resistance was also evaluated, but only marginally and without methodology description. All mechanical properties with the exception of impact resistance showed a strong relationship with density [1]. Effect of particle size and geometry on mechanical properties of the particleboards and chipboards were investigated by Badejo [2], who concluded that the chip geometry is highly correlated with board key properties, including modulus of elasticity and strength. The longer and thinner the particle, the stronger and stiffer the boards. The study [3] confirms this statement. Compressive strength correlates with density and decreases regularly in correspondence with the rising wood aggregate content. Correlation between density and mechanical properties can be attributed to enhanced wood densification, elimination of gaps, and improved connection between matrix and fibre [4]. Stiffness characteristics are also a function of cement-wood ratio. This relationship is based on the fact that cement is inherently a more rigid material than wood. Therefore, greater cement-wood ratios result in higher modulus of elasticity values, as these values are dependent on the total amount of the stiff and incompressible cement matrix [4]. Anyway, most of the studies are focused on particle boards [5–8] and complex information about the cement bonded wood chip material characteristics, with emphasis on properties at impact load, are still missing, as the usual current applications do not require them. On contrary, such a data is essential for using of cement bonded wood chip material as a part of safety road barriers. Based on the results of the physico-mechanical tests, the numerical model was created and validated and used for further design and numerical simulation of the novel safety road barrier.

## 2. Experimental program

### 2.1. Materials and procedures

*2.1.1. Specimens.* Three variants of cement bonded wood chip material (designation WSR50, WSO80 and WSW85, see Figure 1 left) were investigated in two moisture states – at operational humidity and soaked (water content 10 wt.%). The individual materials differ in cement ratio in the whole composite – the WSR50 specimen contains the highest amount of cement, the specimen WSW85 the lowest. The materials were produced by VELOX Werk GesmbH, test specimens were prepared by cutting from the 1000 × 500 mm boards. The thickness of the slab is given by its surface structure and varies from 50 to 85 mm. The profile was cut off before preparation of specimens. The dimensions and number of the specimens for each particular test are given in Table 1.

**Table 1.** Dimensions and number of the specimens to be tested.

| Test | Number of specimens per state and type | Dimensions |
|---|---|---|
| Tensile strength test | 10 | 250 mm free length |
| Compressive strength test – perpendicular to the plane of the slab | 10 | 50 × 50 × 25 mm |
| Compressive strength test – in the plane of the slab | 10 | 100 × 40 × 200 mm |
| Flexural strength test | 48 | 50 × 25 × 500 mm |
| IZOD test | 20 | 100 × 50 × 25 mm |
| Drop test (Newton Cradle) | 20 | 50 mm diameter, 20 mm length |

*2.1.2. Mechanical properties at quasistatic load and bulk density.* The mechanical parameters were obtained using strength testing machine TIRAtest 2710, R58/02. The compressive, tensile and flexural load was applied in quasi-static conditions at speed of 5 mm/min. The tests were performed according to the standards ČSN EN 310 (flexural strength, bulk density), ČSN EN 789 (compressive strength in the plane of the slab, tensile strength) and ČSN EN 826 (compressive strength perpendicular to the plane of the slab).

*2.1.3. Mechanical properties at impact load.* IZOD test was performed to determine the impact toughness of woodchip materials according to the PZN – VUSTAH 0200 02:2012 standard. Specimen is fixed to the device in form of cantilever beam. A pivoting arm is raised to a specific height (constant potential energy) and then released (see Figure 1 right). The arm swings down hitting and breaking the specimen. The energy absorbed by the specimen is calculated from the height the arm swings to after hitting the sample. The results are expressed in energy lost per unit cross-sectional area:

$$a_i^{IZOD} = \frac{E_i}{h \cdot b} \times 10^3 \ J.m^{-2} \qquad (1)$$

where:
$E_i$... lost energy in Joules; h...thickness of the specimen in mm; b... width of the specimen in mm.

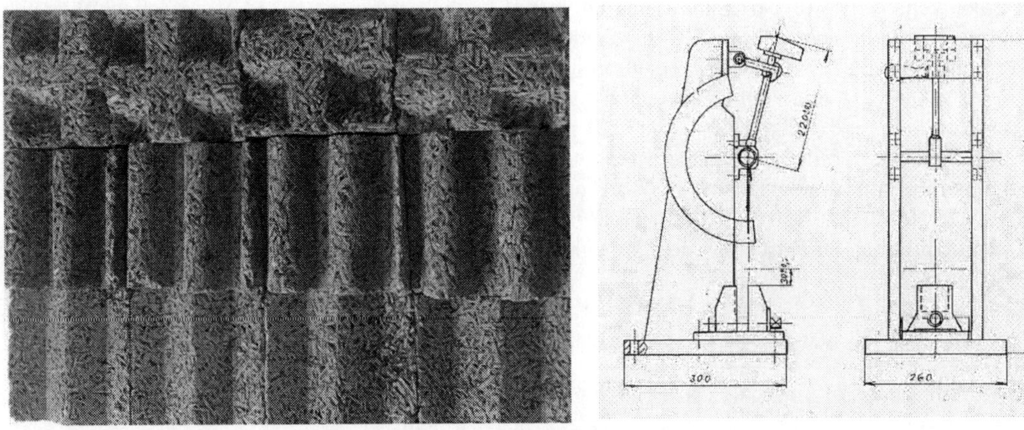


**Figure 1.** Specimens of wood-chip materials (left) and scheme of the IZOD test rig (right).

For drop test and energy absorption evaluation, the Newton cradle principle was used. The specimen is fixed to the steel rod and loaded with the impact rod raised to a specific height and released, see Figure 2. The energy absorbed by each specimen can be simply evaluated by residual potential energy of the impact rod. The impact velocity obtained in this test is 3.5 $m \cdot s^{-1}$, which is relatively low, but still this test methodology enables to map significantly different process of deformation regarding to static compressive tests.

**Figure 2.** Drop test device based on Newton cradle principle.

*2.1.4. Numerical simulation.* Numerical model of WSO80 material was prepared. This material was selected because of its best impact resistance expressed as the highest impact toughness. The creation of the numerical model is described in detail in the section 2.2.2. Created numerical model was used in numerical simulation of the road safety barrier crash test according to the standard ČSN EN 1317.

### *2.2. Results and discussion*

*2.2.1. Result of the mechanical properties at quasi-static and impact load.* The results of determined physico-mechanical properties at both moisture stages (at quasi-static load) are summarized in Table 2 and 3. The results of IZOD and Newton Cradle drop test are summarized in Table 4. All mechanical properties showed a strong relationship with density and cement ratio – the lower the cement ratio and thus density, the lower mechanical properties, which corresponds with the findings in [1]. The relative absorbed energy was the only exception, it showed opposite trend.

The moisture content affects the behaviour of the material. Under quasi-static load, the soaked specimens showed lower strength values compared to the specimens tested at operational state. In the higher moisture content state, 26–33% reduction in flexural strength was observed. Similar results were published in [9], where 28% reduction in the flexural strength was reported on specimens being exposed to the 85% relative humidity (RH) environment compared to the specimens exposed to RH 65%. The compressive strength in the slab plain was reduced by 27–42% and tensile strength by 19–30%. At dynamic load, the opposite trend was observed, as the moisture content increases toughness of the main part of the composite - wood. The impact toughness of soaked specimens was by 16.1% (WSW85), 30.0% (WSR50) and 38.2% (WSO80) higher than the impact toughness of specimens tested at operational state. Similarly, the moisture content positively influences the values of relative absorbed energy.

**Table 2.** Physico-mechanical properties of specimens at quasi-static load (operational state).

| Designation | Bulk density ($kg \cdot m^{-3}$) | Compressive strength (MPa) | Compressive stress (perpendicular)* (MPa) | Flexural strength (MPa) | Tensile strength (MPa) |
|---|---|---|---|---|---|
| WSR50 | 792 | 3.3 | 2.3 | 2.84 | 0.47 |
| WSO80 | 634 | 2.1 | 2.2 | 2.08 | 0.36 |
| WSW85 | 550 | 1.1 | 0.8 | 0.56 | 0.13 |

**Table 3.** Physico-mechanical properties of specimens at quasi-static load (soaked).

| Designation | Bulk density ($kg \cdot m^{-3}$) | Compressive strength (MPa) | Compressive stress (perpendicular)* (MPa) | Flexural strength (MPa) | Tensile strength (MPa) |
|---|---|---|---|---|---|
| WSR50 | 1,108 | 1.9 | 1.1 | 1.96 | 0.37 |
| WSO80 | 926 | 1.1 | 0.9 | 1.32 | 0.29 |
| WSW85 | 740 | 0.8 | 0.4 | 0.42 | 0.09 |

*) at 10% compression

**Table 4.** IZOD test and drop test results.

| Designation | Impact toughness ($kJ \cdot m^{-2}$) | Impact toughness – soaked ($kJ \cdot m^{-2}$) | Relative absorbed energy (%) | Relative absorbed energy – soaked (%) |
|---|---|---|---|---|
| WSR50 | 3.77 | 4.90 | 54.4 | 62.1 |
| WSO80 | 3.80 | 5.25 | 57.6 | 63.9 |
| WSW85 | 2.73 | 3.17 | 60.1 | 62.2 |

*2.2.2. Numerical model and simulations.* The numerical model of wood-chip material WSO80 (see Figure 3) was prepared with regards to real production procedure of the wood-chip board. The outer shape of the board element was prepared as a rigid empty mould. A large number of wood beam elements were simulated as parts falling to mould by a standard gravity load. A rigid surface pressed all randomly placed wooden elements to the final shape. An explicit numerical code was used for this simulation.

IOP Conf. Series: Materials Science and Engineering **379** (2018) 012025 doi:10.1088/1757-899X/379/1/012025

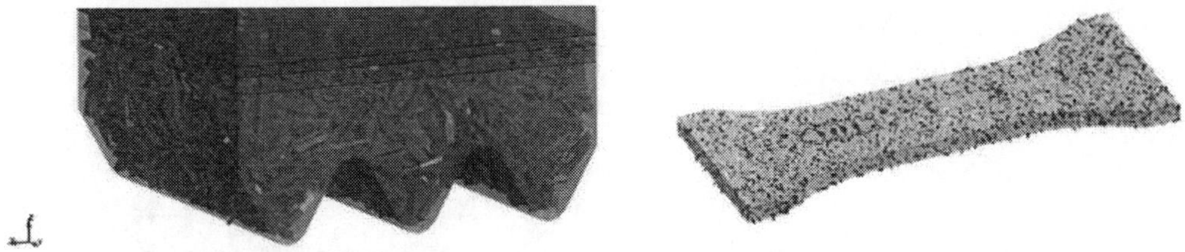

**Figure 3**. Cement-bonded wood-chip numerical material model, basic board and tensile test sample.

The cement binder was modelled by solid FE elements with required outer shape. Beam elements of wood chips were modelled with standard material properties for spruce wood with longitudinal orientation (bilinear material model). A special condition *CONSTRAINED_LAGRANGE_IN_SOLID from LS-DYNA explicit numerical code was used for modelling of joint forces between beam elements of wood and solid elements of cement binder.

Analysed wood-chip material exhibits significantly different behaviour at tensile, flexural and compressive load. A material model with different stiffness curves in tensile and compression was selected in the first step (LS-DYNA code *MAT_PLASTICITY_COMPRESSION_TENSION). Using this model, good correlation between the real measurement and numerical simulation was detected only in the case of tensile and compressive test. The flexural strength values obtained from the real test were almost two times higher than those calculated by numerical simulation. The parameter correlation study had to be performed to optimise the material model. Input parameters were defined by all material constants (Young modulus, compressive and flexural load-displacement curves scale coefficients). Output parameters were adjusted using the results of real measurements – load-displacement curves and (in the case of IZOD test) scalar values. It was necessary to extend a material model in more precise erosion criteria, especially maximum strain at failure. The presented adjustments were followed by an optimization study, which has found an optimal set-up of material model constants. Optimization study was performed using the numerical system optiSLang. The description of final cement-bonded wood-chip material model for explicit code depicts Figure 4.

```
*MAT_PLASTICITY_COMPRESSION_TENSION
$ WS80_Velox_material
$#     mid        ro         e        pr         c         p      fail      tdel
        50 634.0E-09    0.2100     0.149       0.0       0.01.00000E20       0.0
$#   lcidc     lcidt     lcsrc     lcsrt    srflag    lcfail        ec      rpct
         3         4         0         0       0.0         0   0.00000       0.9
$#      pc        pt     pcutc     pcutt     pcutf
   0.00003  0.000010       0.0       0.0       0.0
$#       k
       0.0
$
*MAT_ADD_EROSION
$#     mid      excl    mxpres     mneps    effeps    voleps    numfip       ncs
        50     0.000                                                 1         2
$#  mnpres     sigp1     sigvm     mxeps     epssh     sigth   impulse    failtm
 -1.500e-4 0.000e00     0.000  2.000e-3     0.000     0.000     0.000     0.000
```

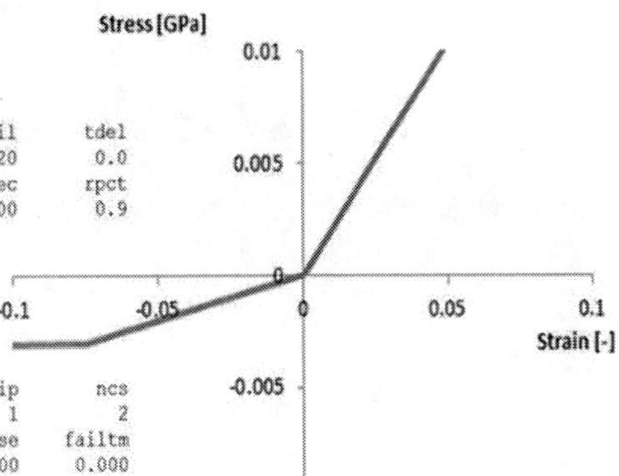


**Figure 4.** Cement-bonded wood-chip material model for explicit code.

The created model was verified using the results of Newton cradle test. Relative absorbed energy value obtained from the numerical simulation (65,27%) was closed to the value obtained from the real experiment (60,13%). From the point of view of absorbed energy, the correspondence of the obtained values is sufficient to confirm the accuracy of the created model.

IOP Conf. Series: Materials Science and Engineering 379 (2018) 012025 doi:10.1088/1757-899X/379/1/012025

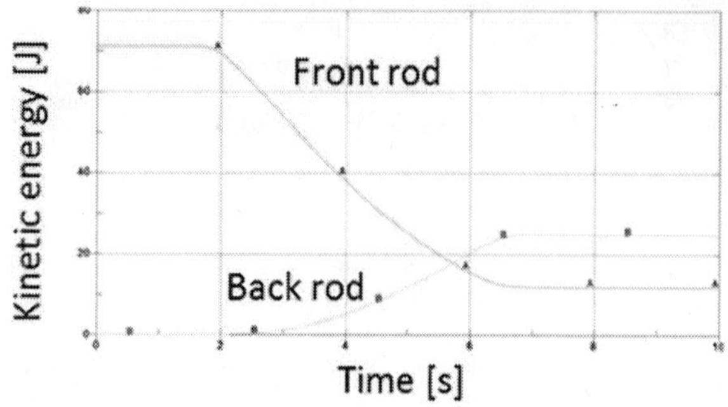

**Figure 5.** Numerical simulation of Newton cradle test – the course of kinetic energy of rods in time (left), numerical model of the test (right).

Created numerical model was used in numerical simulation of the newly designed road safety barrier. The CAE model was created with respect to its use in complex, explicit impact analysis according to ČSN EN 1317. The construction of the barrier consists of concrete foot and beam part with implemented noise protective slab construction. I-profile beam elements are embedded in concrete foots. Wood-chip noise protective slabs are inserted into the I-profile. The construction is easy and quick to erect and disassemble.

The development of the CAE model is complicated because of the high variability of the mechanical properties of noise protective wood-chip slabs caused by the irregularity of individual wood particles and the inhomogeneous distribution of the binder. Another problem is the free positioning of noise protective slabs, which leads to a contact analysis, which in combination with a fast dynamic factor leads to a difficult converging task. This fact is connected with the necessity of choosing a small time computing step, which is time consuming, but provides accurate results.

**Figure 6.** Crash test numerical simulation results – an example of the impact of 38 t vehicle.

An example of numerical simulation of road safety barrier crash test (using a 38 t vehicle) is depicted in Figure 6. Within the next steps, the results of the numerical simulations will be verified by the real crash tests. With reliable models and methodology, the numerical simulation can partly replace real crash tests, which makes it the great tool in development process of road safety barriers and other protective elements as it can dramatically decrease the associated costs.

## 3. Conclusions

The response of CBWC material under static and dynamic load was investigated both experimentally and numerically. Three types of materials with different amount of the binder were investigated in two moisture content states. The following conclusions can be drawn:

-Moisture content can seriously affect the mechanical properties of wood-chip panels - soaked specimens showed significantly lower quasi-static load strength values compared to the specimens tested at operational state. 26–33% reduction in flexural strength, 27–42% in compressive strength and 19–30% in tensile strength was observed. The lower density panels experienced greater decreases in their flexural and tensile strength.

-Opposite trend was observed in the case of dynamic loading. The impact toughness of soaked specimens was by 16.1% (WSW85), 30.0% (WSR50) and 38.2% (WSO80) higher than the impact toughness of specimens tested at operational state. Similarly, the moisture content positively influences the values of relative absorbed energy.

-All mechanical properties showed a strong relationship with density and cement ratio – the lower the cement ratio and thus density, the lower mechanical properties, except for the relative absorbed energy, which showed opposite trend.

-Adjusted *MAT_PLASTICITY_COMPRESSION_TENSION model was found suitable for numerical simulation of chip-board material. After adjustments, good correlation of results of numerical simulations and all performed real mechanical tests was achieved.

**Acknowledgments**

The authors wish to express their gratitude and sincere appreciation to the authority of the Ministry of Industry of the Czech Republic, project No. FV10120, for financial support.

## References

[1] Dinwoodie J M and Paxton B H 1989 *Constr. Build. Mater.* **3**
[2] Badejo S O J *Wood Sci. Technol.* **22** pp 357-370
[3] Semple K E and Evans P D 2007 *Wood Fibre Sci.* **39** pp 120–131
[4] Frybort S, Mauritz R, Teishinger A and Muller U *Bio resources* **3** pp 602–626
[5] Quiroga A, Rintoul I 2015 *Constr. Build. Mater.* **84** pp 253–260
[6] Jorge F C, Pereira C and Ferreira J M F 2004 *Holz Roh Werkst.* pp 370–377
[7] Kondrup C 1990 *Holz Roh Werkst.* **48** pp 31–35
[8] Fan M Z, Diwoodie J M, Bonfield P W and Breese M C 1999 *J. Mater. Sci.* **34** pp 1729–1740
[9] Pritchard M, Ansell M P, Thompson J H and Bonfield P Q 2001 *Wood Sci. Technol.* **35** p 395

IOP Conf. Series: Materials Science and Engineering **379** (2018) 012026 doi:10.1088/1757-899X/379/1/012026

# Application of the FRP rebar in constructions for reduction of thermal bridges – new insights

**J Prokeš**[1]

[1] PREFA KOMPOZITY, a.s., Kulkova 4231/10, 61500 Brno, Czech Republic

Email: prokes@prefa.cz

**Abstract.** The paper describes usage of composite rebar in applications with reduction of thermal bridges. There are described benefits of the composite solution without metal parts. The composite elements are used for tensile, compressed and also shear reinforcement. Technical solution of assembly and the basic mechanical properties and thermal insulating characteristics are described.

## 1. Introduction

### *1.1. The issue of breaking the thermal bridges in the reinforced concrete constructions*

Thermal bridge is phenomenon of construction which occurs when materials with different thermal conductivity are used for exterior surfaces or parts. It means that the thermal bridges are weaknesses within a building's construction where heat and/or cold is transferred at a significantly higher rate than through the surrounding envelope parts. There are basically two types of this phenomenon: geometric thermal bridges where part of the structure projects through the building envelope, and material thermal bridges where materials with different conductivity are used in combination. In practice, these effects are often combined. A classic example are balcony slabs, where problems occur when no serious consideration is given to the connections.

The thermo-photograph (Figure 1) bellow shows the situation, when thermal bridges at balconies were not taken care of, the balconies act as the cooler ribs; the heat is conducted off the building and the rooms adjacent to the balconies are cooled [1].

### *1.2. The effect of thermal bridges*

*1.2.1. Higher energy consumption.* Due to the thermal loss at the balcony connection, heat is drawn from inside, which results in a significant increase of heating costs and energy consumption.

*1.2.2. Mould formation.* Interior temperatures of the adjacent rooms can drop easily below the dew point. This leads to condensation, deterioration of plaster and paintwork and also mould formation. Long term exposure to condensation can lead in serious deterioration of the building.

*1.2.3. Uncomfortable living space.* Cold surface temperatures cause uncomfortable living space for occupants.

**Figure 1.** Thermograph photograph of building with balcony with significant thermal bridge [1].

### *1.3. Current methods of solution*

Current methods include creating of a gap between ceiling slab and balcony slab, and usage of materials with lower thermal conductivity.

Typically, load-bearing thermal insulation elements made from austenitic steel and thermal insulation [2] are used. These elements form a thermal break between the balcony and the interior floor while transferring load and maintaining full structural integrity. Also, the warmer interior temperature is maintained decreasing condensation, and preventing the formation of mould. Compare Figure 2 and Figure 3 or Figure 4.

Application of only FRP rebar is quite new. No other studies dealing with only FRP rebar are known to the author of this paper. State of the art is described in patents DE 4 102 332 (A1), EP 2 138 641 (A2), DE 1 9711 813 (A1), DE 19508292 (A1), DE 4 040 433 (A1). These patents describe usage of FRP and steel rebar together. At least, shear reinforcement is made on steel. No patent claims usage of only FRP rebar.

## 2. Properties of load-bearing thermal insulation elements made from FRP rebar

### *2.1. Types of rebar*

There are several types of FRP (Fiber Reinforced Polymer) rebar: glass fiber FRP (GFRP), carbon fiber FRP (CFRP), glass & carbon fiber FRP (C-GFRP) [2] or other types (e.g. basalt FRP). The used fibers have an effect on mechanical properties, environmental resistivity [3] and price. The basic mechanical properties of FRP rebar are described in Table 1. The structure of glass fiber reinforced rebar is illustrated in Figure 5. As shown in Table 1, the FRP rebars have higher tensile strength and lower modulus of elasticity and density.

**Table 1.** Typical mechanical properties of FRP rebars (volume content of fibers is 50–75%), comparison with steel rebar [3] and PREFA KOMPOZITY FRP rebar.

| **Characteristic/rebar material** | Steel | GFRP | C-GFRP | CFRP |
|---|---|---|---|---|
| **Module of elasticity longitudinal [GPa]** | 200 | 50 | 80 | 140 |
| **Tensile strength longitudinal [MPa]** | 550 | 1000 | 1150 | 1870 |
| **Density [kg/m3]** | 7850 | 2100 | 2000 | 1600 |

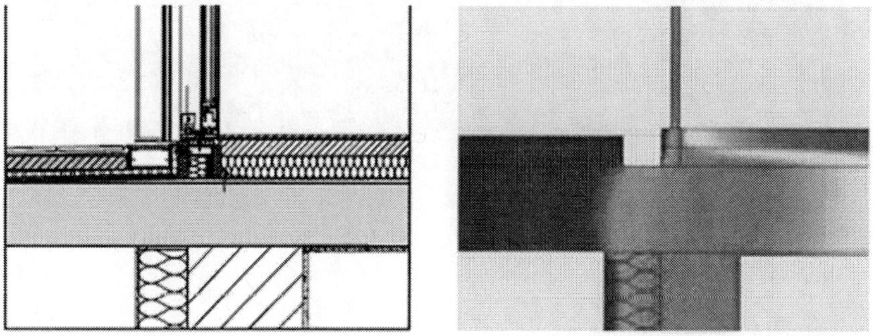

**Figure 2.** Construction where thermal bridges are not solved. The ceiling and balcony slabs are connected through the concrete and reinforcing steel [2].

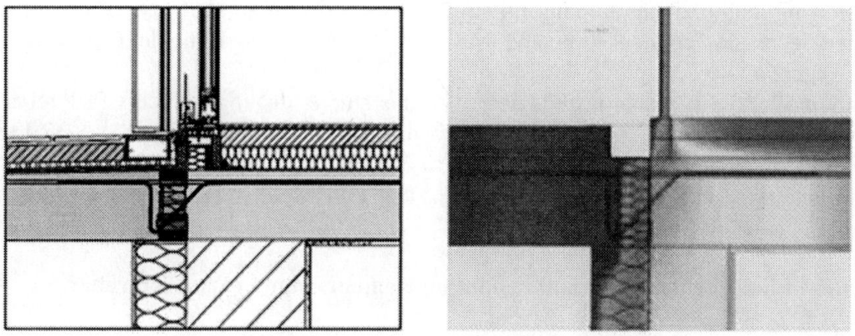

**Figure 3.** Construction where load-bearing thermal insulation elements from austenitic steel [2] are used. The thermal bridge is restricted by insulation layer, but heat can flow through the austenitic steel reinforcement.

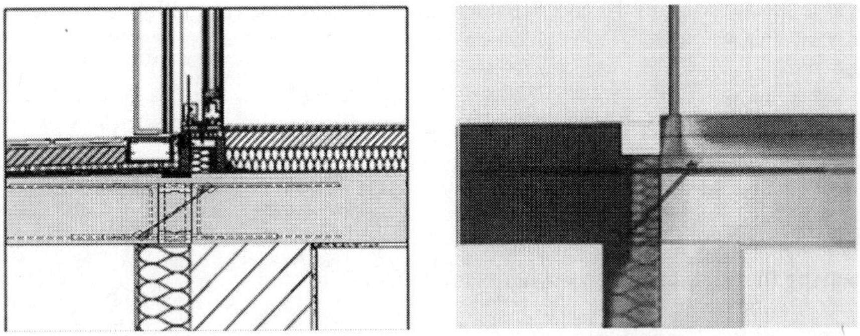

**Figure 4.** Construction where load-bearing thermal insulation elements from FRP rebar are used.

**Figure 5.** FRP rebars. (From left to right GFRP, C-GFRP a CFRP) [4]

*2.2. Thermal conductivity*

The characteristic feature of the composite reinforcements is the low coefficient of the thermal conductivity. Table 1 shows the values of the coefficient of commonly used steel reinforcement (different types of steels), the experimentally determined value of this coefficient for the individual reinforcement element, thermal coefficient of extruded polystyrene (data supplied by the extruded polystyrene manufacturer) and the average thermal coefficient of load bearing thermal insulation element made from FRP (Figure 6).

It is clear from Table 1 that the coefficient of thermal conductivity of composite reinforcements is about 2 orders of magnitude lower than in case of steel reinforcements. The resultant, experimentally determined, ratio of 0.066 $W \cdot m^{-1} \cdot K^{-1}$ is thus in the same order of magnitude as the coefficient of the insulating element.

**Table 2.** Overview of the coefficients of thermal conductivity of individual materials.

| Material | Coefficient of thermal conductivity $[W \cdot m^{-1} \cdot K^{-1}]$ |
|---|---|
| **Steel** | 50 – 100 |
| **GFRP Rebar** | 0.49 |
| **Extruded Polystyrene** | 0.032 |
| **Load-bearing thermal insulation elements made from FRP** | 0.066 |

*2.3. Technical solution*

The load-bearing thermal insulation element made from FRP was used according to the patent [5]. The basic structure is shown in Figure 6 and 7. This element contains:

- Pulled reinforcement (Figure 6, position 1),
- Compressed reinforcement with auxiliary element (Figure 6, position 2 and 6),
- Shear reinforcement with auxiliary element (Figure 6, position 3 and 5),
- Load distribution reinforcement (Figure 6, position 4),
- Thermal insulation material (Figure 6, position 7).

These basic parts can differ in number, diameter, angle, etc. The desired load capacity and stiffness can be achieved by changing of mentioned parameters.

There are several parts of FRP load-bearing element that are not listed in patent [5] and aforementioned list. FRP rebars are flammable and above critical temperature (glass transition temperature) lose mechanical properties. The critical for usage of this element are fire-proof properties. The necessary part of this element is a fire insulation, which should be used at the bottom of the element at least. All parts must be assembled into one piece, which enables easy handling at construction site.

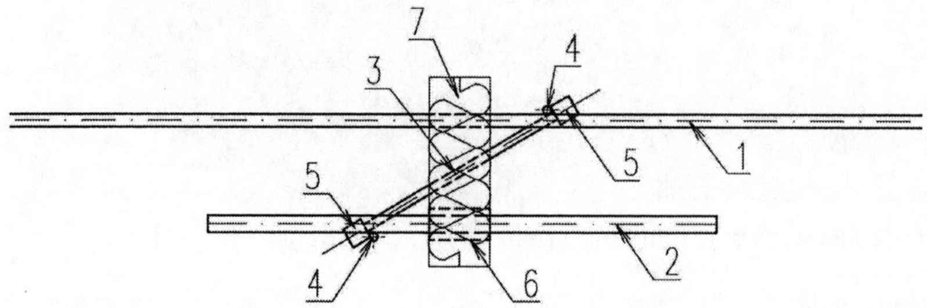


**Figure 6.** Load-bearing thermal insulation element made from FRP. Symbols: 1 – pulled rebar, 2 – compressed element, 3 – shear element, 4 – auxiliary reinforcement for load distribution, 5 – auxiliary element for increasing of cohesion strength of shear rebar, 6 – auxiliary element for increasing of compression strength of compressed rebar, 7 – insulation material.

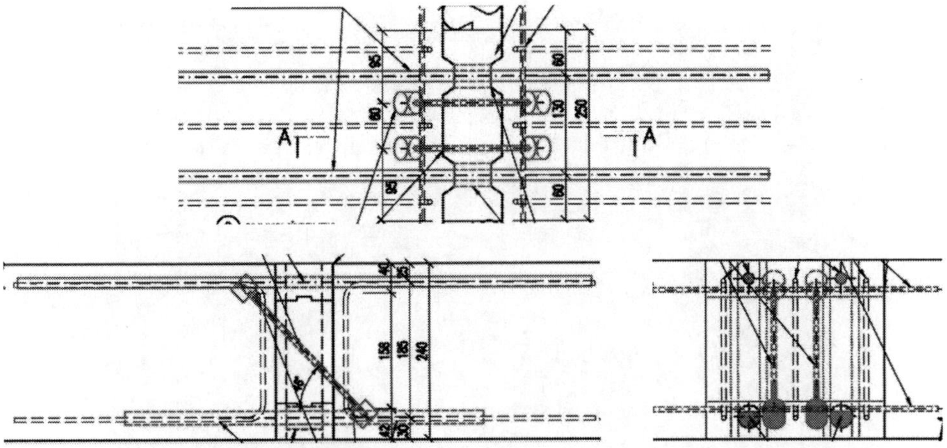

**Figure 7.** Load-bearing thermal insulation element made from FRP in construction. FRP reinforcement – red colour, steel reinforcement – black colour.

*2.4. Testing of mechanical properties of element*

The designed FRP load-bearing thermal insulation element was tested experimentally in order to determine the ultimate load bearing capacity of the assembly and its rigidity [6]. The thickness of the thermal insulation layer was 60 mm. The FRP reinforcement arrangement is shown in Figure 8. The load was applied until sample failed at the end of the slabs. The test samples were subjected to a flexural static loading test (see Figure 9). Than the feasibility of this solution under normal construction conditions was critically evaluated. The dependence of the applied force and the measured deflection of the end of the balcony slab is shown in Figure 10. It can be stated that both the behavior of the elements and failure mode were expected in the course of the load test.

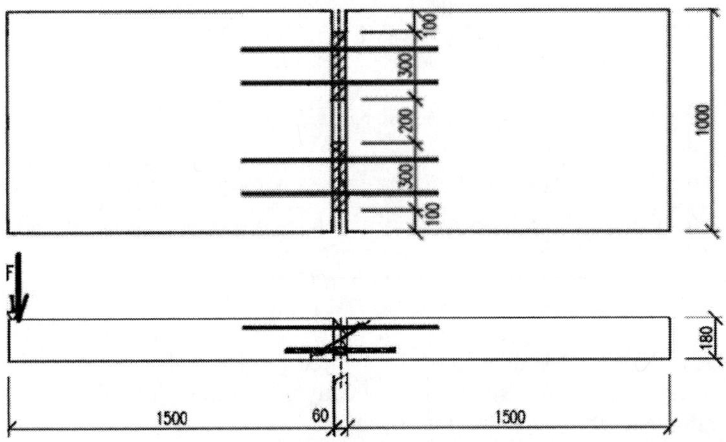


**Figure 8.** Arrangement of FRP reinforcement in the concrete sample [6].

**Figure 9.** Arrangement of the experiment [6].

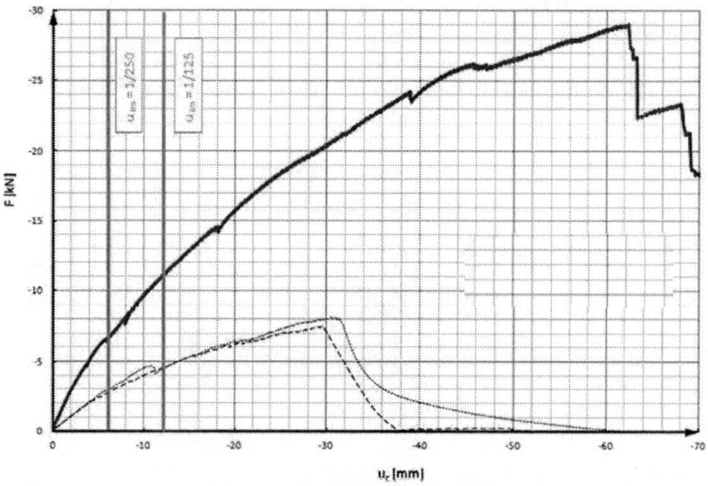


**Figure 10.** The dependence of the applied force F and the measured deflection $u_c$ of the end of the balcony slab (blue line). Without the influence of its own weight. [6]

## *2.5. Calculations of the long-term load-bearing capacity of the element*

Technical feasibility study [7] was carried out according above-mentioned experiments and parameters. The influence of the reinforcement and height of the load-bearing insulation element on the resulting bend and shear bearing capacity was calculated. Five reinforcement variants and nine heights (see Figure 11 and Table 3) were calculated and assessed. The minimum technical possible height is 160 mm and there is downward trend of the shear bearing capacity above 300 mm. Calculation was carried out according to the standard EN 1990 [8].

Variants A to E (Table 3) for heights H 160 to 300 mm (Figure 11) were calculated. The resulting long-term (50 years) bend and shear bearing capacity is shown in Table 4 and Figures 12 and 13.

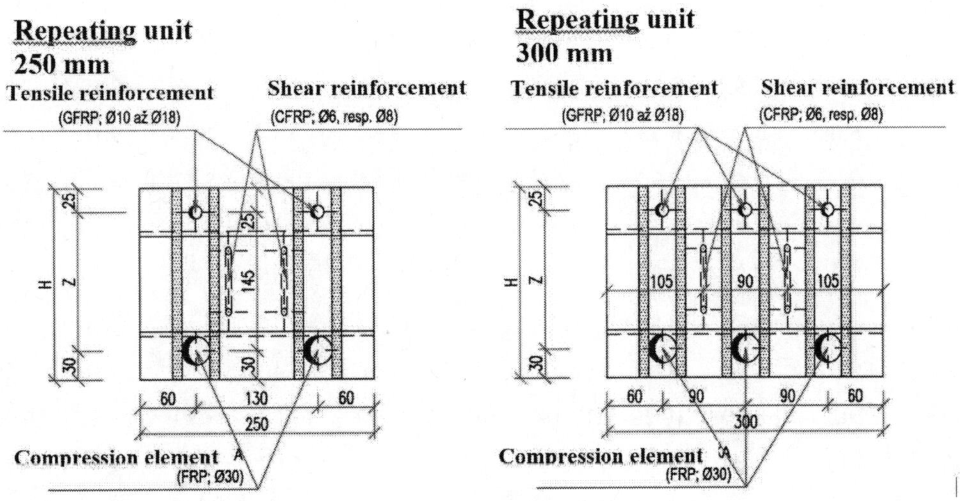


**Figure 11.** The design of reinforcement for the repeating unit 250 mm and 300 mm. The height H is parameter of the calculation. [7]

**Table 3.** Calculated variants of FRP load-bearing unit. [7]

| **Variant** | Tensile Reinforcement | Compression element | Shear reinforcement |
|---|---|---|---|
| **A** | 2 pc $\phi$14, type GFRP | 2 pc $\phi$30, FRP composite with short glass fiber | 2 pc $\phi$6, type CFRP |
| **B** | 3 pc $\phi$14, type GFRP | 3 pc $\phi$30, FRP composite with short glass fiber | 2 pc $\phi$6, type CFRP |
| **C** | 3 pc $\phi$18, type GFRP | 3 pc $\phi$30, FRP composite with short glass fiber | 2 pc $\phi$6, type CFRP |
| **D** | 2 pc $\phi$14, type GFRP | 2 pc $\phi$30, FRP composite with short glass fiber | 2 pc $\phi$8, type CFRP |
| **E** | 1 pc $\phi$14, type GFRP | 1 pc $\phi$30, FRP composite with short glass fiber | 2 pc $\phi$8, type CFRP |

**Table 4.** Calculated bend $M_{rd}^{LT}$ and shear $V_{Rd}^{LT}$ bearing capacity. [7]

| Height H [mm] | A | | B | | C | | D | | E | |
|---|---|---|---|---|---|---|---|---|---|---|
| | $V_{Rd}^{LT}$ | $M_{rd}^{LT}$ | $V_{Rd}^{LT}$ | $M_{rd}^{LT}$ | $V_{Rd}^{LT}$ | $M_{rd}^{LT}$ | $V_{Rd}^{LT}$ | $M_{rd}^{LT}$ | $V_{Rd}^{LT}$ | $M_{rd}^{LT}$ |
| **160** | 5.44 | 3.93 | 5.44 | 6.11 | 5.44 | 6.11 | 7.26 | 3.78 | 7.26 | 1.60 |
| **180** | 6.83 | 4.75 | 6.83 | 7.37 | 6.83 | 7.37 | 9.10 | 4.59 | 9.10 | 1.97 |
| **200** | 7.75 | 5.59 | 7.75 | 8.64 | 7.75 | 8.64 | 10.34 | 5.42 | 10.34 | 2.37 |
| **220** | 8.62 | 6.43 | 8.62 | 9.92 | 8.62 | 9.92 | 11.49 | 6.25 | 11.49 | 2.76 |
| **240** | 9.27 | 7.27 | 9.27 | 11.20 | 9.27 | 11.20 | 12.35 | 7.08 | 12.35 | 3.16 |
| **250** | 9.57 | 7.70 | 9.57 | 11.85 | 9.57 | 11.85 | 12.76 | 7.51 | 12.76 | 3.37 |
| **260** | 9.87 | 8.14 | 9.87 | 12.50 | 9.87 | 12.50 | 13.16 | 7.95 | 13.16 | 3.59 |
| **280** | 10.29 | 8.99 | 10.29 | 13.79 | 10.29 | 13.79 | 13.72 | 8.79 | 13.72 | 4.00 |
| **300** | 10.55 | 9.85 | 10.55 | 15.08 | 10.55 | 15.08 | 14.07 | 9.64 | 14.07 | 4.41 |

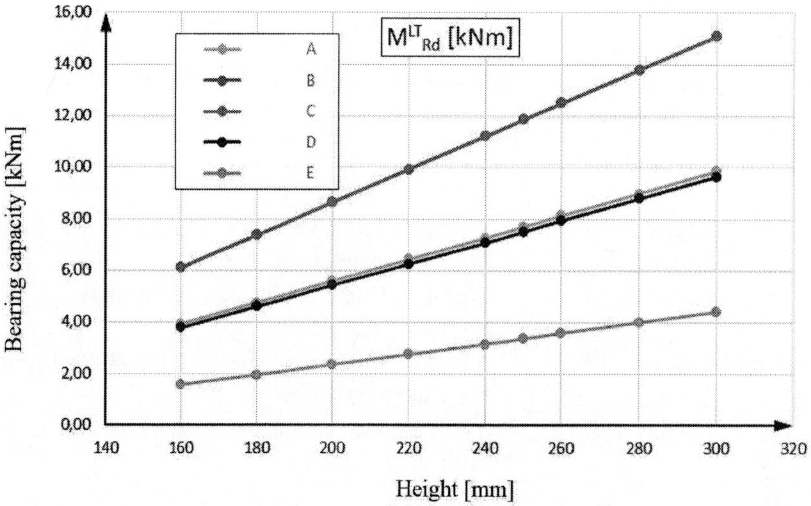


**Figure 12.** The dependence of bend bearing capacity and height of element. The line B is overlapped by C. [7]

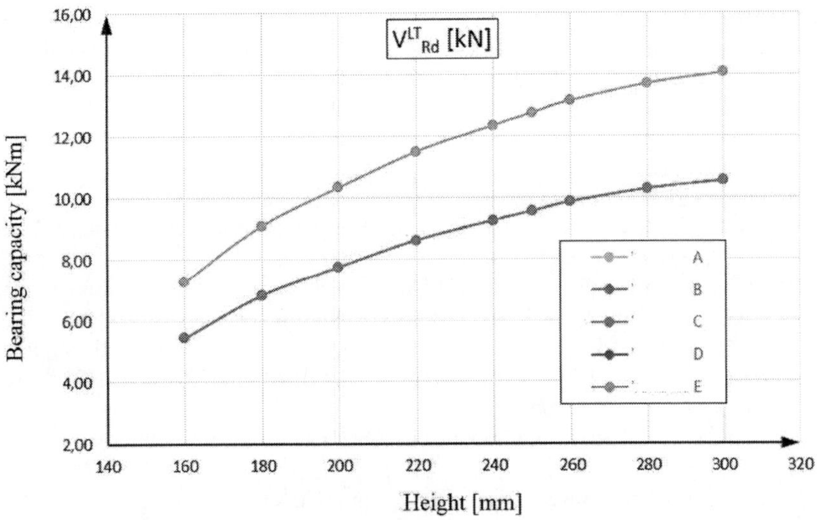


**Figure 13.** The dependence of shear bearing capacity and height of element. The lines A and B are overlapped by C and D is overlapped by E. [7]

The linear dependence of bend bearing capacity and height of element is shown in Figure 12. The breakage of tension and compression reinforcement is checked in the calculation.

It can be seen that despite the increase of the tension reinforcement in the variant C compared to the variant B, there was no increase in bearing capacity – the compression bearing is limiting in both cases (Variants B and C are overlapped). However, when the deflection of the end of the slab is calculated, increasing the cross-section of the reinforcement in variant C will increase the bending stiffness and achieve a lower deflection.

The change in shear reinforcement has a negligible effect on the bending resistance, compare variant A to variant D in Figure 12.

The comparison of two different shear reinforcements shows dependence of shear bearing capacity on the angle between shear compression reinforcement (see Figure 13).

The shear bearing capacity is increased when the cross-section of shear reinforcement is increasing, however, when the reinforcement area is increased by about two times (2 pieces of $\phi 6$ in variant A or 2 pieces of $\phi 8$ in variant D), the load bearing capacity is increased only about one third.

## 3. Conclusions

The calculations have shown that the FRP load bearing thermal insulating element exhibits good bending and shear resistance and its behavior has an expected trend. The parametric studies were carried out and the following conclusions were drawn:

Bending capacity of the element is determined by load-bearing capacity of implemented compression reinforcement. To optimize the technical solution, it is advantageous to increase the mechanical characteristics of compressed elements.

The shear capacity of the element is primarily determined by the ability to transfer securely the force generated in the shear reinforcement. To optimize the technical solution, it is advantageous to strengthen the element anchoring, as only to increase the cross-section of shear reinforcement is not sufficient and cannot be fully utilized.

There is also necessary to verify calculated dependencies by means of long-term tests, especially tests of compression elements.

IOP Conf. Series: Materials Science and Engineering **379** (2018) 012026 doi:10.1088/1757-899X/379/1/012026

**Acknowledgments**
This work was published with support of Ministry of Industry and Trade of Czech Republic. The project number FV10588 and CZ.01.1.02/0.0/0.0/15_019/0005058.

**References**

[1] Technical information (*Online* www.tzb-info.cz)
[2] Technical information Schoeck (*Online* www.schoeck.com/)
[3] ACI 440.1R-06: Guide for the Design and Construction of Structural Concrete Reinforced with FRP Bars (American Concrete Institute )
[4] Technical information PREFA KOMPOZITY, a.s. (*Online* www.prefa-kompozity.cz)
[5] Štěpánek P, Girgle F and Prokeš J 2016 Sestava prvků pro kotvení *Czech Patent* CZ305718
[6] Prokeš J and Girgle F 2016 Application of the composite rebar for construction elements with the interruption of thermal bridges *ICEBMP - 20. MEZINÁRODNÍ KONFERENCE, Ekologie a nové stavební hmoty a výrobky*
[7] Štěpánek P and Girgle F 2017 Kompozitní i-KORB Návrh prvku a parametrická studie (Brno: VUT Brno)
[8] EN 1990:2002 Eurocode - Basis of structural design (CEN)

IOP Conf. Series: Materials Science and Engineering **379** (2018) 012027 doi:10.1088/1757-899X/379/1/012027

# Fracture response of alkali-activated slag mortars reinforced by carbon fibres

**H Šimonová[1], L Topolář[1], P Schmid[1], Z Keršner[1] and P Rovnaník[1]**

[1] Brno University of Technology, Faculty of Civil Engineering, Veveří 331/95, 602 00 Brno, Czech Republic

E-mail: simonova.h@vutbr.cz

**Abstract**. The paper reports a study of the effect of carbon fibres on the mechanical fracture parameters of alkali-activated slag mortars. The carbon fibres were added in the amount of 1, 2 and 3%, respectively with respect to the mass of the slag. The mechanical fracture parameters were determined using evaluation of fracture tests carried out on 40 × 40 × 160 mm beam specimens with an initial central edge notch. The monitored parameters were compressive strength, modulus of elasticity, effective fracture toughness and specific fracture energy. The specimen response during fracture tests was also monitored by means of acoustic emission. It was shown that as the addition of carbon fibres increased the value of compressive strength and modulus of elasticity of alkali-activated slag dropped to 50% in case of the highest amount of fibres. The effective fracture toughness is not significantly influenced by addition of carbon fibres. On the other hand, the fracture energy value gradually increases with addition of carbon fibres, up to more than twofold increase in case of the highest amount of fibres.

## 1. Introduction

Nowadays, the binders based on ordinary Portland cement (OPC) are the most commonly used in manufacturing of building materials like concrete, mortars and fine-grained composites. On the other hand, the great emphasis is placed on environmental protection through reductions of $CO_2$ emissions of which large amount is produced by cement industries. There are two feasible ways how to reduce the $CO_2$ emissions in production of above mentioned building materials. The first one is application of secondary raw materials as supplementary cementing materials [1]. The other way is use the alkali-activated binder, this type of material is even more effective in reducing $CO_2$ emissions and energy consumption [2, 3]. These binders are formed by mixing of some aluminosilicate based material, such as blast furnace slag or fly ash with an alkaline activator and water [4, 5]. The mechanical properties and application possibilities of materials based of alkali-activated binders are very similar to those based on OPC. Their major disadvantage is an increased shrinkage during hardening period, caused by both autogenous and drying shrinkage, which finally results in volume contraction, microcracking and deterioration of tensile and bending properties [6]. The addition of different types of fibres into alkali-activated matrix might lead to reduced cracking tendency and improvement tensile properties of these materials as in case of OPC based materials.

Only few papers concerning mechanical properties of alkali-activated slag composites with carbon fibres have been published so far [7–10]. They are focused mainly on determination of compressive and flexural strengths but unfortunately no information on the fracture properties of these composites is available in the literature. Considerable discrepancies in the influence of carbon fibres addition on the compressive strength were reported. While Vilaplana or Cui [9, 10] observed an improvement of mechanical properties with increasing amount of fibres, Alcaide et al. [8] reported strength deterioration of alkali-activated slag paste with 1 and 3% of carbon fibres.

Since no fracture tests on alkali-activated slag mortars with carbon fibres have been performed yet the main objective of this work was to determine the basic mechanical fracture parameters of these composites and analyse the effect of carbon fibres on fracture behaviour.

## 2. Material and specimens

The alkali-activated slag was composed of granulated blast furnace slag supplied by Kotouč, s.r.o. (CZ) finely ground to the specific surface of 383 $m^2/kg$ with the mean particle size of 15.5 μm and solid sodium silicate Susil MP 2.0 (Vodní sklo, CZ) as activator. The molar $Na_2O/SiO_2$ ratio of the silicate activator is 2.0 and the $SiO_2$ content is 52.4%. Quartz sand with maximum grain size of 2.5 mm was used as aggregate in order to prepare AAS mortar. The carbon fibres Tenax A HT C124 with the average length of 3 mm were added in the amount of 1, 2 and 3% respectively with respect to the mass of the slag. Since carbon fibres have nonpolar character 2% solution of Triton X-100 was used for easier dispersion of fibres in water. The designation of individual mixtures is based on carbon fibres content: CF0, CF1, CF2, and CF3. The mixture CF0 is reference without carbon fibres content.

For the preparation of the specimens, the following procedure was applied. At first, carbon fibres were dispersed in part of mixing water together with Triton X-100. Then, sodium silicate activator, slag and quartz aggregate was added and the mixture was stirred in a planetary mixer for about 7 min to prepare a fresh mixture. Finally, defoaming agent Lukosan S was added in order to detrain air bubbles from the fresh mixture. The amount of activator added was 20% of the mass of slag and an aggregate/slag ratio was 3.0 and water content varied with the amount of carbon fibres to achieve an accurate consistency. The composition of mixtures is presented in Table 1.

**Table 1.** Mixture proportions of alkali-activated slag mortars.

| Mixture component | Unit | CF0 | CF1 | CF2 | CF3 |
|---|---|---|---|---|---|
| Slag | g | 450 | 450 | 450 | 450 |
| Sodium silicate | g | 90 | 90 | 90 | 90 |
| Aggregate | g | 1350 | 1350 | 1350 | 1350 |
| Carbon fibres | g | – | 4.5 | 9.0 | 13.5 |
| 2% Triton X-100 | g | – | 5 | 10 | 15 |
| 1% Lukosan S | g | – | 5 | 10 | 15 |
| Water | ml | 180 | 210 | 235 | 270 |

The mixtures were cast into prismatic moulds of the size 40 × 40 × 160 mm. After 24 hrs the hardened specimens were immersed in a water bath at 20°C for another 27 days. Before the fracture tests were performed, all specimens were pulled out of the water and allowed to dry spontaneously under ambient conditions for 24 hrs.

## 3. Fracture tests

The three-point bending fracture tests were conducted on beam specimens with initial stress concentrator made by a diamond blade saw before testing. The depth of initial edge notch on the bottom side of specimen was approximately 1/3 of specimen depth. The span length was 120 mm. Fracture experiments were carried out on a very stiff LabTest 6-1000.1.10 multi-purpose mechanical testing machine with displacement control. If displacement increment loading of specimen is performed, it is possible record the load vs. displacement (*F–d*) diagrams using induction sensors connected in HBM SPIDER 8 device during the testing.

At the beginning of the specimen loading, there are quite often recorded small fluctuations in the measured values. These are caused by the contact of specimen with the supports, pushing it into the supports, and crushing of small protrusions on the specimen's surface. These phenomena usually soon stop, and the diagram continues with a linear part. Therefore, it is advisable to correct the beginning part of the measured diagram, to obtain the proper estimation of the modulus of elasticity value. The first step is approximation of the linear part of diagram with a straight line, follow by determination of the intersection of this line with the horizontal axis, and then all points of the diagram are shifted equidistantly, thus the intersection becomes the origin of the coordinate system. The processing of measured diagrams including the above-mentioned phenomena was performed in GTDiPS software [11], which uses advanced transformation methods to process extensive point sequences. The aim of the corrections of the measured diagrams was primarily to shift the origin of the coordinate system, the smoothing of the diagram and the reduction of the number of points. For this purpose, a chain of transformation steps has been put together in the software. The selected *F–d* diagrams, processed in above mentioned way, with well recorded descending part are shown in Figure 1.

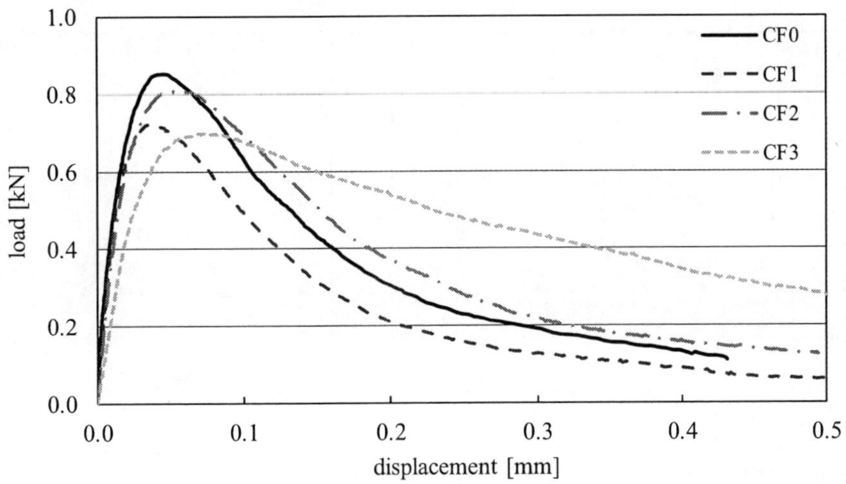


**Figure 1.** Selected *F–d* diagrams of alkali-activated mortar specimens tested at three-point bending.

After the above described corrections of measured *F–d* diagrams, their initial parts were used to estimate the modulus of elasticity values. Furthermore, the effective crack elongation and effective fracture toughness values were determined using the Effective Crack Model [12], which combines the linear elastic fracture mechanics and crack length approaches. The work of fracture and specific fracture energy values were obtained from the whole *F–d* diagrams according to the RILEM method [13]. The informative compressive strength values were also determined on the fragments remaining after the fracture experiments had been performed.

To describe acoustic emission signals formed during the three-point bending test, the attention was focused on the number, duration, amplitude and energy of AE signals [14]. The AE signals were recorded by measuring equipment DAKEL XEDO with four acoustic emission sensors IDK-09 with 35 dB preamplifier, which were attached to the specimen surface by beeswax. The number of AE events corresponds to the material ability to withstand the fracture. The duration of an AE signal is the time difference between the crossing of the first and last threshold. The amplitude is the highest measured voltage in a waveform. This is an important parameter in AE inspection because it determines the detectability of the signal. Signals with amplitudes below the operator-defined limit are not recorded.

Energy of AE signals is defined as an area under the envelope of the rectified linear voltage time signal from the transducer [15].

## 4. Results and discussion

The mean values (obtained from 3 independent measurements) and standard deviations (presented as error bars) of the informative compressive strength values of the tested alkali-activated slag mortars are depicted in Figure 2 (left). The compressive strength value decreased with addition of 1, 2 and 3% of carbon fibres by 5, 23 and 56%, respectively, which is in contradiction with some data reported in literature [9, 10]. It can be assumed that lower compressive strength value of alkali-activated slag mortars reinforced by carbon fibres is due to matrix higher porosity compared to denser matrix of non-reinforced mortar. However, the main difference is probably caused by the level of fibre dispersion, i.e. the way of pre-treatment, mixing and the type of dispersing agent. Ultrasonic treatment applied by Vilaplana [9] probably led to very well dispersed fibre mixture which resulted in the improvement of mechanical properties, Alcaide [8] used methylcellulose as dispersing agent which probably led to insufficient dispersion, thus lower strengths.

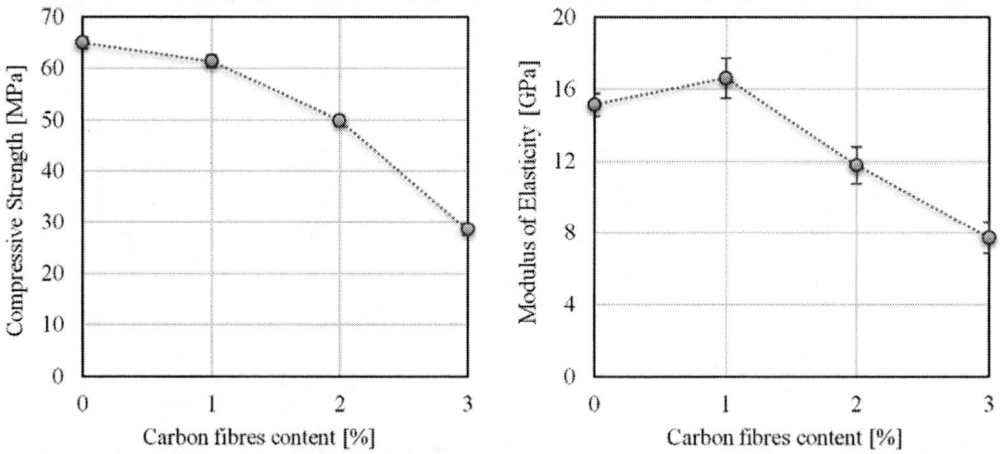


**Figure 2.** Compressive strength (left) and modulus of elasticity values of alkali-activated mortars with different amount of carbons fibres.

The mean values and standard deviations of selected mechanical fracture parameters obtained from recorded *F–d* diagrams in above described way [12, 13], i.e. modulus of elasticity, effective fracture toughness and specific fracture energy, are summarized in Figure 2 (right) and Figure 3. Modulus of elasticity values increased with addition of 1% of carbon fibres by 10%. For higher amount of carbon fibres, the similar decrease of modulus of elasticity value was achieved as in case of compressive strength. In comparison to other parameters, the effective fracture toughness value is not significantly influenced by addition of carbon fibres, see Figure 3 (left). The difference with addition of carbon fibres is about 10% in comparison with reference mortar. On the other hand, the fracture energy value gradually increases with addition of carbon fibres, up to more than twofold increase in case of the highest amount of fibres. It follows that addition of carbon fibres into alkali-activated matrix does not have any significant effect on stable crack propagation. On the other hand, after the maximum load is reached more energy is needed to unstable crack propagation with addition of carbon fibres into the matrix, which can be connected with bridging of the crack by fibres.

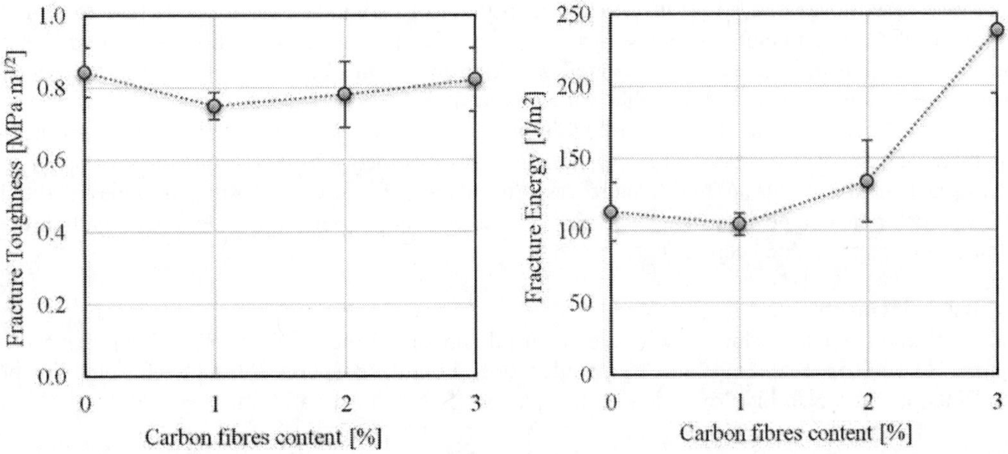


**Figure 3.** Effective fracture toughness (left) and specific fracture energy values of alkali-activated mortars with different amount of carbons fibres.

**Table 2.** Mean values of selected AE parameters (coefficients of variation in %).

| Parameter | Unit | CF0 | CF1 | CF2 | CF3 |
|---|---|---|---|---|---|
| Number of AE events | – | 264 (0.6) | 308 (0.4) | 224 (1.2) | 171 (0.8) |
| Duration of AE signals | μs | 1316 (0.1) | 1177 (0.1) | 1171 (0.1) | 1171 (0.4) |
| Amplitude of AE signals | mV | 2861 (0.2) | 2103 (0.3) | 2183 (0.6) | 2249 (0.2) |
| Energy of AE signals $10^{-5}$ | V·s | 252.2 (1.4) | 221.8 (0.2) | 134.4 (0.4) | 94.6 (0.6) |

The mean values and coefficient of variations of selected parameters of AE signals obtained from AE measurements during the fracture tests are summarized in Table 2. The number of AE events values increased with the addition of carbon fibres in amount 1% by 17%. For the higher amount of carbon fibres, the decrease of the number of AE events value was recorded, the observed trend was similar as in case of modulus of elasticity values. The duration of AE signals values decreased with the addition of carbon fibres. The signal damping was 11% in comparison with reference specimens. The values of AE signal amplitude decreased with the addition of carbon fibres. Lower values of amplitude probably represent the formation of smaller microcracks. The energy of AE signal values gradually decreased with increasing amount of carbon fibres addition.

## 5. Conclusions

The aim of this study was to quantify the effect of carbon fibres as conductive filler on the fracture properties of alkali-activated slag mortar. Although carbon fibres were primarily added in order to enhance the electrical properties of the building material, it should not deteriorate its mechanical fracture properties. On the contrary, the addition of fibres into alkali-activated matrix should lead to a reduction in cracking tendency and improvement in tensile properties of these materials. The specimen response during fracture tests was also monitored by means of acoustic emission.

It was shown that as the addition of carbon fibres increased the value of compressive strength and modulus of elasticity of alkali-activated slag dropped to 50% in case of the highest amount of fibres.

This might have been caused by insufficient dispersion of carbon fibres in the alkali-activated slag matrix. The effective fracture toughness is not significantly influenced by addition of carbon fibres. On the other hand, the fracture energy value gradually increased with addition of carbon fibres, up to more than twofold increase in case of the highest amount of fibres.

The processed AE data showed that addition of carbon fibres affects mechanical properties; in a number of cases the decrease of these parameters was recorded. The work confirms that analysis of AE signals obtained during the fracture tests of alkali-activated composites provides additional information about composite damage and can be used to better understand the effects of carbon fibres addition to alkali-activated composites.

**Acknowledgements**
This outcome has been achieved with the financial support of the Czech Science Foundation under project No 16-00567S and the Ministry of Education, Youth and Sports of the Czech Republic under the "National Sustainability Programme I" (project No. LO1408 AdMaS UP).

**References**

[1] Keun-Hyeok Y, Yeon-Back J, Myung-Sug C and Sung-Ho T 2015 Effect of supplementary cementitious materials on reduction of CO2 emissions from concrete *J Cleaner Prod* **103** pp 774–783

[2] Keun-Hyeok Y, Jin-Kyu S and Keum-Il S 2013 Assessment of $CO_2$ reduction of alkali-activated concrete *J Cleaner Prod* **39** pp 265–272

[3] McLellan B C, Williams R P, Lay J, van Riessen A and Corder G D 2011 Costs and carbon emissions for geopolymer pastes in comparison to ordinary Portland cement *J Cleaner Prod* **19(10)** pp 1080–1090

[4] Provis J L and van Deventer J S (eds) 2014 *Alkali activated materials: state-of-the-art report, RILEM TC 224-AAM* (Dordrecht: Springer)

[5] Shi C, Krivenko P V and Roy D 2006 *Alkali-activated cements and concretes* (London: Taylor & Francis)

[6] Cincotto M A, Melo A A and Repette W L 2003 Effect of different activators type and dosages and relation with autogenous shrinkage of activated blast furnace slag cement *Proc. of the 11th International Congress on the Chemistry of Cement* (Durban, South Africa) pp 1878–1888

[7] Natali A, Manzi S and Bignozzi M C 2011 Novelfiber-reinforced composite materilasbased on sustainable geopolymer matrix *Procedia Engineering* **21** pp 1124–1131

[8] Alcaide J S, Alcocel E G, Puertas F, Lapuente R and Garces P 2007 Carbon fibre-reinforced, alkali-activated slag mortars *Materiales de Construccion* **57** pp 33–48

[9] Vilaplana J L, Bazea F J, Galao O., Alcocel E G, Zornoza Z and Garces P 2016 *Con Build Mat* **116** pp 63–71

[10] Cui H, Feng T, Yang H, Bao X, Tang W and Fi J 2018 *Con Build Mat* **161** pp 442–451

[11] Frantík P and Mašek J 2015 GTDiPS software http://gtdips.kitnarf.cz/

[12] Karihaloo B L 1995 *Fracture Mechanics and Structural Concrete* (New York: Longman Scientific & Technical)

[13] RILEM TC-50 FMC Recommendation 1985 Determination of the fracture energy of mortar and concrete by means of three-point bend test on notched beams *Materials & Structures* **18** pp 285–290

[14] Grosse C U and Ohtsu M 2008 *Acoustic Emission Testing: Basics for Research – Applications in Civil Engineering; with contributions by numerous experts* (Heidelberg: Springer)

[15] Sagar R V 2009 An experimental Study on Acoustic Emission Energy and Fracture Energy of Concrete *Proc. of the National Seminar & Exhibition on Non-Destructive Evaluation* (Tiruchirappalli, India) pp 225–228

# Development of conductive composite coatings with specific functional properties

**B Nešpor**[1], **M Nejedlík**[1] **and L Vavrušková**[1]

[1] Research Institute for Building Materials, Hněvkovského 30/65, 617 00 Brno, Czech Republic

Email: nespor@vustah.cz

**Abstract**. The article presents the development results of advanced composite coatings with enhanced electric conductivity intended for detection of disruptions in the base material structure. Conductive coatings were prepared from a low viscosity epoxy resin and three types of admixtures - amorphous carbon black, micronized graphite and multi-walled carbon nanotubes (MWCNTs). Such multicomponent coatings were tested together with hand-applied graphite layers and commercial spray-applied colloidal graphite and copper-based electroconductive coatings, for comparison. The influence of the admixtures on electrical conductivity was evaluated. The course of impedance response of polymer coatings, graphite layers and copper layers was investigated during bending in press, drilling and high-speed loading by rifle projectiles. Conductive coatings showed a distinctive change in impedance under mechanical damaging by drilling and during dynamic testing. These functional coatings represent innovative direction for the development of new progressive ballistic and anti-blast shield systems.

## 1. Introduction

The number of terrorist attacks and their victims is growing dramatically in the world. Ensuring security of the population and protecting critical infrastructure against terrorist threats is one of the key challenges of current era. One of the protection options are progressive mobile barrier systems that are resistant to dynamic threats (projectiles, explosions). These are, for example, ballistic and counter-explosive shield systems with damage detection function. The aim of presented research is to develop a segment of ballistic and anti-blast shield systems, which, in conjunction with the monitoring system, will detect damage in the structural integrity of the material.

In previous research [1–8], it was found that conductive concrete with an appropriate dosage of electrically conductive particles and/or fibers exhibit the properties necessary for strain monitoring and detecting its own material structure damage during both static and dynamic loading. However, the complexity of the cement composites behavior and above all, difficult reproducibility of the results in changing climatic conditions quite complicates their possible use. Therefore, easy-to-apply thin-film coating sensors were investigated to offset the disadvantages of the cementitious composites.

Generally, conductive coatings contain conductive components such as carbon black, metallic components (copper, silver) or carbon nanoparticles. Several articles comparing these functional fillers have been published recently. Berkei et al. [9] investigated the electrical properties and percolation threshold of carbon nanotubes. Low concentrations from 0.5% to 1% phr (parts per hundred rubber) carbon nanotubes result in anti-static properties, whereas electrical conductivity is reached in the range of 2–4% phr. In contrast, in some coatings even high concentrations of 8% phr carbon nanotubes show no effect at all. Tinthoff et al. [10] have confirmed in their research that MWCNTs can be used as an alternative to the classical conductive pigments like carbon black or metallic particles. Senesac and Thundat [11] investigated the piezoresistive properties of experimental nano-cantilever beam sensors. Single-walled carbon nanotubes (SWCNTs) could be used for explosion detection.

## 2. Materials and methods

### *2.1. Technology of preparation and application of coatings*

The increased electrical conductivity of the coatings was achieved by the appropriate choice of matrix, functional micro/nano admixtures, their optimum amount and also the preparation technology itself. Initially, 25 samples of conductive coatings with varied composition were prepared from a low viscosity epoxy resin and styrene-butadiene dispersion with following fillers incorporated (see Figures 1a-c):

- amorphous carbon black (up to 6% w/w),
- micronized graphite (up to 20% w/w),
- multi-walled carbon nanotubes (MWCNTs, 0.5% w/w).

To reduce system viscosity and ease the application, volatile substances (acetone, technical alcohol and gasoline, isopropyl alcohol, etc.) were added into the mixture. Components were stirred by Silverson L5M-A high shear laboratory mixer at 5000 rpm for 5 minutes. This procedure provided a high degree of mixture dispersion but only for epoxy resin samples. Unfortunately, styrene-butadiene polymer samples could not be adequately mixed with functional fillers so the workability was considerably worse, thus only a limited number of testing bodies was prepared. All prepared materials were consecutively applied onto the surface of glass fiber reinforced concrete (GFRC) with a plastic spatula, a brush or a roller. All these ways created a thin film.

The mentioned coatings were supplemented with hand-applied graphite lead layers and commercially available spray-applied electroconductive lacquers based on colloidal graphite and copper, see Figure 2. All these coatings were also applied on the surface of GFRC.

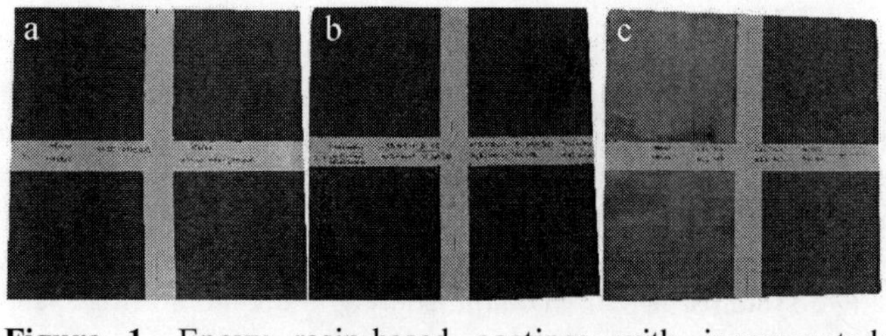


**Figure 1.** Epoxy resin-based coatings with incorporated: (a) carbon black; (b) MWCNT; (c) micronized graphite.

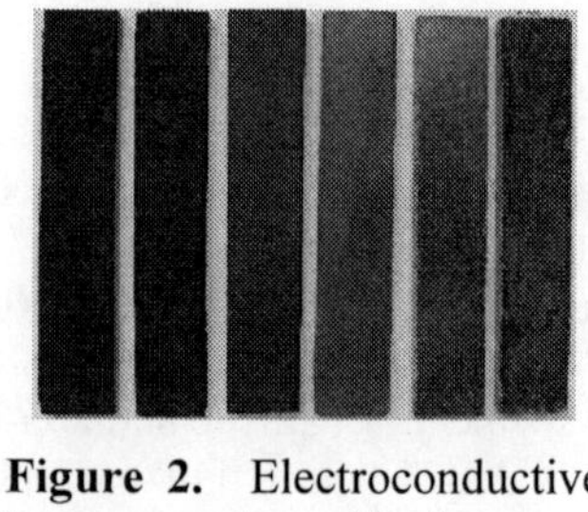

**Figure 2.** Electroconductive lacquers and graphite layers.

### *2.2. Dispersion analysis of conductive components in coatings*

In the first phase of testing, the dispersion of conductive components in individual matrices was investigated. Microscopic analysis and visual assessment of the mirror gloss of the surface were performed on the hardened epoxy coatings, see Figures 3a-c. The next step was to verify the degree of electrical conductivity and its uniformity over the entire surface of the coating, see Figure 4.

Alternating current was used to measure the electrical properties of conductive coatings. The resistance to this alternating current is expressed by the electrical impedance in ohms Ω and measured by a hand-held LCR meter (and impedance analyzer-recorder) by a direct two-point method at the corners of the test specimens during positive half-waves of excited alternating current with a frequency of 1 kHz at voltage of 1 V. The impedance of the composite coating depends on the distance between copper electrodes. The length of the element (l) is directly proportional to impedance.

In contrast, the impedance is in inverse proportion to the active surface (S):

$$R = \frac{r \times l}{S} \tag{1}$$

where r is specific impedance of the material and R is impedance of the element.

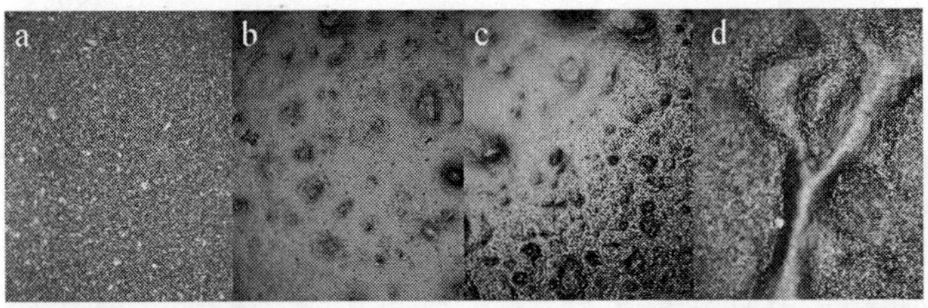


**Figure 3.** Microscopic analysis of the conductive coatings: (a) matt surface; (b) glossy surface; (c) matt/gloss transition; (d) wrinkle structure of MWCNTs-enriched coating.

**Figure 4.** The measurement of impedance.

*2.3. Impedance response testing under mechanical damage*

Impedance measurements of the conductive coatings under mechanical stress was performed by a test ripping machine TIRA test 2710. Contact parts of the press were separated by electrical insulation. Individual samples of conductive polymer coatings on GFRC boards (250 × 250 × 10 mm), and graphite and copper layers on GFRC boards (250 × 50 × 10 mm) were statically loaded with a three-point bend. The coatings were directly connected to a digital impedance measurement device by clamped contacts.

Promising samples were subsequently damaged by hand drilling device. The hole was drilled through GFRC base boards and the drill penetrated the conductive coating on the other side. The conductive coatings were directly connected to the impedance meter via clamped contacts.

*2.4. The testing of impedance response during ballistic loading*

Three types of promising conductive coatings were prepared for dynamic testing, which were applied onto dried and cleaned GFRC surfaces with glued copper contacts (i.e. grounding tape). The coatings were: epoxy coating with amorphous carbon black incorporated (6% w/w), surface-applied graphite lead and spray-applied electroconductive graphite lacquer. The contact parts of the coatings were provided with an electrically insulating tape after curing, see Figures 5a-c. For comparison, a second series of identical coatings were provided with self-adhesive conductive tape (instead of glued tape) and covered with a transparent non-conductive epoxy gel, protecting against mechanical damage, humidity and weather, see Figures 5d-f.

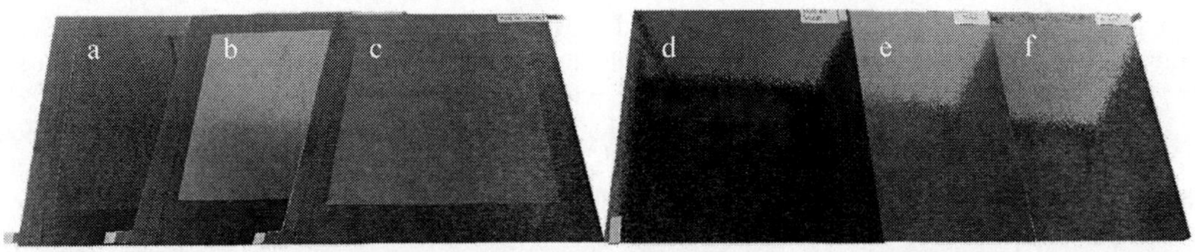


**Figure 5.** (a) Epoxy resin/carbon black coating; (b) graphite coating, (c) graphite lacquer; Epoxy-gel encapsulated samples of: (d) epoxy resin/carbon black coating; (e) graphite coating; (f) graphite lacquer.

Both nondestructive and destructive tests were carried out in a ballistic laboratory test tunnel. The impedance of the conductive coatings was measured during a perpendicular impact of the 7.62 × 51 mm NATO ball ammunition. The main measuring device was an impedance analyzer with a recording computer. The test specimens were connected to the apparatus via a two-conductor copper wire and crocodile clips. The test samples were loaded with rifle projectiles with initial velocity of 830 $m \cdot s^{-1}$. Individual shots were aimed at a minimum pitch of 12 cm, see Figure 6.

**Figure 6.** (a) Table with a clamped-fixed sample, a protective shield and an optical gate for measuring bullet speed; (b) firing rest; (c) sandwich configuration for non-destructive testing; (d) sample after bullet penetration.

Non-destructive testing was ensured by ballistically-resistant cover plates made of densely steel fiber reinforced concrete (SFRC). During destructive experiments, the samples were only covered with GFRC, see Figures 7–8. Each sample was subjected to four shots, two non-destructive and two destructive.

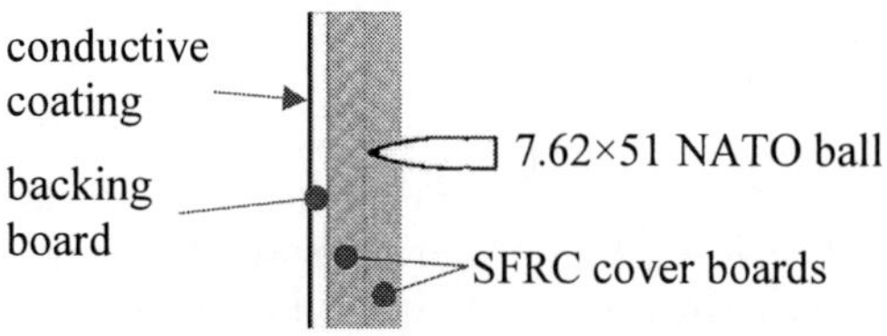


**Figure 7.** Destructive test configuration.

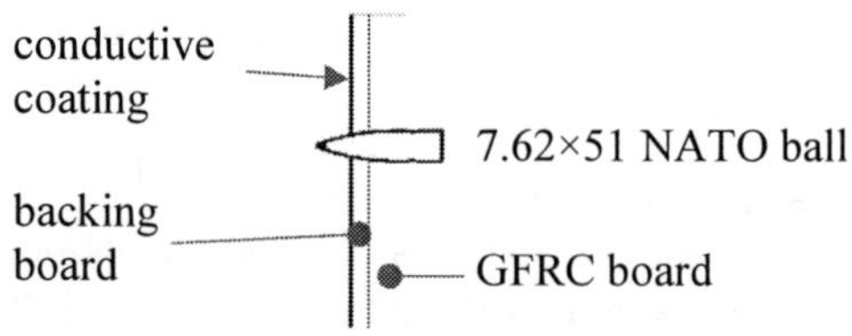


**Figure 8.** Non-destructive test configuration.

## 3. Results and discussion

### *3.1. Preparation technology and application methods*

In the first phase of the tests, the preparation technology and the application method for coatings with functional micro/nano admixtures incorporated was verified. Based on evaluation of application and functional properties, a selection of suitable binders and additives was conducted. Doses of the active ingredients, the time of homogenization and the mode of application were optimized.

The dispersion of functional fillers has considerable effect on the final properties of the epoxy coating. Carbon epoxides need to be applied in one very thin film layer (e.g. with a roller) onto adsorbent substrates to spread the coating uniformly and without creating non-conductive resin layers. Insufficient dispersion of the filler in the surface layer is visible to the naked eye. Microscopic coating analyses corresponded with the results of gloss intensity evaluation. Epoxy coatings with evenly dispersed carbon filler (with low and stable impedance) show lower values of mirror gloss.

### *3.2. Impedance measurements of non-loaded samples*

In the second phase of the project, samples with the most beneficial properties, i.e. the lowest impedance, were identified. Among the low-viscosity epoxy resins, the sample enriched with amorphous carbon black (3% w/w) achieved stable impedance below 5 kΩ. MWCNTs-doped samples were less applicable as they exhibited tendency to create wrinkled structure (see Figure 3d) and showed a little higher impedance, about 12 kΩ. Micronized graphite, unlike nanotubes, was easily dosed in large quantities up to 20% w/w. However, nor better neither more stable coating impedance (50–100 kΩ) was achieved with such high dosage.

The styrene-butadiene matrix dispersions were less workable and did not create conducting chains, i. e. showed extremely high impedance, almost beyond measurability.

The most satisfactory results of impedance measurements were achieved with:

- epoxy coatings enriched with conductive carbon black (3% w/w),
- surface-applied graphite lead layer,
- electroconductive lacquer based on colloidal graphite.

These three samples also exhibited the best results under static and dynamic load, as shown in 3.3.

### *3.3. Impedance measurements of statically and dynamically loaded samples*

The most promising samples after impedance measurements with no load were subjected to impedance measurements during static loading, i. e. bending. Almost none of the conductive coatings transforms the bending load into an evaluable impedance change. The only exception was the hand-applied graphite. During the bending test, the impedance began to increase after reaching 2/3 of the maximum applied force. In a drilling test, three variants of conductive coatings exhibited a measurable increase in impedance (see Figure 9) and were subjected to dynamic loading tests.

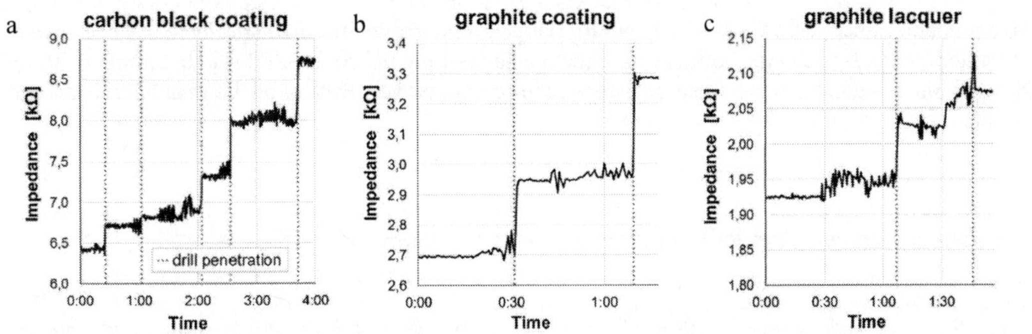


**Figure 9.** Course of impedance under mechanical damage by drilling (a) carbon black coating; (b) graphite coating; (c) graphite lacquer.

During dynamic processes, all three tested materials exhibited a significant change of impedance. The individual impacts of projectiles were well recognizable in impedance course graphs. The impedance of the samples increased during both destructive and non-destructive testing methods, see Figure 10.

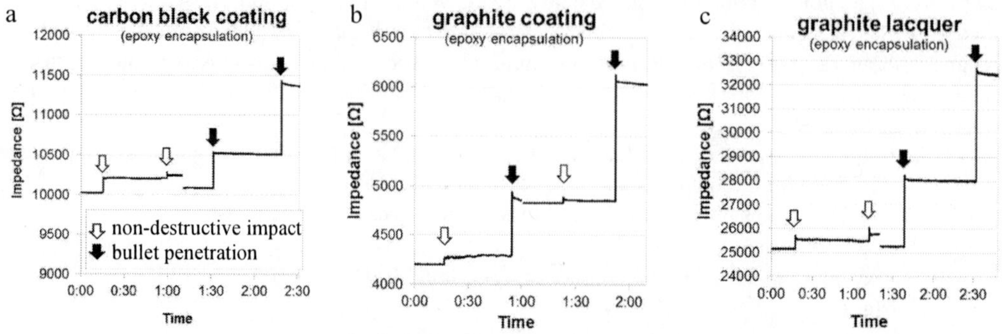


**Figure 10.** Course of impedance under ballistic loading (a) carbon black coating; (b) graphite coating; (c) graphite lacquer.

## 4. Conclusion

The thin-film sensor coatings with enhanced electric conductivity represent an innovative direction for the development of 'smart' blast-resistant and ballistic-resistant barriers with failure detection ability. Easy-to-apply conductive coatings can be prepared from low viscosity epoxy resin with amorphous carbon black incorporated. Similar results were achieved using surface-applied graphite lead and spray-applied electroconductive graphite lacquer. Sensor coatings with non-conductive epoxy gel encapsulation (protecting against mechanical damage, humidity and weather) showed similar detection properties to the identical coatings without encapsulation. The conductive epoxy coating and graphite layers enable detection of material structure damage under dynamic ballistic loading. The materials exhibited impedance change both during direct damage of the matrix, and after indirect impact through a cover plate. Before practical use of these sensor coatings, verification of the results reproducibility in changing climatic conditions is advised.

## Acknowledgements

This paper was financially supported by the TA CR project TE02000162 'CAMPT - Centre of Advanced Materials and Technologies for Protection and Safety Enhancement'.

## References

[1] Sihai W and Chung D 2006 *The role of electronic and ionic conduction in the electrical conductivity of carbon fiber reinforced cement* (*Carbon 44.11*) pp 2130–2138
[2] Chen P-W and Chung D 1993 Concrete reinforced concrete as a smart material capable of non-destructive flaw detection *Smart Mater. Struct.* **2** pp 22–30
[3] Chen P-W and Chung D 1996 Carbon fiber reinforced concrete as an intrinsically smart concrete for damage assessment during static and dynamic loading *ACI Mater. J.* 93.4 pp 341–350
[4] Fu X and Chung D 1996 Self-monitoring of fatigue damage in carbon fiber reinforced cement *Cem. Concr. Res.* 26.1 pp 15–20
[5] Fu X, Lu W and Chung D 1998 *Ozone treatment of carbon fiber for reinforcing cement* (Carbon 36.9 ) pp 1337–1345
[6] Qizhao M, Binyuan Z, Darong S, Zhuoqiu L 1996 *Resistance changement of compression sensible cement specimen under different stresses* (Univ. Technol. 11.3) pp 41–45
[7] Shi Z-Q and Chung D 1999 Carbon fiber reinforced concrete for traffic monitoring and weighing in motion *Cem. Concr. Res.* 29.33 pp 435–439
[8] Nešpor B and Nejedlík M 2018 Development of Electrically Conductive Composite Sensors with the Addition of Functional Fillers *Solid State Phenomena* **272** pp 34–40
[9] Berkei M 2011 Conductive coatings: Using carbon nanotube *CHEManager Europe* vol 7-8/2011

IOP Conf. Series: Materials Science and Engineering **379** (2018) 012028 doi:10.1088/1757-899X/379/1/012028

p 10
[10] Tinthoff T, Hanitzsch N, Pickave M, Tecklenburg J, Willing N and Berkei M 2012 Conductive coatings Using carbon nanotubes *Paint & Coatings Industry* vol 28.1 pp 16–18
[11] Senesac L and Thundat T G 2008 Nanosensors for trace explosive detection *Materials Today* vol 11.3 pp 28–36

# Non-standard experimental tests of sandstone and its pre-cracking for fracture testing

**M Šperl [1] and M Drdácký[1]**

[1] Institute of Theoretical and Applied Mechanics of the Czech Academy of Sciences, rosecká 809/76, 190 00 Prague 9, Czech Republic

Email: sperl@itam.cas.cz

**Abstract.** The paper presents a methodology for investigating fracture phenomena in sandstone. It shows the preparation of test specimens with a cyclic loading generated crack, control of the test specimen preparation and verification by means of x-ray microCT and digital image correlation (DIC) techniques. There are also presented tests studying tensile and fatigue characteristics of arkose sandstone used i. a. for the construction of the medieval Charles Bridge in Prague where weather induced tensile damage and failure occurred. Finally, the new methodology for pre-cracking of the specimens was verified experimentally and fracture toughness was computed.

## 1. Introduction

The presence of cracks in many historical structures indicates the action of external forces accompanied by internal strain gradients. This is usually a repetitive process, and damage cumulation may occur in porous brittle or quasi-brittle materials. The authors have performed a study of environmental fatigue effects for which they developed a suitable methodology for testing historical stone subjected to repeated tension strains [1].

Cumulative tensile damage may occur in stone monuments due to mainly two reasons. The stone objects are subjected to repeated change of temperature from solar irradiation when the heat is transferred unevenly inside the material. The accompanied different thermal dilation may cause dangerous stress and cause cumulative crack propagation, typically initiated in usually present interior defects. Similarly, negative effect has been observed in composed materials, namely masonry made of stone and mortar with different thermal dilation characteristics. Recent examples are presented in Figure 1 (detritic limestone sculpture in Eisgarn - Austria) and Figure 2 (sandstone masonry block of the Charles bridge parapet wall). In the second case, a very strong mortar has been used for repair of the bridge and the block cracked across the whole profile during a very severe winter when temperatures dropped under -20°C. Such adverse effects called for studies of behaviour of similar stone structures under extreme climate effects.

This paper takes advantage of the experimental assessment of changes in the mechanical characteristics of sandstone from Božanov (one of the typical medieval rock materials used in Charles Bridge) due to the accumulation of damage. The Young modulus and the Poisson number were investigated, using a verified methodology for testing stone in simple tension and in cycling simple tension/compression loading. The results have shown that the first tension load-displacement can be approximated very satisfactorily by a power function, and the optical DIC method again demonstrated its capacity and suitability for measuring the complex deformation field on porous surfaces and on

IOP Conf. Series: Materials Science and Engineering **379** (2018) 012029 doi:10.1088/1757-899X/379/1/012029

naturally well-structured surfaces. Nevertheless, the methodology that had been adopted required painstaking specimen preparation and highly-skilled staff and can be recommended only for special small-series tests.

**Figure 1.** Ruptured limestone sculpture.

**Figure 2.** Ruptured bridge rail stone.

This paper also needed a special methodology for the preparation of test specimens with defined cracks for a series of tests to study the effects of consolidation on fracture behaviour, crack propagation and ultimate loads. Experimental investigations on specimens with cracks have rarely been published. Two such studies were published in [2] and [3] but they were focused a bit differently.

## 2. Experimental works

Božanov stone is a greyish beige gross grain strong arkose sandstone without marked layering, Figure 3 (medium grained arkose sandstone by geologists). The material used for two types of tests was extracted from the eleventh arch of the Charles Bridge parapet wall, which had to be replaced during recent repairs. The material can be characterized as a quasi-brittle inelastic silicate composite.

**Figure 3.** Composition of the Božanov sandstone.

### *2.1. Tensile, compression and fatigue tests*

Testing specimens (cut out from the stone block) of dimension of 50 mm × 50 mm × 200 mm were fixed into steel rectangular tubes with axially welded flat steel hangers which served to fix the set into hydraulic grips (by the help of the special-self procedure with safety loop) of the loading frame Instron 1343 – Figure 4. For gluing a two component Sikadur-31 CF RAP resin was used. The fixtures had to be prepared with careful precision in order to ensure perfect alignment and perpendicular arrangement for tension and combined loading. The specimens in prismatic form were cut without geometrical imperfections.

**Figure 4.** Test specimens and test arrangement.

Four pilot experiments were completed on the specimens from Figure 4. The tests involved i) compression, ii) very low cycle fatigue (altering unsymmetrical cycle), iii) tension, iv) fatigue (altering unsymmetrical cycle). Compression force was applied in three steps: -1.5 MPa, -26 MPa and -37 MPa. During the test the deformation was measured in order to study change of Young modulus and Poisson number in relation to the applied load – Figure 5. Hysteresis and irreversible deformation was observed (measured by DIC) and the Poisson number reached a value of 0.23.

The tension tests were performed with a very low velocity of the crosshead movement at about 5 μm/min. A crack initiated at one interior defect and due to very low loading velocity it was possible to evaluate the overall energy absorbed during the crack origin and propagation - Figure 6. The load-displacement diagram shows the overall amount of energy absorbed during the test (0.4 J) and for the crack growth (0.16 J) which corresponds to a very brittle material behaviour even though the strength was decreased by a defect inside the material. The course of stress and deformation can be well described up to the fracture by means of a suitable power function.

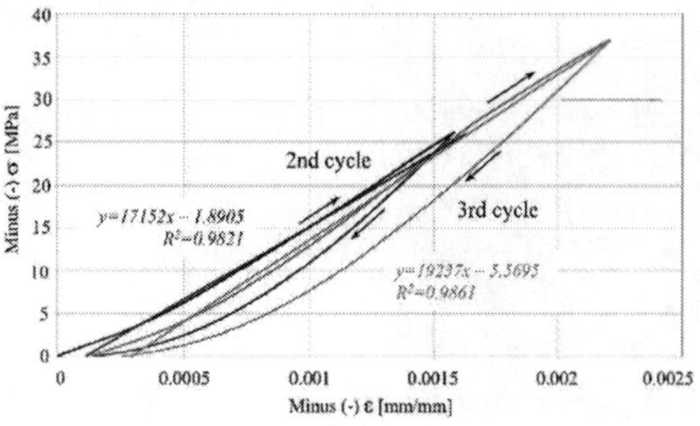


**Figure 5.** Load-deformation diagram – compression.

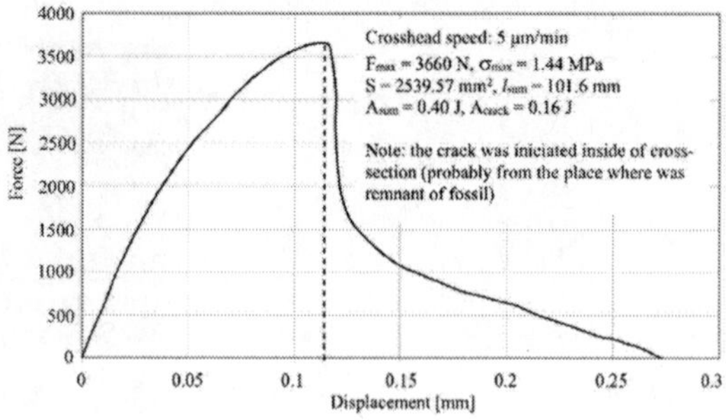


**Figure 6.** Load-displacement - tension test.

IOP Conf. Series: Materials Science and Engineering **379** (2018) 012029 doi:10.1088/1757-899X/379/1/012029

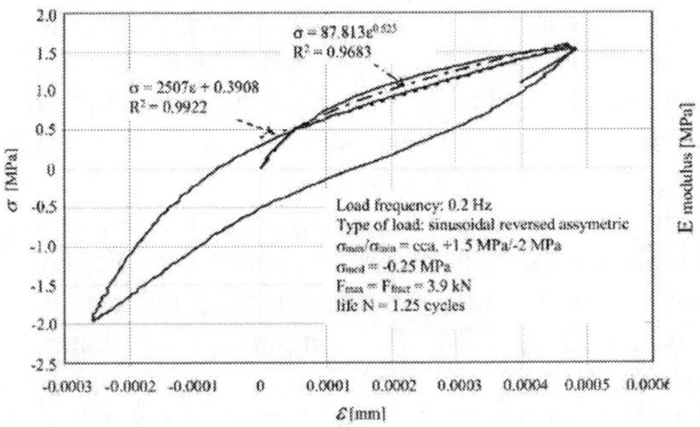


**Figure 7.** Low-cycle fatigue diagram.

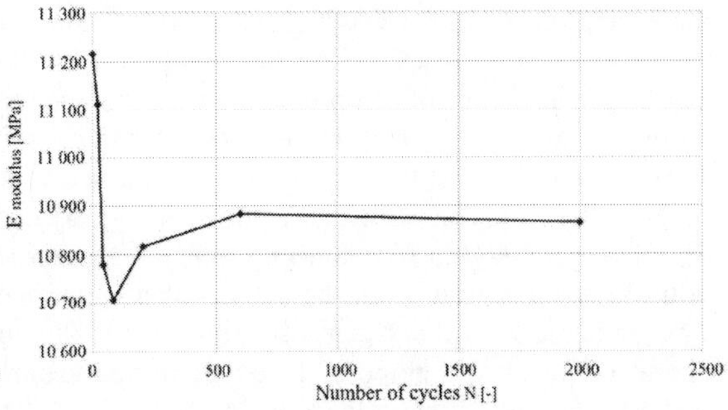


**Figure 8.** Change of E modulus during cycling.

Sinusoidal altering non-symmetrical loading cycle with the stress limits of +1.5 MPa and -2 MPa was applied during the very low fatigue test. The stress mean value was -0.25 MPa, the stress double amplitude of 3.5 MPa and frequency of loading of 0.2 Hz, Figure 7.

Fatigue tests used the same sinusoidal altering non-symmetrical loading cycle, here with the stress limits of +0.265 MPa and -0.354 MPa. Two modes of cycling velocity were used - a frequency of 0.25 Hz for the main loading sequences and at certain stages 0.01 Hz for a group of three cycles with the measurement of deformations from which the second cycle was evaluated, and the E modulus calculated. Such a detailed measurement was accomplished up to 2000 cycles. Then the specimen was loaded up to failure. The overall lifespan reached 224 908 loading cycles. The modulus of elasticity E changed during the cycling. Its value dropped from the initial figure of 11 214 MPa to 10 705 MPa after 82 loading cycles (about 5% decrease). Then its value stabilized around 10 865 MPa after 2002 cycles, see Figure 8.

### 2.2. *Three point bending tests with cycling and fracture tests*

Test specimens with dimensions of 20 mm × 20 mm × 100 mm were placed on one side in the center with an initiation notch 1 mm thick and 2 mm deep. Then the specimens were cyclically loaded in three-point bending generated by a resonance pulsating machine - Rumul Mikrotron. The initiation notch was placed on the side in tension and the load pulsated in a mode with the asymmetry of the cycle being R = 0.0385 – i.e., with a nearly vanishing load cycle. The resonance loading frame started cycling with an initial force impulse and then the loaded specimen acted as an elastic member in the system. The cycling at a given force was controlled by a resonance frequency which was dependent on the stiffness of the test specimen [4]. Then, a change in stiffness of the specimen was followed up during the loading, which signaled the crack initiation and propagation, and enabled control of the crack depth. To this end, a change in loading frequency was measured. In this way it was possible to prepare a series of test specimens with approximately the same damage in front of the initiation notch. However, a series of preliminary tests had to be carried out in order to identify a suitable level for the loading force. They included a static three-point bending test (3PB) and a series of fatigue tests on various low-force levels to determine the approximate fatigue behaviour of the test material.

Then, in order to ascertain the depth of the initiated cracks X-ray micro-tomography was applied. A special table-top loading device allowing simultaneous X-ray imaging of the test specimen during loading was employed. A specimen was loaded in a cylindrical chamber made of a high-strength composite which had a low attenuation for X-rays and allowed observation from all directions. Both simple 2D transmission images and CT acquisitions could be acquired in individual loading steps. The focal spot of the X-ray source was approximately 50 μm with this setting, which was sufficient with respect to the desired resolution. A scintillation flat panel detector of 2048 × 2048 px resolution and 200 μm single pixel size was used. The resulting imaging geometry allowed 4× magnification, which entailed the resolution of 50 μm at one pixel. The standard three-point bending test was performed with the specimen. The load was applied only until the moment when the crack was observable by simple transmission radiography. In this position CT data (1200 projections per 360°) was acquired. The acquisition time of one projection was 2 s. The images obtained were corrected with regard to beam hardening effect by a set of alumina filters and subsequently a final CT reconstruction was carried out. A 3D visualization of the reconstruction obtained was performed in a VG studio.

Now these specimens with the same damage are consolidated using the most typical agents: elastified Steinfestiger 300 (cca 30% concentration of the active substance), Paraloid (2% concentration), Funcosil 100 (10% concentration of the active substance), and will be tested by standard three-point bending. The specimens without consolidation were already tested in 3PB. By comparing the acquired characteristics it will be possible to determine the consolidation effects.

The acquired ultimate loads were used to calculate $K_{IC}$ toughness [5], which attained the value of 0.331 for the notched beam, 0.349 for the notched beam with the crack in $MPam^{0.5}$. The toughness values are approximate estimates based on an engineering approach to the crack depth assessment from the above mentioned X-ray CT scans and the DIC reconstructions and analytical equation (1).

$$K_I = \sigma\sqrt{\pi a} \cdot F(a/b) \tag{1}$$

where:

$$\sigma = \frac{6M}{b^2} \quad \Rightarrow M = \frac{F \cdot s}{4}$$

$$F(a/b) = 1{,}090 - 1{,}735(a/b) + 8{,}20(a/b)^2 - 14{,}18(a/b)^3 + 14{,}57(a/b)^4$$

$a$... ... total depth of crack (mm),
$b$......specimen height (mm),
$s$......span length (mm),
$\sigma$......stress per unit thickness of the specimen (N/mm²).

## 3. Results and Discussion

Cycling the specimens in the high frequency resonance loading frame helped to understand fatigue behaviour of the Božanov sandstone. At a given force level (3PB arrangement under point load cycling from -10 N to -260 N) the majority of specimens exhibited fatigue life in a range between 90 000 and 140 000 cycles. However, there were specimens with one order lower or higher fatigue life due to the heterogeneity of the sandstone. To prepare the test specimens, a lower number of loading cycles was applied (between 32 000 and 40 000), and only specimens exhibiting behaviour similar to those of the above mentioned majority were accepted for further experiments.

A typical record of the change of resonance frequency is presented in Figure 9. Visible steps are the results of the regulation loop reaction which keeps the mean loading value within an interval of +/- 2 N.

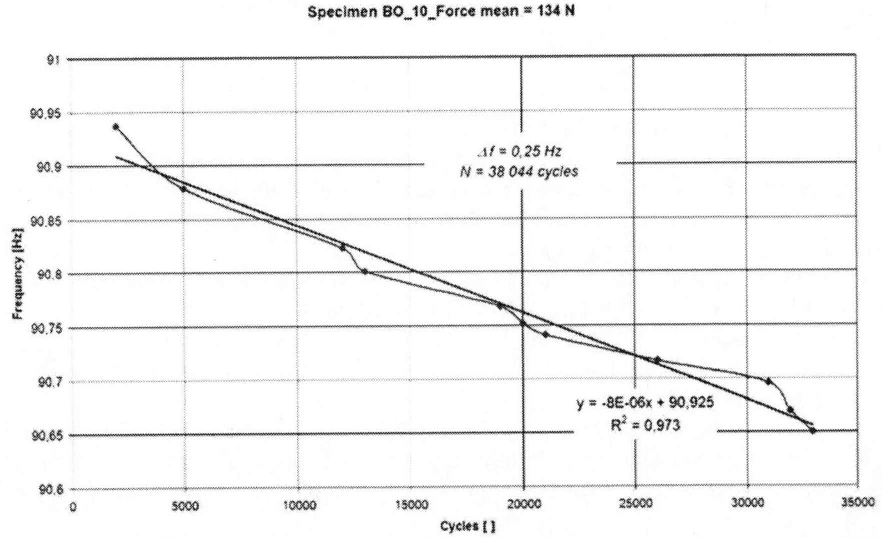


**Figure 9.** Decrease of loading frequency during cycling of the specimen bo_10 (for 134 N).

The character of the damage after cycling was determined by means of X-ray CT. It was not possible to identify the crack originating in the granular structure of the sandstone without applying a slight 3PB loading to the specimen and opening the crack. Thus, the specimen under investigation (BO_5 after 140 300 cycles) was loaded with 200 N and under such loading scanned with the X-ray CT. Then the DIC method was used to identify similar situations within the sections in the specimen depth. DIC proved to be a useful tool for crack propagation studies, e.g. [6] and [7]. A comparison of crack visualization for loaded and unloaded specimens is shown in Figure 10.

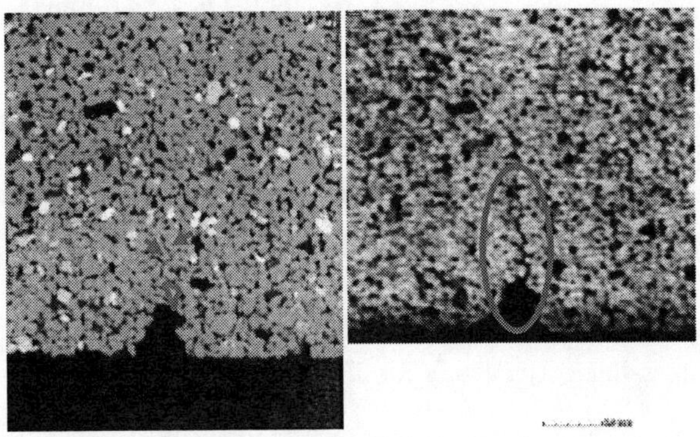

**Figure 10.** X-ray pictures of unloaded (left) and loaded (right) sandstone specimen BO_5.

The results of uniaxial tests (Figure 7) also shows that the first loading course can be well approximated by a power function. However, the parameters in the constant differ from the previous case. This may have been influenced by the interior defect in the first case, or just by the natural dispersion of the characteristics of the material. The change in the tension branch of the fatigue loop in the second cycle should be noted. Compared with the first cycle, it is remarkably flatter and straighter. It can even be approximated by a linear function. It signals major damage to the material developed in the first loading cycle, which corresponds to the applied stress. This specimen exhibits rather high hysteresis. The compression component is steeper than the tension component. Some results were in good agreement with a previous work [8].

## 4. Conclusions

The methodology of tension tests using specimens inserted and glued in tubular fixtures is functional and has been verified. However, it is very demanding from the point of view of specimen preparation, and it requires very skilled staff. It can be recommended mainly for special small-series tests. However, the pilot tests were successfully completed, and the energies for crack initiation and propagation in the given sandstone have been determined.

The optical DIC method again proved its capacity and its suitability for measurements of a complex deformation field on porous and naturally well-structured surfaces.

The described methodology involving the use of the resonance loading frame to generate cycled cracks in stone was demonstrated. It enables the successful creation of well-defined and controlled crack damage in stone.

The computer X-ray micro-tomography is a useful tool for visualizing and measuring generated cracks in combination with the digital image correlation technique. The crack is usually close and to identify its depth the specimen must be loaded slightly in order to open the crack and make it more visible.

The fracture toughness of untreated notched and cracked specimens reached almost same values, which means that with such heterogeneous material we cannot expect stress concentration factors similar to those of metals. This magnitude will be good indicator for research on consolidation effects.

## Acknowledgement

The authors acknowledge kind support from Czech Science Foundation Project GAČR P105/12/G059.

## References

[1] Beran P and Drdácký M 2007 Influence of Temperature Changes on Stresses in the Triforium Tracery of St Vitus' Cathedral in Prague" *Proc. Computational Methods for Coupled Problems in Science and Engineering II Coupled Problems 2007* ed E Oñate, M Papadrakakis and B Schrefler pp 433-436

[2] Feng X, Zhang N, Zheng X and Pan D 2015 Strength restoration of cracked sandstone and coal under a uniaxial compression test and correlated damage source location based on acoustic emission *PlOS ONE* 10 (12): e0145757 (doi: 10.1371/journal.pone.0145757) p 20

[3] Ferestade I, Hosseini P and Heidary R 2017 Fracture toughness estimation of ballast stone used in Iranian railway *J. of Rock Mech. And Geotech. Eng. 9/2017* pp 892-899

[4] Vavřík D, Jandejsek I, Fíla T and Veselý V 2013 Radiographic observation and semi-analytical reconstruction of fracture process zone silicate composite specimen *Acta Technica CSAV* vol 58 No 3 pp 315–326

[5] Gross 1965 Stress-Intensity Factors for 3PB Specimens by Boundary Collocation *NASA TN D-3092*

[6] Lin Q and Labuz J F 2013 Fracture of sandstone characteristics by digital image correlation *Int. J. of Rock Mechanics & Mining Sci.* **60** pp 235–245

[7] Nath F, Salvati P, Mokhtari M, Seibi A and Hayatdavoudi A 2017, *doi.org/10.2118/187515-MS SPE Eastern Regional Meeting* (USA: Society of Petroleum Engineers) ID: SPE-187515-MS

[8] Čechová E, Drdácký M, Frankeová D, Lesák J, Slížková Z, Valach J, Vála O, Zeman A and Zíma P 2010 Research report of ITAM AS CR (in Czech) *Technology guidelines for restoration of the XIth arch of the Charles Bridge in Prague* (Prague: ITAM) p 139

# The assessment of mechanical properties of thermally stressed composite materials using radio frequency electrical field

**I Kusak[1], M Lunak[1], R Dvorak[1] and Z Chobola[1]**

[1] Institute of Physics, Faculty of Civil Engineering, Brno University of Technology, Veveří 95, 602 00 Brno, Czech Republic

Email: kusak.i@fce.vutbr.cz

**Abstract.** The main topic of this paper is the testing of cement-based composite materials that were stressed by high temperature. Impedance spectroscopy has been used as a non-destructive measurement method. The specimens were made from mortar mixture of CEM I type Portland cement, silica sand and fine gravel of various size classes for two sets of specimens. The sets, A and B, differed from each other also by the plasticizer used. The specimens were exposed to high temperatures in the range 23–1200°C. Studying the structural changes in the process of heat build-up is an efficient way to assess the reliability of whole construction. Using the impedance spectra the relative permittivity, electrical conductivity and other characteristics have been obtained. In accordance with Debye's theory of dielectrics the changes of impedance parameters have been put into context with structural changes. For higher range of frequencies, 100 MHz up to 3 GHz, R&S ZNC vector analyzer with DAK 12 coaxial probe manufactured by Speag has been used. In this frequency spectrum the electrical conductivity and relative permittivity as a function of frequency have been measured.

## 1. Introduction

If concrete is exposed to elevated temperatures, its mechanical and chemical properties, such as the allowable tensile and compressive stresses, the modulus of elasticity, the Poisson's ratio and some other quantities, will undergo a change. These changes depend on the temperature change rate, temperature extreme values, fire duration and concrete type [1, 2].

Thanks to its excellent resistance to compressive forces and to its high durability, concrete is widely used as the main structural material in advanced structures. However, concrete exhibits some shortcomings. One of them consists in its low toughness, or, in other words, its brittleness, due to which concrete structures may collapse suddenly or even in an explosive way. The seriousness of this problem arises with the concrete strength and the structure size. If the ultimate strength is exceeded, even locally only, a disaster may occur without previous warning in the form of a plastic deformation preceding the breakdown. High temperatures enhance the magnitude of this problem. There is a number of large concrete structures, such as TV towers, dams, bridges, nuclear power plant containments, which may be exposed to elevated temperatures, for example, due to fire, a nuclear disaster, or an act of terrorism. The elevated temperature exposure may be either shock-type (quick heating or cooling) or long-time. Concrete structures may be exposed to elevated temperatures in consequence of fire, explosions or chemical reactions. When concrete is heated to a certain temperature, C-S-H gel dehydration, thermal incompatibility between aggregates and cement grout, and porous pressure in the cement grout will take place, being the main harmful factors [3, 4, 5].

Thermal stress induced structural change monitoring is of high importance for the determination of the overall structural reliability.
The present paper focuses on the study and applicability of the impedance spectroscopy to the testing of cement-based composites of cement mortar, aggregates and quartz sand mixes, which have been realized in the framework of this research and were degraded while being exposed to elevated temperatures (200°C – 1 200°C).

## 2. Experiment setup

For our research we have prepared concrete specimens with dimensions 0.1 m × 0.1 m × 0.4 m.

Designed mixtures are at Table 1. Consistency 550 mm flow a cone. The specimens were soaked in water for 28 days and then dried at first at the laboratory temperature and then for 48 hours in a ceramic furnace at a temperature of 110°C. The concrete specimens were heated in Rhode KE 130B programmable laboratory furnace at the heating rate of 5°C/min. Selected temperatures T = 200°C, 400°C, 600°C, 800°C, 1 000°C and 1 200°C were maintained for 60 minutes.

The principal difference in the composition of B-type specimen consists in the use of a lower grading, namely, 11/22, and, at the same time, in the plasticizer used.

For higher range of frequencies, 100 MHz up to 3 GHz, the vector analyzer R&S ZNC has been used with the connected coaxial probe DAK 12 manufactured by Speag. In this frequency spectrum the electrical conductivity and relative permittivity as a function of frequency has been measured.

**Table 1.** Designed mixtures A and B.

| Compounds | Amount of each compound for 1 $m^3$ [kg] | |
|---|---|---|
| | Mixture A | Mixture B |
| Cement CEM I 42.5 R | 345 | 345 |
| Fine aggregate Žabčice 0/4 mm | 848 | 896 |
| Coarse aggregate Olbramovice 8/16 mm | 980 | 521 |
| Coarse aggregate Olbramovice 11/22 mm | - | 391 |
| Superplastizer Sica Viscocrete 2030 | 2.8 | 2.5 |
| Mix water | 160 | 173 |

The compressive strength was measured in compliance with EN 12390-3:2009 Testing hardened concrete - Part 3: Compressive strength of test specimens. The compressive strength test was carried out on broken ends of the beam, on which a flexural tensile strength test had been previously performed. Steel fixtures for the upper and lower pressure application areas were used to ensure an accurately defined area of 100 × 100 mm. The compressive force was applied perpendicularly to the compaction direction.

The tensile strength was measured in compliance with EN 12390-5:2009 Testing hardened concrete testing - Part 5: Flexural tensile strength of test specimens. The tensile strength test was carried out on beams by means of a four-point bending. The bending force was applied perpendicularly to the compaction direction of the test specimens [6, 7].

## 3. Results

The electric conductivity of A and B-formula specimens was measured by means of a vector analyser in the frequency range from 100 MHz to 3 GHz. The readings ranged from $10^{-6}$ S/m to 0.15 S/m. Growth trends were established, however, they were not uniformly distributed over all of the formulas and frequencies. Slopes of the lines were calculated, however, they were not uniformly distributed over all of the formulas and frequencies.

For the A-formula specimen set Figure 1, a steep growth takes place up to the frequency of 500 MHz. Subsequently, a linear rise follows up to 2 350 MHz, where a steeper increase sets on again. The readings are oscillating around the level, which shows the linear trend with the mentioned nickpoint.

The highest electrical conductivity values were found on the specimen which has not been loaded thermally. Lower values correspond to the specimen which was dried up at 200°C. Its conductivity gets close to that of the 20°C specimen at the highest frequency. A noticeable decrease in value occurred at 400, 600, 800°C, where the differences in conductivity values narrow. A further drop was shown – as expected – by the specimens, which had been heated to 1 000°C and 1 200°C. Their conductivities differed from each other at 2 356 MHz only. The above mentioned frequency value is approximate only and cannot be specified accurately. It is rather a region.

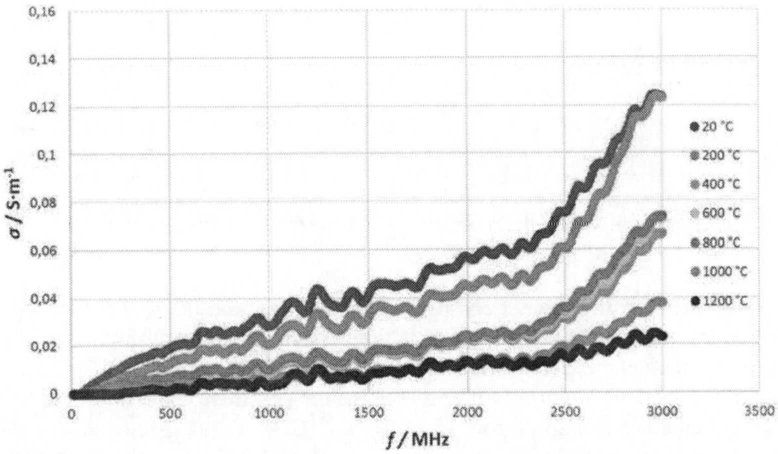


**Figure 1.** Electrical conductivity of B-formula concrete specimens, as measured by ZNC vector analyser.

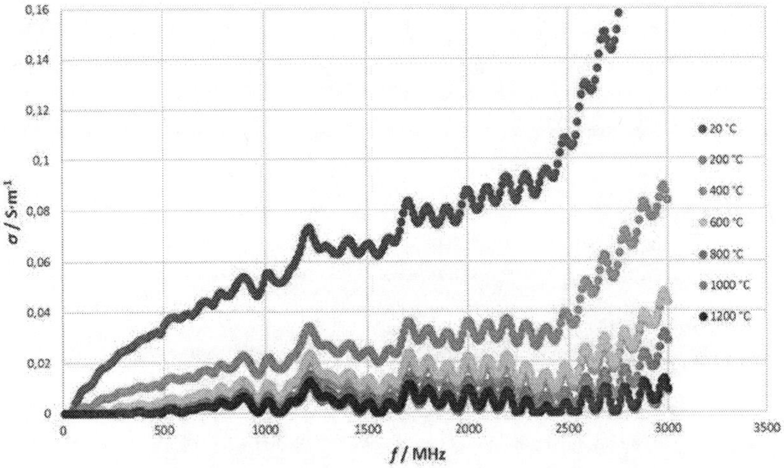


**Figure 2.** Electrical conductivity of B-formula concrete specimens, as measured by ZNC vector analyzer.

The conductivity plots of the B formula specimens Figure 2 are similar in shape. We also observe the rapid change of slope, however at a lower frequency than above – at about 2 450 MHz. The highest conductivity values are reached by no-thermal-stress subjected specimens. However, the conductivities of other specimens are below those of the A-formula specimens. The B-formula specimens contain higher grading aggregates (11/22 Olbramovice), the same amount of cement but a lower amount of water. Due to the thermal stress, water is released on a bigger scale, because of the lower binder amount and, consequently, lower number of conducting paths from the C-S-H gel. The latter disintegrates in a larger extent, too. At higher temperatures, the conductivity change is caused by the concrete structure transformation to the vitreous phase.

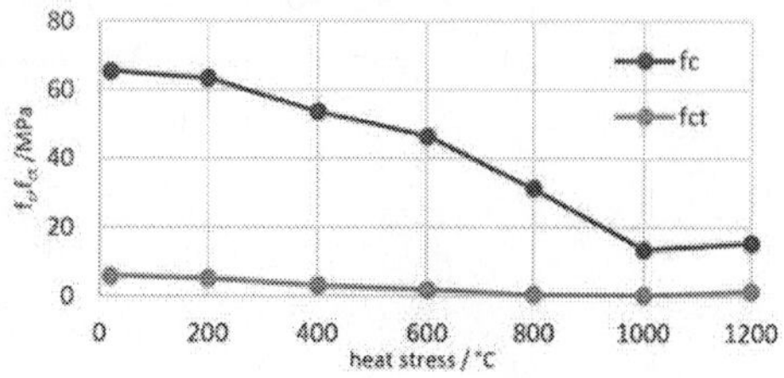


**Figure 3.** Compressive and flexural tensile strengths for A-formula specimens under stress.

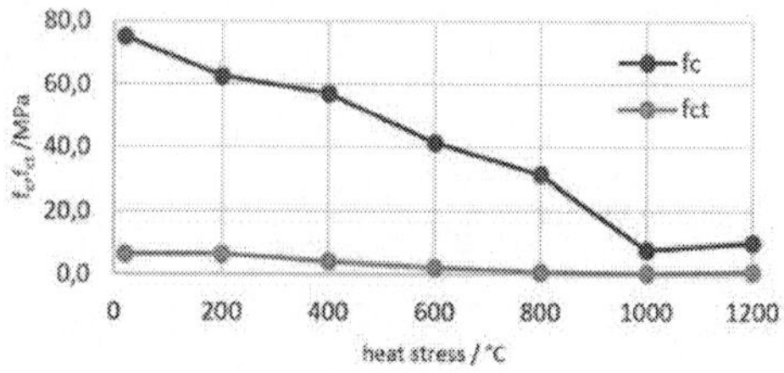


**Figure 4.** Compressive and flexural tensile strengths for B-formula specimens under stress.

The B type specimens exhibit higher compressive strength $f_c$ and flexural tensile strength $f_{ct}$ values for the mentioned thermal stress variants. The B-type specimens contain higher grading aggregates. Lower conductivities than at the A-formula are only observed at 1 000°C and 1 200°C.

Lower permittivity values Figure 5 of the A-formula specimens are observed in the dielectric constant spectra at frequencies ranging from 100 MHz to 3 GHz. These spectra show a decreasing trend up to 1 GHz to oscillate around a single value. They show a decline in the vicinity of 3 GHz. The higher degree of firing results in a lower value of the permittivity, which is the same as with the B-formula specimens, where, however, the values corresponding to 400°C and 600°C have been unexpectedly interchanged. At 1 000°C and 1 200°C, the permittivity values have been interchanged, too. Only these, last-mentioned changes and their trend are in accordance with the increase of $f_c$, $f_{ct}$ for 1 200°C as against 1 000°C.

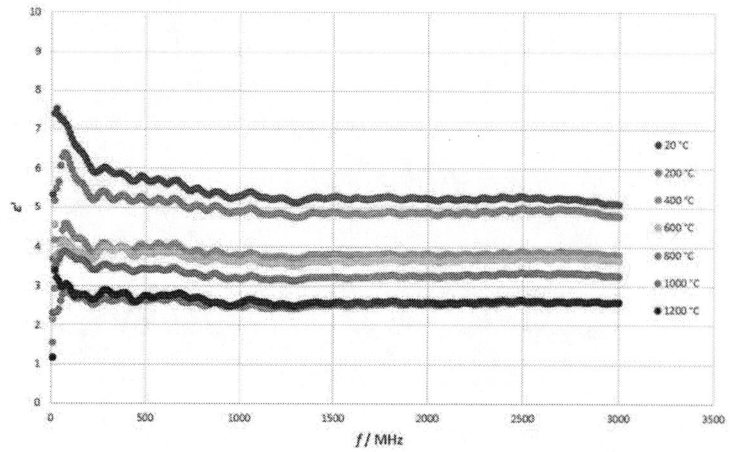


**Figure 5.** Relative permittivity of A-formula specimens under stress.

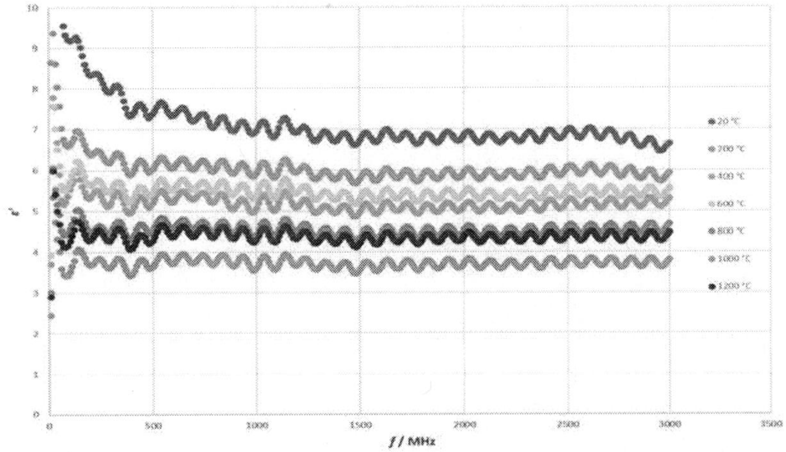


**Figure 6**. Relative permittivity of B-formula specimens under stress.

When comparing the spectra of Figures 5 and 6, we observe a larger scattering (apparently waviness) of the measured points for B-formula specimens, which contain coarser aggregates. The waviness of the spectrum shape for the B-formula specimens is due to the behaviour of the granite grade, which is present in the higher grading broken stone from Olbramovice. The granite grade used has a higher content of polarizable particles, whereas the cementing compound contributes to the conductivity.

## 4. Conclusion

In the present paper, two sets of concrete beams, which had been subjected to thermal stresses of various intensities, were successfully tested by means of a vector analyzer in the frequency range from 100 MHz to 3 GHz. This unique measurement method is non-destructive. In spite of that, the DAK contact probe measurements provided us with conductivity and permittivity spectra.

The conductivity and permittivity values decreased when the annealing degree increased, which is in agreement with the concrete structure metamorphosis into a vitreous phase. For comparison, spectra showing the decreasing concrete strength versus the growing annealing degree were presented. Downward trends were compared with a slight increase resulting from the specimens' transition to the vitreous phase.

In this experiment, the analyzer, being equipped with a Speag probe, proved to be a fast-operating, non-destructive tool, which has been underutilized in the NDT practice thus far. Monitoring of impedance parameters being in correlation with material changes is very important for evaluation of structural reliability.

## Acknowledgments

This paper has been worked out under the project GAČR No.16-02261S supported by Czech Science Foundation and project No. LO1408 "AdMaS UP - Advanced Materials, Structures and Technologies", supported by Ministry of Education, Youth and Sports under the „National Sustainability Programme I".

## References

[1] Monica J H and Harris A T 2013 *Progress in Materials Science* **58** pp 1056–1102
[2] Saafi M, Andrew K, Mcghon D, Taylor S, Rahman M, Yang S, Zhou X and Tang P L 2013 *Construction and Building Materials* **49** pp 46–55
[3] Lhoták P 2014 http://www.uochb.cas.cz/Zpravy/PostGrad2004/7_Lhotak.pdf
[4] Matsutani S, Shimosako Y and Wang Y 2012 *Physica A* **391** 5802
[5] Fernandez-Jimenez A 1999 *Cement Concrete Res.* **29** 1313

[6] Kusak I, Lunak M and Schauer P 2013 *Appl. Mech. Mater*. **248** 370
[7] Tupy M, Sotiriadis K, Kusak I, Lunak M, Stefkova D and Petranek V 2015 *J. of Mater in Civil. Eng.* **28** 04015171

IOP Conf. Series: Materials Science and Engineering **379** (2018) 012031 doi:10.1088/1757-899X/379/1/012031

# Release of ammonia from conventional power plant fly ash after the introduction of SNCR process

**D Beranová[1], T Opravil[1], P Ptáček[1] and R Snop[2]**

[1] Brno university of Technology, Faculty of Chemistry, Materials Research Centre, Purkyňova 464/118, Brno 612 00, Czech Republic

[2] ČEZ Energetické produkty, s.r.o., Komenského 534, Hostivice 253 01, Czech Republic

E-mail: xcberanovad@fch.vut.cz

**Abstract.** The paper deals with a study of various factors affecting the release of ammonia from conventional power plant fly ash after introduction of SNCR (Selective non-catalytic reduction). SNCR is a method of flue gas denitrification based on the chemical reduction of nitrogen oxide into nitrogen and water vapour. Reduction has many by-products, various ammonia salts, which could contaminate fly ash and deteriorate its properties and application. The chemical and phase composition of fly ash was determined by X-Ray Fluorescence and X-Ray Diffraction. To predict the contaminated fly ash behaviour, the samples of fly ash with known ammonia content was prepared by mixing fly ash with exact amount of ammonia hydrogensulfate. Tests were performed on the samples prepared from contaminated fly ash as it was, contaminated fly ash with deionized water and as a mixture of contaminated fly ash, lime, gypsum and water (further stabilizate). Effects of various storage and mixing conditions on prepared samples properties were studied.

## 1. Introduction

As the main source of energy in thermal power plants is used coal. During the processes, the solid residues as fly ash and slag are produced. Coal combustion fly ash, energetic by-product, is commonly used in many branches i.e. for coal mines recultivation, terrain work or in construction industry [1].

Coal combustion produces emissions, carbon, sulfur and nitrogen oxides [2]. There are some of the processes, which can decrease the amount of emitted sulfur and nitrogen oxides. To reduce sulfur oxides emissions, the wet flue gas desulfurization is most commonly used [3]. Reduction methods as SCR (Selective catalytic reduction) and SNCR (Selective non-catalytic reduction) are the main used methods for decrease $NO_x$ emissions. The reactions between nitrogen oxides and ammonium in the high temperature have been overall studied [4, 5]. Products of these reductions are nitrogen and water vapour. SCR process takes a place at the temperature 80–420°C and with the presence of a catalyst. SNCR process takes place at the higher temperatures, 800–900°C and without the presence of the catalyst [2]. The application of these methods in manufacturing technology could bring some technical problems. It is known, that a small amount of injected ammonia doesn't react with $NO_x$ and results in ammonia emissions, so called ammonia slip. It typically represents less than 1% of injected ammonia, but it reacts with sulfur to form ammonium sulfate or bisulfate [6], which probable remains into a surface of fly ash or downstream equipment [7, 8, 9, 10].

Amount of ammonium salts remained on the fly ash surface depend on $SO_3$ content, fly ash sulfur content, an alkalinity of the fly ash, the ammonia concentration, and the ash loading in the flue gas [11].

The paper studied the properties and utilization of fly ash coming from plant where the SNCR technology is used and where the presence of unfavorable by-products is predictable.

## 2. Experimental

### *2.1. Methods*

The amount of ammonia in samples was determined by a spectrophotometric method with Nessler's reagent. The first step was an extraction of ammonia to the soluble form which can be analyzed. Samples was mixed with deionized water with the ratio - 1/10 (solid/water) and the leaching procedure took place for 2 hours. Solid particles were filtered by Buchner funnel. A solution for measurement was prepared by mixing of 2.89 ml sample solution with 0.01 ml 5% solution of potassium-sodium tartrate and 0.1 ml Nessler's reagent. Absorption was measured after 10 minutes of mixing. Spectrometer HACH-LANGE DR 2800 and 1 cm cuvettes were used. The absorption wavelength was 425 nm. The concentration of ammonia was determined on the basis of the calibration primer.

### *2.2. Materials*

Coal combustion fly ash which doesn't contain SNCR by-products (ammonia concentration lower than 5%), industrial gypsum ($CaSO_4 \cdot 2H_2O$), lime ($CaO$) and ammonium bisulfate ($NH_4HSO_4$) were used for preparing samples.

The chemical composition of used fly ash was examined by X-Ray fluorescence on the device Xenemetric EX-6600 SSD and is given in Table 1. The phase composition of used fly ash was examined by X-Ray diffraction on the device Empyrean (Panalytical) and is given in Figure 1.

**Table 1.** The chemical composition of fly ash (major and minor oxides) [%].

| $SiO_2$ | $Al_2O_3$ | $CaO$ | $Na_2O$ | $K_2O$ | $MgO$ | $SO_3$ | $Fe_2O_3$ | $TiO_2$ | $P_2O_5$ | | | | |
|---|---|---|---|---|---|---|---|---|---|---|---|---|---|
| 47.3 | 29.8 | 4.31 | 0.88 | 1.51 | 1.29 | 1.06 | 19.2 | 1.43 | 0.27 | | | | |
| $Cr_2O_3$ | $MnO$ | $V_2O_5$ | $CoO$ | $NiO$ | $CuO$ | $ZnO$ | $As_2O_3$ | $Rb_2O_3$ | $ZrO_2$ | $SrO$ | $BaO$ | $CeO_2$ | $Tb_4O_7$ |
| 0.02 | 0.18 | 0.07 | 0.02 | 0.02 | 0.03 | 0.07 | 0.08 | 0.02 | 0.03 | 0.04 | 0.10 | 0.04 | 0.02 |

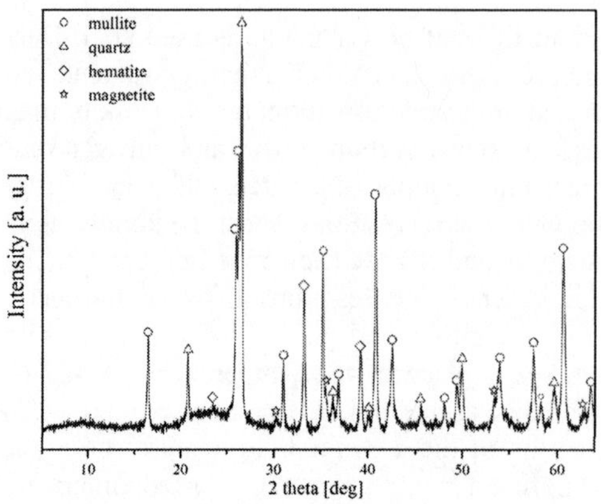


**Figure 1.** The phase composition of fly ash.

### *2.3. Tested samples*

The samples were prepared from coal combustion fly ash doped by ammonium bisulfate. Fly ash, was mixed with small amount of water and ammonium bisulfate solution. The quantity of $NH_4HSO_4$ was chosen to correspond 50, 100, 150 and 200 ppm of $NH_3$ (ppm $NH_3$ = mg $NH_3$ per kg of fly ash). The wet doped fly ash was dried in a dryer at 60°C to constant weight. The real ammonia concentration of dried samples was determined by spectrophometric method with Nessler's reagent.

In Table 2 the experiments are summarized. The three types of mixtures were tested – FA, FAW and FALG. All components (see Table 2) were put in a laboratory mixer and mixed for three minutes. Prepared samples were stored in containers without cover under the dry laboratory conditions (temperature 24°C, humidity 30%).

The effect of a mixing time and a velocity on ammonia release was examined on FAW samples. The samples were mixed for 0.5, 1, 2, 3, 5 and 10 minutes at the velocity of 58, 74, 138 and 220 rpm. The effect of the mixing time was also examined on FALG samples. First one was mixed in the laboratory mixer for 3 minutes, and then was stayed in container without cover under the dry laboratory conditions for 57 minutes. The second one was mixed for 60 minutes. The mixing velocity for both samples was 74 rpm.

The fly ash pre-heated on the temperature of 60, 80, 100 and 160°C was also used for preparing FALG samples. The fly ash was maintained at the temperature for 2 hours in the dryer and then immediately mixed with other components.

The laboratory depot simulation was performed in big container. A big container with regularly spaced small holes in the bottom was used for simulating depot in a laboratory. The container was filled with 50 kg of FAW in a layer about 40 cm. It was put in another container, where was captured water. Every 24 hours, 10 l of demineralized water was poured onto the FAW. Immediately before pouring, samples from three different levels were collected.

**Table 2.** Samples contain.

| | FA | FAW | FALG |
|---|---|---|---|
| Fly ash [wt.%] | 100 | 60–85 | 60–85 |
| Lime [wt.%] | × | × | 1-10 |
| Industrial gypsum [wt.%] | × | × | 1-10 |
| Water [wt.%] | × | 15–25 | 15–25 |

## 3. Results and discussion

The release of ammonia was studied on samples FA and FAW during six months. Time dependences of ammonia content are presented in Figure 2 and 3. The number of samples corresponds to the quantity of ammonia donated to fly ash by $NH_4HSO_4$.

The concentration of ammonia in all samples gradually decreased with the time. After six months, the ammonia concentration was about 65% of the starting concentration. The storage of coal combustion fly ash contaminated by SNCR by-products and its using for re-cultivating purposes should have an air-polluting effect with ammonia or some ammonium compounds.

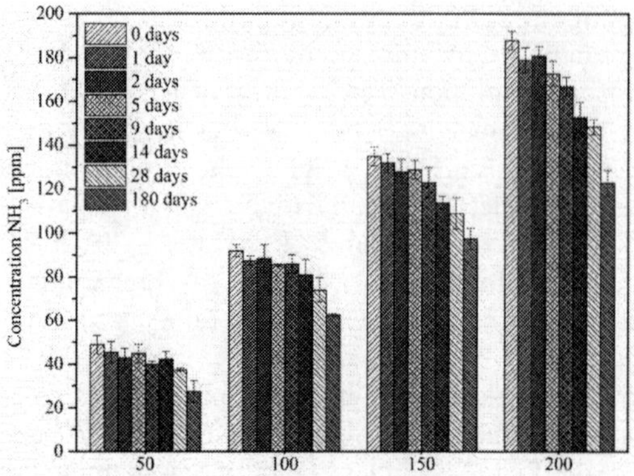


**Figure 2.** The time dependence of ammonia content in FA.

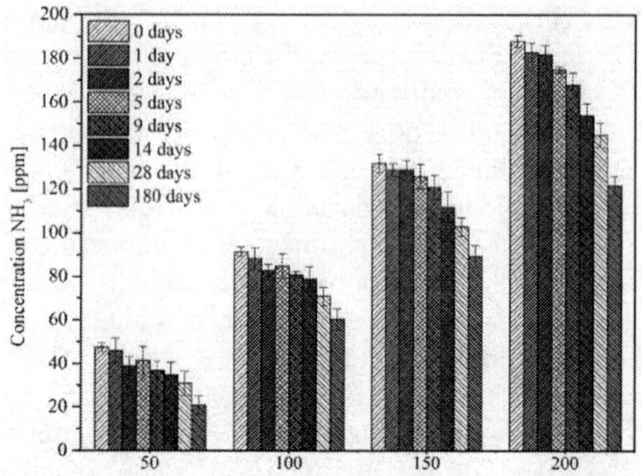


**Figure 3.** The time dependence of ammonia content in FAW.

The coal combustion fly ash cannot be used for storage and re-cultivation as it is, but it has to be modified. There are several different ways, how can be fly ash utilized and further used for recultivation purposes. Contaminated fly ash can be mixed with water to reduction dustiness and improve handling. This form can be used for repositories. Samples FAW simulate this way of utilization. Further samples called FALG simulate the second way of possible utilization. The way is to mix the fly ash with lime and industrial gypsum and produce a stabilized compound, which is suitable for terrain work. The process of mixing is affected by SNCR by-products. The release of ammonia during mixing process was studied and the results are presented in Figure 4 and 5.

The concentration of ammonia in samples FAW was during mixing approximately same (Figure 4) during 10 minutes. During experiments the applied mixing velocity was also studied (Figure 4) and it was observe that the increasing or decreasing velocity have not a significant effect on ammonia release. So the velocity of 74 rpm was used as optimal for experimental procedure.

Addition of lime to fly ash in aqueous conditions causes the increase of pH. Ammonium salts react in aqueous solution with high pH to create gas ammonium, which releases from the sample in very short time. Shou et al. [12] also study releases of ammonia from concrete composites. According to their results, the highest concentration of ammonia in the air above the concrete surface was after 20 minutes (around 300 ppm) from starting of mixing of batch components. They measured the concentration of ammonia in gaseous form in situ during mixing and after above the molded samples. Their results showed slightly different behavior that can be partially explained by measuring the ammonia in gaseous form instead of measuring the ammonia converted to soluble salt, as is our work. The formation of soluble salt should be affected with the high pH of used mixing solution, so the part of presented ammonia could evaporate immediately after mixing the components together. Therefore, the release measured finally in the solution could be lower.

The mixing time of FALG samples has a significant effect on the ammonia release (Figure 5). About 50% of ammonia releases in the first 60 seconds of mixing. Fast release of ammonia continues after mixing, too. Dark columns on Figure 5 presented the mixture, which was 3 minutes mixed and then stayed. The mixture could get rid of ammonia almost completely by mixing with prolonged time (Figure 5, light columns). After 60 minutes of mixing, the residual amount of ammonia was lower than 10 ppm. The resulting amount of ammonia in the mixture, which was mixed for 60 minutes, was about 15 times lower than in the mixture, which was mixed just 3 minutes.

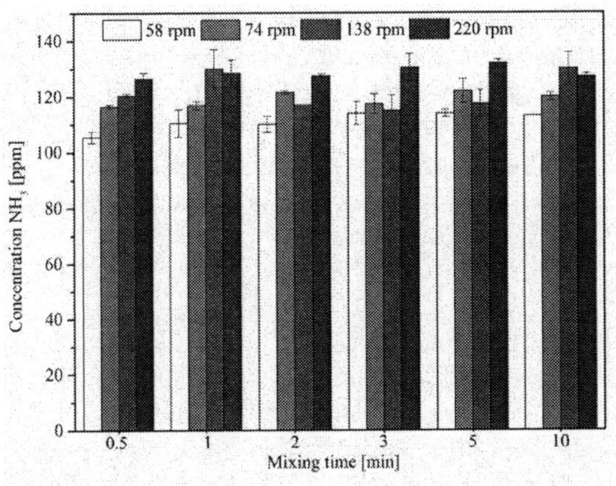


**Figure 4.** The mixing time dependence of ammonia content in FAW.

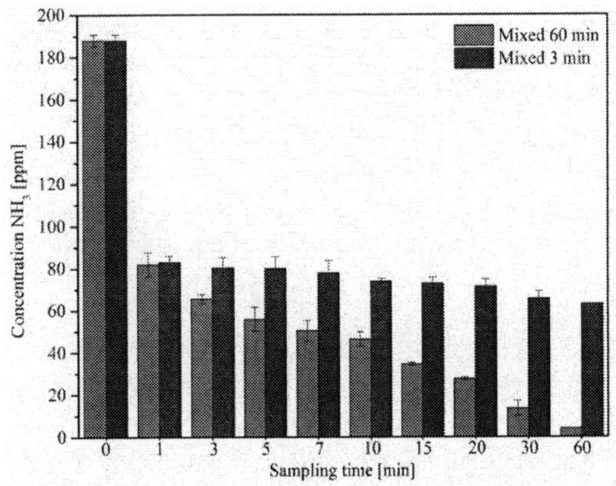


**Figure 5.** The mixing time dependence of ammonia content in FALG.

The fly ash is about 160°C hot when it leaves the boiler. So for the next experiment, the temperature treated fly ash was used for preparing FALG samples. The results are presented in Figure 6.

The high temperature of the fly ash slightly increases the release of ammonia. Immediately after mixing were concentrations of ammonia in samples about three times lower and after 24 hours were in all samples less than 30 ppm. It is a really low concentration which isn't healthy injurious. This ammonia concentration was reduced in next 28 days to value about 15 ppm. Maximum safe level of ammonia in the air is 25 ppm for long-term exposure (8 h) and 35 ppm for short-term exposure (15 min) [13]. So the determined values meet the requirements required for the product application.

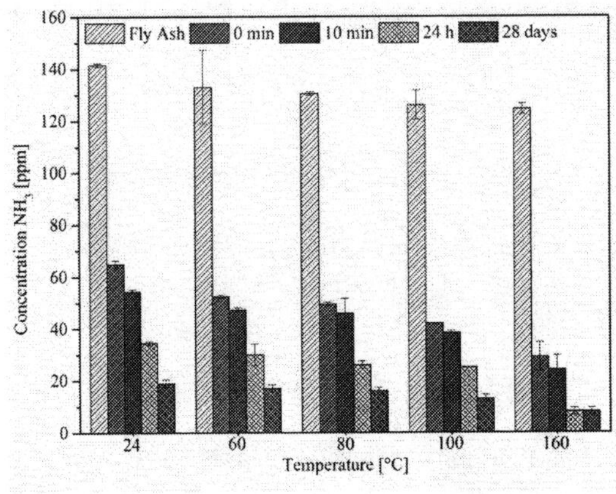


**Figure 6.** The temperature dependence of ammonia content in FALG.

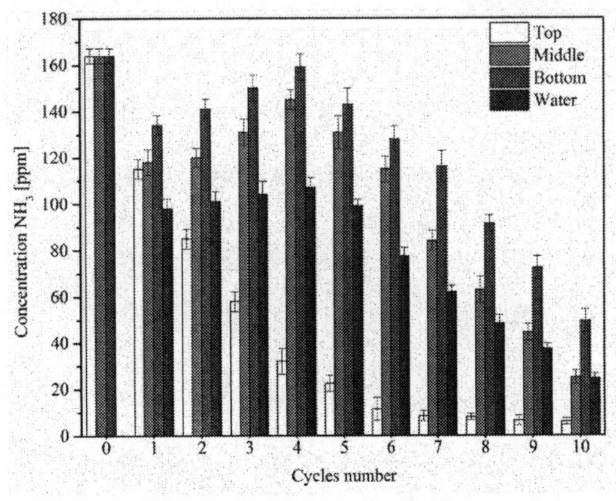


**Figure 7.** Depot simulation.

The results simulated the depot storage under the laboratory conditions are presented in Figure 7. Ammonium bisulfate is well soluble in water so it washes out from FAW samples. During first four cycles, the concentration in the middle and bottom decreased, because of washing out ammonium salts from the top level. The concentration of ammonia in added water, which has flowed during the FAW layer, was in the first five cycles very similar. When the concentration of ammonia in the middle and bottom level of FAW started to decrease the concentration of ammonia in flowed water decrease too. After 10 cycles, the concentration of ammonia in the top and middle levels of FAW and flowed water was lower than 30 ppm. But in the bottom level was still about 50 ppm. In the top level was less than

20 ppm of ammonia after 4 cycles. At the end of the experiment, the concentration in the top level was about 5 ppm. Similar leaching of ammonia salts studied Wang et al. in [8]. Results of their experiments confirmed, that 94% of ammonium bisulfate contained in fly ash released from mixture during first 110 minutes. This experiment show, that using coal combustion fly ash contaminated by SNCR by-products for restoring of the landscape after mining or other terrain adjustments poses a risk to the environment. The decrease of ammonia content in surface water and groundwater can have a negative impact on the life of aquatic organisms and on the quality of drinking water [14].

## 4. Conclusion

The presented work was focused on the release of ammonia from coal combustion fly ash contained SNCR by-products. The mentioned by-products could be various ammonium salts, most represented by ammonium sulfate and bisulfate. The salts probably aren't strongly bonded to the fly ash particles, because of ammonia content in samples storage without cover decrease with time. Such fly ash cannot be used for terrain work, because of the high solubility of ammonium salts in aqueous conditions, what has a negative impact on the environment. In the alkaline environment, ammonium releases in very short time. The mixing of samples with alkaline pH solutions has a significant effect on ammonia release. It was determined that after 60 min of mixing, the residual ammonia concentration is lower than 5% of the starting one. The high concentration of ammonia in the air has a negative impact on human health and industrial equipment. These are the reasons, why fly ash contaminated by ammonium salts cannot be used for concrete production.

## Acknowledgement

This work was supported by the project "Materials Research Centre at FCH BUT - Sustainability and Development, REG LO1211, with financial support from National Programme for Sustainability I (Ministry of Education, Youth and Sports)".

## References

[1] Ahmaruzzaman M 2010 *Prog. Energy Combust. Sci.* **36** pp 327–363
[2] Lee G-W, Shon B-H, Yoo J-G, Jung J-H and Oh K-J 2008 *J. Ind. Eng. Chem.* **14** pp 457–67
[3] Dou B, Pan W, Jin Q, Wang W and Li Y 2009 *Energy Convers. Manage.* **50** pp 2547–53
[4] Kang Z, Yuan Q, Zhao L, Dai YK, Sun B and Wang T 2017 *Appl. Therm. Eng.* **123** pp 635–45
[5] Van Caneghem J, De Greef J, Block C and Vandecasteele C 2016 *J. Cleaner Prod.* **112** pp 4452–60
[6] Rubel A M 2002 *Prepr. Pap. - Am. Chem. Soc., Div. Fuel Chem.* **47** pp 834–35
[7] Larrimore L 2002 *Prepr. Pap. - Am. Chem. Soc., Div. Fuel Chem.* **47** pp 832–33
[8] Wang H, Ban H, Golden D and Ladwig K 2002 *Prepr. Pap. - Am. Chem. Soc., Div. Fuel Chem.* **47** pp 836–38
[9] Mussatti D C, Srivastava R, Hemmer P M and Strait R 2002 *The EPA Air Pollution Control Cost Manual* (North Carolina: Office of Air Quality Planning and Standards)
[10] Liang X, Zhong Z, Jin B, Chen X, Li W, Wei H and Guo H. 2010 *Korean J. Chem. Eng.* **27** pp 1483–91
[11] Bittner J, Gasiorowski S and Hrach F 2001 *Int. Ash Util. Symp.* (University of Kentucky)
[12] Shou L, Hayes J, Cheng W, Wu C-Y, Townsend T, Vinson T and Schert J 2014 *Air Qual. Atmos. Health* **7** pp 505–13
[13] 2011 *EH40/2005 Workplace exposure limits* (Health and Safety Executive)
[14] Roessler J, Cheng W, Hayes J B, Shou L, Ferraro C, Wu C-Y, Bergin M and Townsend T G 2016 Fuel **184** pp 613–19

IOP Conf. Series: Materials Science and Engineering 379 (2018) 012032 doi:10.1088/1757-899X/379/1/012032

# Possibilities of elimination of ammonia slip from technological water in power plants

**J Hajzler[1], T Opravil[1], J Pořízka[1] and P Ptáček[1]**

[1] Brno University of Technology, Faculty of Chemistry, Materials Research Centre, Purkyňova 464/118, Brno Cz-612 00, Czech Republic

Email: xchajzlerj@fch.vutbr.cz

**Abstract**. The purpose of this study is to find new possibilities of elimination or separation of ammonium compounds from technological water produced by powerplants and incinerators, in which the secondary flue gas denitrification methods have been implemented (SCR, SNCR). This paper seeks for a new method of ammonium slip elimination from technological water due to the absence of industrially applicable method and thus fixation of ammonium ions in composite cement eventually in binder containing fly ash. The experiments confirmed the validity of the complex compound utility in given issue. The new method consists in adding a soluble cobalt compound to get an insoluble cobalt complex, the processing conditions were optimized during laboratory testing.

## 1. Introduction

$NO_x$, which is contained in flue gas, is composed of NO (90%) and $NO_2$ (5%). Both gases occur during the combustion process. Primary, relatively harmless, colorless non-drying NO (90–95%) is formed, which reacts with oxygen to form brown, irritating $NO_2$. The remaining nitrogen oxides are present in very small amounts (5%) and often transform between NO and $NO_2$ [1]. $NO_2$ is acid gas that can further react in the atmosphere with OH-radicals to form nitric acid. $NO_x$ contributes to acid rain and eutrophication. To protect human health from the negative effects of $NO_x$ emission limit values (ELV) exist. For waste incineration plants the ELV for $NO_x$ is 200 mg/Nm$^3$. Primary measures usually do not suffice to reach this ELV, therefore secondary measures must be introduced. The most often used method is selective non catalytic reduction (SNCR). This method is relatively simple and cost-efficient [2, 3]. The principle of the method is the thermal conversion of nitrogen oxides by injecting a reducing reagent into the flue gas stream at an appropriate temperature range, which is known as temperature window [4, 5]. Reducing reagent selectively reacts with nitrogen oxides and converts them into harmless molecular nitrogen and water vapour at relatively high temperatures (850–1100°C) at 0 % $O_2$ concentration [6, 7]. Ammonia, urea and cyanuric acid are used as the reagent. Under stoichiometric conditions, the chemical process can be written according to equations (1), (2) and (3) [2, 6].

$$NH_2CONH_2 + H_2O \rightarrow 2\ NH_3 + CO_2 \quad (1)$$

$$4\ NH_3 + 4\ NO + O_2 \rightarrow 4\ N_2 + 6\ H_2O \quad (2)$$

$$4\ NH_3 + 2\ NO_2 + O_2 \rightarrow 3\ N_2 + 6\ H_2O \quad (3)$$

If the flue gas temperature is too low, $NO_x$ conversion decreases, the reaction rates become very slow and ammonia slip occurs [6]. If the flue gas temperature is above 1100°C, the reaction of $NH_3$ oxidation becomes predominant (4) [8].

$$4\,NH_3 + 5\,O_2 \rightarrow 4\,NO + 6\,H_2O \tag{4}$$

The fate of reducing reagent (ammonia) is shown in Figure 1.

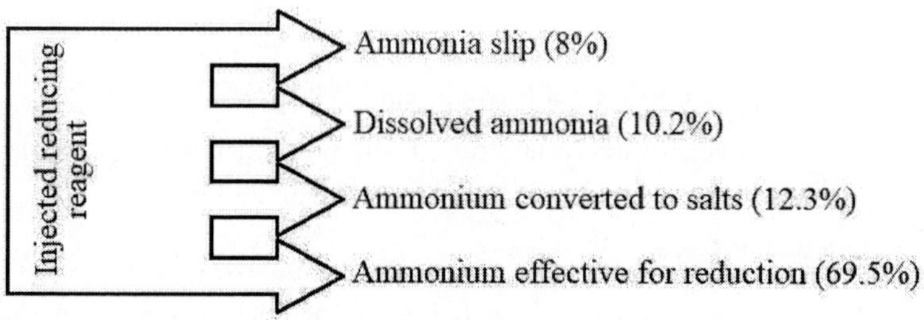


**Figure 1.** The fate of ammonia.

The SNCR method is usually sufficient to meet the requirement of EVL of 200 mg/Nm$^3$. If, however, the ELV is reduced to 100 mg/Nm$^3$, as it is already the case in some regions in Germany [9], the SNCR method might not be suitable to meet this ELV, because the removal efficiency is about 50%. In that case, the method of selective catalytic reduction (SCR) is an interesting alternative, because the removal efficiency is up to 90% and it would be the tail end solution of the problem [2, 10]. In SCR the reduction reagent reacts with nitrogen oxides on a $TiO_2/V_2O_5/WO_3$ catalyst surface, at the temperature range between 220–350°C [1]. However, the installation, operation and maintenance costs are noneffective. The installation of SNCR system makes just 20–35% of the cost of the SCR system [11]. But there is another obstacle regarding the environment. Using the SCR method instead of the SNCR method direct environmental impact (reducing $NO_x$) in the impact categories acidification or eutrophication is reduced. Unfortunately SCR involves higher indirect environmental impacts than SNCR, mainly due to the need to reheat the combustion gas burning fuel oil in already existing facilities with SNCR. When fuel oil is used as an energy source to reheat the flue gas, these indirect impacts are higher than the impacts directly avoided by lower $NO_x$ emissions. Those indirect impacts have an influence on global warming. From that reason, the optimization of SNCR should be preferred to installing the tail end SCR in existing installations. For new installations, where the space to locate the catalyst after the boiler is available, avoiding the need to reheat the combustion gas could be an alternative for SNCR [2].

Even if the temperature window is kept in the SNCR system, the formation of ammonia slip occurs, which is the result of incomplete reaction of $NO_x$ with the reduction reagent. Ammonia slip results in several problems, including ammonium hydrogen sulphate or another ammonia salts formation, which can further cause the corrosion of facility walls. The absorption of ammonia by fly ash may prevent its further use. Apart from ammonium hydrogen sulphate also huge amount of another soluble ammonium salts, such as sulphates, nitrates, chlorides, are contained in ammonia slip [12].

The problem is that almost all ammonium salts are well soluble in water, but an insoluble complex compound ammonium hexanitrocobaltate was found. This ammonium complex is formed very quickly when ammonium ions react with soluble sodium hexanitrocobaltate, which is yellow brown crystal well soluble in water and insoluble in alcohols. The water solution is used in analytical chemistry and is known as Fischer salts [13].

The aim of this study is to find a new possibility of elimination of ammonia residues with complex compounds from technological water in the facilities, where high temperature combustion takes place and where secondary denitrification flue gas was introduced. The study focuses on finding a new method, as there is currently no industrially used method, which would solve this problem.

## 2. Experimental

### *2.1. Materials*

The Portland clinker was ground to 380 $m^2/kg$ (plant Mokrá, CZ). In the experiment, ash (power plant Počerady) was doped with a solution of ammonium hydrogen sulphate (Carl Roth, ≥98,5%, p.a.) in the amount of 30, 50 and 200 ppm and then dried. Sodium hexanitrocobaltate was prepared by the precipitation [14] at Faculty of Chemistry. Gypsum water is water after SNCR. (power plant Počerady).

### *2.2. Sample preparation*

For experiments the agglomerate (mixture of ash and water - transport to the compost) and the stabilizer (ash mixture of lime and energy gypsum) were chosen. According to ČSN EN 197-1, CEM II / B-V paste and mortar were made. The composition of mixtures was designed according to ČSN EN 196-1. The size of beams was 20 × 20 × 100 mm. The composition of each mixture is shown in Table 1. The aqueous extracts were carried out according to ČSN EN 12457-4. Prepared and weighed beams from mixtures after 1, 4 and 7 days were placed in the bottle and then poured with distilled water at the weight ratio 1:10. The bottle was placed in a head-heel shaker for 24 h. The particles were allowed to sediment for 15 minutes. Subsequent filtration was carried out under reduced pressure on a 0.45 μm porous filter.

**Table 1.** The composition of mixtures.

| Mixture | Clinker [g] | Ash ppm [g] | | | Sand [g] | FGD gypsum [g] | CaO [g] | $H_2O$ [g] | $Na_3[Co(NO_2)_6]$ [mg] |
|---|---|---|---|---|---|---|---|---|---|
| | | 30 | 50 | 200 | | | | | |
| A | 0 | 70 | 0 | 0 | 0 | 5 | 3 | 22 | 300 |
| B | 0 | 78 | 0 | 0 | 0 | 0 | 0 | 22 | 300 |
| C | 0 | 78 | 0 | 0 | 0 | 0 | 0 | 22 | 0 |
| D | 455 | 245 | 0 | 0 | 0 | 0 | 0 | 269.0 | 55 |
| E | 455 | 0 | 245 | 0 | 0 | 0 | 0 | 261.7 | 91.5 |
| F | 455 | 0 | 0 | 245 | 0 | 0 | 0 | 206.9 | 365.5 |
| G | 455 | 245 | 0 | 0 | 2 100 | 0 | 0 | 269.0 | 55 |
| H | 455 | 0 | 245 | 0 | 2 100 | 0 | 0 | 261.7 | 91.5 |
| I | 455 | 0 | 0 | 245 | 2 100 | 0 | 0 | 206.9 | 365.5 |
| J | 455 | 245 | 0 | 0 | 0 | 0 | 0 | 280 | 0 |
| K | 455 | 0 | 245 | 0 | 0 | 0 | 0 | 280 | 0 |
| L | 455 | 0 | 0 | 245 | 0 | 0 | 0 | 280 | 0 |
| M | 455 | 245 | 0 | 0 | 2 100 | 0 | 0 | 280 | 0 |
| N | 455 | 0 | 245 | 0 | 2 100 | 0 | 0 | 280 | 0 |
| O | 455 | 0 | 0 | 245 | 2 100 | 0 | 0 | 280 | 0 |

### *2.3. Methods*

The filtered extract was analyzed by means of inductively coupled plasma emission spectroscopy (ICP-OES). In this work the ICP-OES method was performed using the ULTIMA 2 instrument from HORIBA Scientific. An ion chromatography (IC) method was used to determine the ions in analyzed solutions. The IC method was performed using Metrohm's 850 Professional IC. The water to binder ratio (w/b) was set to 0.4. It is the ratio between the total amount of water (g) and the total amount of binder (g), which is the sum of amounts of ash and clinker.

## 3. Results and discussion

Sodium hexanitrocobaltate was used to reduce the ammonium ions contained in the sample of gypsum water. It was found that ammonium hexanitrocobaltate was formed, which precipitated and rapidly settled in the solution. The formation of ammonium hexanitrocobaltate was proceeded according to the equation (5). [14] The first experiment was performed with the sample of gypsum water. The determined contents of selected substances in gypsum water are given in Table 2.

$$3\,NH_4^+ + Na_3[Co(NO_2)_6] \rightarrow \downarrow (NH_4)_3[Co(NO_2)_6] + 3Na^+ \qquad (5)$$

**Table 2.** Analysis of gypsum water using ICP - OES and IC methods.

| Component [mg·l⁻¹] | | | | | |
|---|---|---|---|---|---|
| $SO_4^{2-}$ | sulphates | 16 404.30 | Ni | nickel | 3.94 |
| $Cl^-$ | chlorides | 7 054.56 | Zn | zinc | 3.86 |
| $NO_3^-$ | nitrates | 5 566.86 | Cu | copper | 2.35 |
| $Na^+$ | sodium ion | 5 462.80 | Co | cobalt | 1.78 |
| $\rho_N$ | total ammonium nitrogen | 2 835.45 | Al | aluminium | 0.25 |
| $Mg^{2+}$ | magnesium ion | 1 701.30 | Cr | chrome | 0.21 |
| $K^+$ | potassium ion | 920.9 | Cd | cadmium | 0.11 |
| $Ca^{2+}$ | calcium ion | 696.23 | As | arsenic | 0.03 |
| $F^-$ | fluorides | 80.57 | $NO_2^-$ | nitrite | -* |
| Fe | iron | 8.36 | Pb | lead | -* |

*These components were below the detection limit

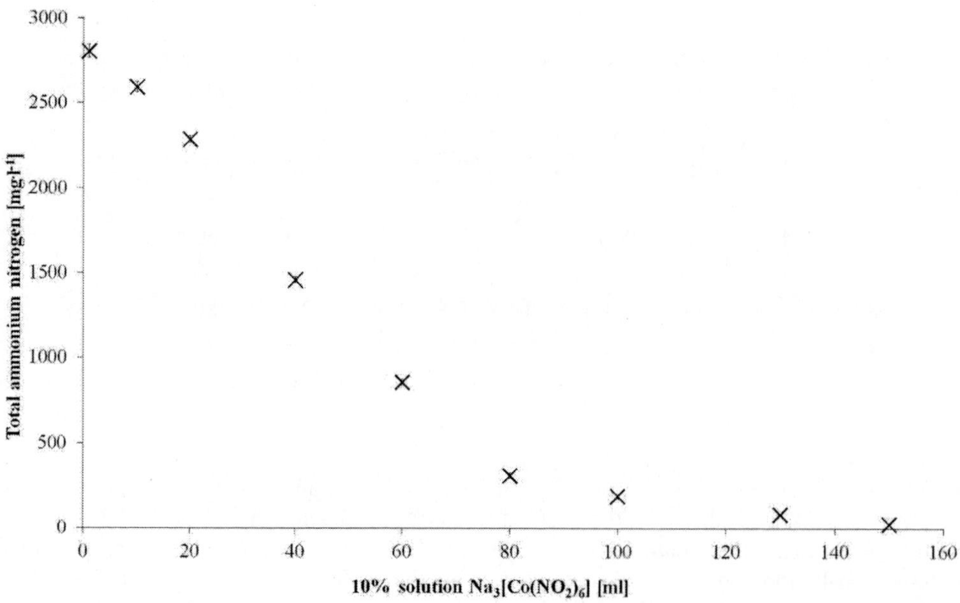


**Figure 2.** Dependency of total ammonium nitrogen on the addition of coordination complex into gypsum water after 10 minutes.

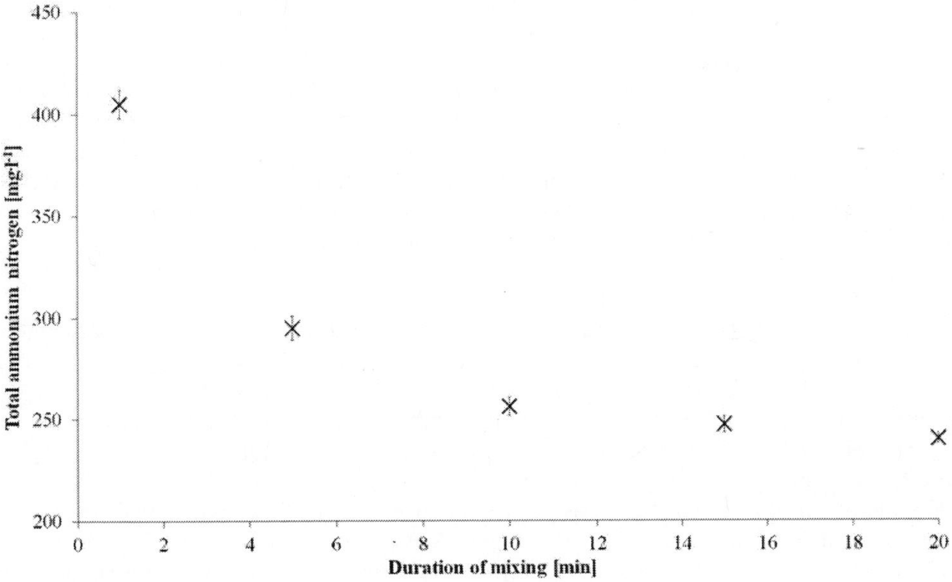


**Figure 3.** Dependency of total ammonium nitrogen the on duration of mixing in gypsum water.

The precipitate was subjected to XRD analysis. The measured spectrum was compared to the database and it was confirmed that the precipitate was indeed ammonium hexanitrocobaltate. Figure 2 shows the trend of total ammonium nitrogen (TAN) in gypsum water decrease with increased addition of 10% water solution of sodium hexanitrocobaltate. Sodium hexanitrocobaltate reduces the ammonium ions present in the sample. The sodium ion is partially substituted by ammonium in the compound. (For detailed description the NMR analysis should be beneficial). This results in the formation of insoluble and stable ammonium hexanitrocobaltate, which precipitates from the solution and quickly settles. The influence of duration of mixing on the TAN decrease in the analyzed sample was further studied and is shown in Figure 3. From the figure it can be seen that with increasing stirring time, the content of TAN in the sample decreases. The amount of nitrogen clearly decreases when being stirred within the first 10 minutes, with longer stirring, the amount of nitrogen in the sample decreases slowly. Hence, 10 minutes of mixing is sufficient time to remove TAN from the sample. On an industrial scale, this time is too long, so it would be advisable to find another way to shorten the mixing interval to the lowest possible time using more appropriate type of agitator, air blowing or heating. From these results the experiment was designed to eliminate the ammonia slip from coal combustion products (CCPs).

Another experiment should test the possibilities of elimination of ammonia sludge from produced solid residues after coal combustion in plants where the second method for SNCR denitrification was installed. Table 3 shows the results of the analysis of mixtures. The device's value limit is about 35 $mg \cdot l^{-1}$.

**Table 3.** Results of the analysis of mixtures by the ICP-OES.

| | **A** | **B** | **C** | **D** | **E** | **F** | **G** | **H** | **I** | **J - L** | **M - O** |
|---|---|---|---|---|---|---|---|---|---|---|---|
| TAN [$mg \cdot l^{-1}$] | 11.401 | 12.439 | 126.545 | 0.028 | 0.043 | 0.058 | 0.016 | 0.030 | 0.031 | 0.065 | 0.031 |
| Co [$mg \cdot l^{-1}$] | 37.5 | 35.6 | * | * | * | * | * | * | * | * | * |

The values of cobalt in mixture C - O were under the limit of detection. From the measured data it can be seen that the addition of the complex in the mixing water reduced the TAN content in the ash 10 times. On the other hand, the cobalt content increased to approximately 35 $mg \cdot l^{-1}$. This value exceeds the determined limit by one third. However, it is clear from the results that the addition of the complex results in a significant reduction of ammonia in the water extracts. Mixtures D to I did not result in increased cobalt content in aqueous extracts after the addition of complex compound to the mixing water. The cobalt values are under the limit of detection in the case of mixtures J to K where no complex is added. The elimination is based on the formation of an insoluble ammonium complex from which the ions are no longer released into the solution.

## 4. Conclusion

The work was focused on the possibilities of elimination of ammonia slip from coal combustion products and from technological water. Sodium hexanitrocobaltate was added to gypsum water and to beams from mortars and pastes. It was demonstrated from the laboratory experiments that insoluble ammonium hexanitrocobaltate is released by the addition of soluble sodium hexanitrocobaltate in gypsum water. After the addition of sodium hexanitrocobaltate, total ammonium nitrogen was reduced in the samples.

The validity of this assumption was confirmed when sodium hexanitrocobaltate was added to the mixture in mixing water. If the complex compound is added to the mixing water, it is incorporated into the product volume and it does not leach into water. Which was confirmed by the aqueous extracts from mixtures, where no elevated cobalt values were measured.

**Acknowledgements**

This work was supported by the project "Materials Research Centre at FCH BUT – Sustainability and Development, REG LO1211, with financial support from National Programme for Sustainability I (Ministry of Education, Youth and Sports)."

## References

[1] Gómez-Garcia M A, Pitchon V and Kiennemann A 2005 Pollution by nitrogen oxides: an approach to NO x abatement by using sorbing catalytic materials *Environment Int.* **31** pp 445–467

[2] Van Caneghem J, De Greef J, Block Ch and Vandecasteele C 2016 NOx reduction in waste incinerators by selective catalytic reduction (SCR) instead of selective non catalytic reduction (SNCR) compared from a life cycle perspective: a case study *J Clean Prod.* **112** pp 4452–4460

[3] De Greef J, Villani K, Goethals J, Van Belle H, Van Caneghem J and Vandecastelle C 2013 Optimising energy recovery and use of chemicals, resources and materials in modern waste-to-energy plants *Waste Management* **33** pp 2416–2424

[4] Morrison G F 1980 Nitrogen Oxides from Coal Combustion – Abatement and Control (London: IEA Coal Research)

[5] Miller J A and Bowman G T 1989 Mechanism and modeling of nitrogen chemistry in combustion *Prog Energ Combust.* **15** pp 289–338

[6] Zandaryaa S, Gavasci R, Lombardi F and Fiore A 2001 Nitrogen oxides from waste incineration: control by selective non-catalytic reduction *Chemosphere* **42** pp 491–497

[7] Görner K and Lentjes A G 1992 *Industrial Air Pollution* (Berlin: Springer)

[8] Rosenberg H S, Curran L M, Slack A V, Ando J and Oxley J H 1980 Post combustion methods for control of NOx emissions *Prog Energ Combust.* **6** pp 287–302

[9] Gohlke O 2010 A new process for $NO_x$ reduction in combustion systems for the generation of energy from waste *Waste Management* **30** pp 1348–1354

[10] Goemans M, Clatysse P, Joannes J, De Clercq P, Lenaerts S, Matthys K and Boels K 2004 Catalytic NOx reduction with simultaneous dioxin and furan oxidation *Chemosphere* **54** pp

1357–1365
[11] Lyon R K 1987 Thermal DeNOx Controlling nitrogen oxides emissions by a noncatalytic process *Environ Sci Technol.* **21** pp 231–236
[12] Institute of Clean Air Companies 2008 Selective non-catalytic reduction (SNCR) for controlling Nox emmisions USA
[13] Klikorka J, Hájek B and Votinsky J 1985 *Obecná a anorganická chemie* (Praha: SNTL)
[14] Hajzler J 2016 *Možnosti eliminace čpavkového skluzu v technologických vodách elektráren* (Brno: Vysoké učení technické v Brně) p 96

IOP Conf. Series: Materials Science and Engineering **379** (2018) 012033 doi:10.1088/1757-899X/379/1/012033

# The possibilities of analysis of limestone chemical composition

**P Šiler[1], I Kolářová[1], J Bednárek[1], M Janča[1], P Musil[2] and T Opravil[1]**

[1] Brno University of Technology, Faculty of Chemistry, Materials Research Centre, Purkyňova 118, CZ-61200 Brno, Czech Republic
[2] Kalcit, s.r.o., Třískalova 902/10A, Lesná, CZ-638 00 Brno, Czech Republic

Email: siler@cfh.vut.cz

**Abstract.** The aim of this work is to show commonly used procedures and methods for limestone analysis. The samples were measured with Inductively Coupled Plasma with Optical Emission Spectrometry (ICP-OES) for chemical compositions determination. For these measurements, the samples were decomposed by dissolving in acid, melting into solution and decomposed by microwave. Due to the time and financial demands of these analyses, there were an effort how to find a satisfactory method providing sufficient results with less time and financial demands. The X-ray diffraction (XRD) method was first tested, followed by X-ray fluorescence (XRF). Samples were measured in powder and melted into pearls. The evaluation was carried out by using of the Fundamental parameters method and by means of instrument calibration. The method of combustion elemental analysis was selected to determine of the sulfur content.

## 1. Introduction

The limestone has been used by humans for several thousand years, and there still grow the possibilities of processing and using of this raw material. Limestone was formed mainly as sedimentary rocks. Most limestones originated in the seas and their genesis was associated with rock-forming organisms. Limestones were also created in fresh waters, lakes, rivers and caves. The main conditions for the formation of limestone were the climate and the absence of clay or sandy material. Calcium carbonate has several polymorphic modifications. In nature, it is predominantly as minerals of calcite and aragonite. Calcite is usually sediment from organic origin. Aragonite secreted from solutions at higher temperatures or in the presence of sulfates. It may also have a biogenic origin from the shells of some molluscs [1].

Limestones are rocks predominantly made by minerals of calcite (calcium carbonate with trigonal structure, $CaCO_3$). They are solid and grained sedimentary rocks of organic or chemical origin. The calcium carbonate content is often above 95%. Most of the limestones were formed by the settling of limestone canals of animals and plants, mainly in the sedimentary basins of the sea. These limestones are called organogenic. In a small amount, limestones were excreted from aqueous solutions in karst lands. Conversion of limestones at high temperature and pressure resulted in crystalline limestones (marbles) in which calcite recrystallized so that the calcareous shells of animals did not appear in marble. Limes only rarely occur clean. Mostly they have admixtures of clay, carbonates, dolomite, sand, or phosphates or iron-containing minerals [1].

Limestone analysis is one of the basic chemical analyzes needed in every company focusing on the production and processing of this raw material. The aim of this work is to introduce commonly used methods and procedures for this analysis and to show what errors can occur in these analyses.

## 2. Sample preparation

All samples were dried at 110°C for 4h. If water content was required, they were weighed before and after drying. To determine the chemical and phase composition, the dried samples were first milled on a vibratory mill.

### *2.1. Acid decomposition*

Determination by decomposition with hydrochloric acid left a large amount of non-degradable fraction after decomposition. The process has been progressively modified. The smallest insoluble fraction was obtained by using aqua regia ($HCl/HNO_3$) = 3/1) followed by addition of HF. After 1 hour of decomposition, the mixture was filtered, diluted with water and then analyzed. Despite the multiple workflow adjustment, satisfactory acid decomposition results have not been achieved since a small amount of undiluted portion of silicon was left in the samples.

### *2.2. Melting into liquid*

Although the standard ČSN 721216 [2] allows the decomposition of samples in acids, the preferred method of decomposition for rocks and minerals is alkaline or alkaline oxidation melting, especially the decomposition of the sample by melting with sodium carbonate. This standard also allows degradation in borate fluxes which are more expensive but have many advantages over carbonates. The use of this mixture resulted in the complete decomposition of all components, and therefore it was not necessary to use HF fumigation, which is problematic for many reasons. In this type of decomposition, the sample was totally distributed. Determined samples were transferred to the solution by picking in platinum crucibles on a Fluxana Vulcan melt. A series of experiments found that the optimal melt ratio is 3 parts of lithium tetraborate ($Li_2B_4O_7$) and 1 part of lithium metaborate ($LiBO_2$). LiBr as a wetting agent was added to this mixture. The whole mixture was melted using propan-butane + air heating gas. The melt thus formed was poured into dilute $HNO_3$ and boiled at 170°C with constant stirring. It was then cooled in air, quantitatively transferred to a volumetric flask, supplemented on 500 ml with distilled water and then determined by ICP-OES.

### *2.3. Pearl preparation*

The specimens were melted into a pearl by drawing in platinum crucibles on a Fluxana Vulcan device. A series of experiments showed that the optimal melt ratio is 3 parts lithium tetraborate ($Li_2B_4O_7$) to 1 part lithium metaborate ($LiBO_2$) as well as samples for melting into solution. The resulting melt was mixed several times automatically and poured hot into a preheated dish, where it was controlled by software programmed refrigeration program.

### *2.4. Distribution by microwave decomposition*

Samples were distributed on the CEM MARS instrument. The samples were decomposed into $HNO_3$. In such an environment, silicon is only decomposed to a minimum. The analysis was performed using ICP-OES.

### *2.5. Powder form measurement*

For samples in the form of powder, samples were placed in special containers supplied by manufacturers of measuring instruments on these instruments. For measurements on the air atmosphere on a handheld XRF spectrometer, they were overlaid with the polyethylene film over which the measurements were made. In the case of vacuum measurements, the samples were equipped with a special foil for vacuum measurement.

### *2.6. Sulfur content determination*
To determine the total sulfur content, samples in powder form were placed in special ceramic crucibles for measurement using a combustion elemental analyser.

## 3. Description of the methods used

### *3.1. ICP-OES (inductively coupled plasma with optical emission spectrometry)*
ICP-OES or inductively coupled plasma with optical emission spectrometry is a trace analytical method for determining trace and significant concentrations of individual elements in the analysed sample. This technique allows analysing almost all elements of a periodic table that can be converted into a solution with sensitivity from ppb to hundreds of ppm. The analyses were carried out on the HORIBA ULTIMA 2 instrument.

### *3.2. XRD (x-ray diffraction)*
X-ray diffraction analysis is a method of determining the structure of crystalline substances. The method is based on the fact that the dimensions of the crystal lattice are comparable to the wavelength of X-rays. Thus, the diffraction (bending) of the radiation can occur on the crystal lattice. Since the pattern of the crystal is periodic, the diffraction pattern of the radiation, after passage through the crystal, has a relatively good relationship with the crystal structure. The method is used to determine the structure understood as a spatial arrangement of crystalline substances. Because of the low detection limit, it is not suitable for the determination of impurities below 1%.

### *3.3. XRF (x-ray fluorescence)*
X-ray fluorescence (XRF) uses the emission of characteristic secondary or X-rays from material that has been excited by high energy X-rays. This phenomenon is widely used for elemental analysis and chemical analysis. Samples were measured primarily on the XENEMETRICS EX-6600 SSD energy dispersion device. The measurements were carried out in vacuum both in pearls and in powder. The evaluation was performed using both the Fundamental Parameter and the instrument calibration using the ICP-OES samples. From the measured values it can be said that this analysis can be used for rapid analysis of the sample for regular quality control. For comparison, measurements were also performed on the Olympus VANTA handheld XRF Spectrometer.

### *3.4. Combustion sulfur analysis in solids*
This analysis is mainly used in the metals, ceramics, minerals, cement industry and other industries. The sample is placed in a ceramic crucible along with a so-called accelerator that promotes the combustion process. Next, the sample is burned in a high-frequency furnace in an oxygen stream. The resulting C, S containing flue gases are converted to $CO_2$ and $SO_2$ by reagents and detected by IR detectors. Samples were measured on a Brooker G4 ICARUS CS instrument.

## 4. Results and discussion
Two samples were observed during analyses. One very pure limestone from the Forest Quarry in Líšeň and the second sample was a polluted limestone used as a quarry stone in the FCH BUT laboratory. The results are shown in Tables 1–5 and Figure 1. Acid decomposition was performed only on a pure limestone sample, as a large amount of non-decomposed portion containing mostly Mg and Si was found in contaminated limestone.

### *4.1. The comparison of XRF and ICP-OES analysis*
The values from comparison between XRF and ICP-OES techniques are present in wt. % in Tables 1 and 2.

**Table 1.** The results of analysis of limestone from Forest Quarry in Líšeň.

| **Pure limestone** | **$Fe_2O_3$** | **$Al_2O_3$** | **$SiO_2$** | **$MgCO_3$** | **$CaCO_3$** |
|---|---|---|---|---|---|
| Powder without calibration | 0.73 | 3.08 | 2.94 | 1.68 | 91.57 |
| Powder with calibration | 0.15 | 0.18 | 0.56 | 0.56 | 98.55 |
| Pearl without calibration | 0.72 | 0.14 | 0.17 | 0.21 | 98.77 |
| Pearl with calibration | 0.08 | 0.30 | 0.57 | 0.56 | 98.50 |
| XRF handed analyzis | 0.10 | 0.19 | 0.67 | 2.12 | 96.92 |
| ICP-OES acid | 0.09 | 0.22 | 0.59 | 0.87 | 98.23 |
| ICP-OES fusion | 0.09 | 0.25 | 0.68 | 0.81 | 98.17 |
| ICP-OES microwave decomposition | 0.12 | LOD | 0.02 | 1.37 | 98.49 |

**Table 2.** The results af analysis of Quarry stone.

| **Quarry stone** | **$Fe_2O_3$** | **$Al_2O_3$** | **$SiO_2$** | **$MgCO_3$** | **$CaCO_3$** |
|---|---|---|---|---|---|
| Powder without calibration | 0.95 | 5.38 | 23.9 | 0.45 | 69.33 |
| Powder with calibration | 0.54 | 0.70 | 5.69 | 0.45 | 92.62 |
| Pearl without calibration | 2.00 | 0.25 | 2.10 | 0.21 | 95.43 |
| Pearl with calibration | 0.38 | 1.08 | 7.95 | 0.98 | 89.61 |
| XRF handed analyzis | 0.42 | 1.21 | 7.61 | 1.25 | 89.51 |
| ICP-OES acid | 0.27 | 0.58 | 1.78 | 1.20 | 96.17 |
| ICP-OES fusion | 0.38 | 1.09 | 8.31 | 1.18 | 89.04 |
| ICP-OES microwave decomposition | 0.28 | LOD | 0.02 | 1.53 | 98.10 |

### *4.2. XRD analysis results*

Another option is XRD (X-ray diffraction) analysis. The evaluation was performed by semi-quantitative analysis. This analysis was made for a pure limestone sample. Since this sample is quite clean, it is possible to determine the sensitivity of this method for limestone analysis. From this analysis the data in Figure 1 and Table 3 were obtained.

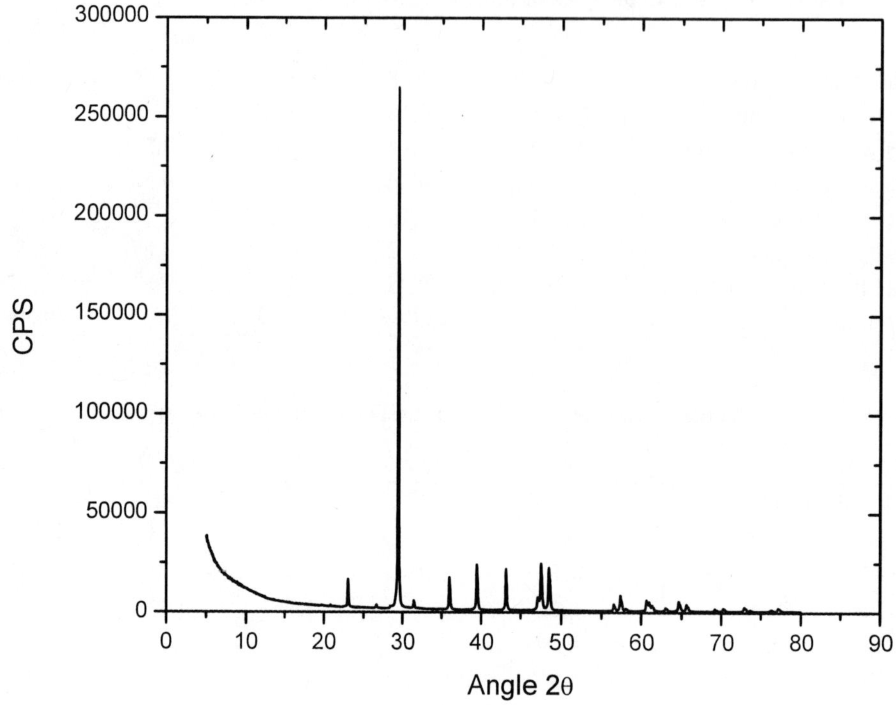


**Figure 1.** XRD spectrum of pure limestone.

**Table 3.** The results of limestone analysis by XRD.

| Compound | wt. % |
|---|---|
| Calcite | 99.2 |
| Quartz | 0.7 |
| Muscovite | 0.1 |

*4.3. Sulfur content measurement*

The purpose of these tests was to find out the differences when using individual calibrations. Calibration was performed on various standards. First, a one-point calibration on the dolomite stone certified material SX09-11 ($SO_3$ = 0.064%) was used because the limestone certified materials mostly do not contain sulfur data. Subsequently, the multi-element calibration was performed on steel standards CRM CZ 2004 A, CRM CZ 2005 A, CRM CZ 2007, CRM CZ 2015 A, CRM CZ 2021 A, CRM CZ 2024 A (with amount of S = 0.0091 – 0.0464) and blast furnace slag NCS HC 15803, NCS HC 13825, NCS HC 13824 with amount of S = 0.535 – 0.98%. Samples were measured five times. The results are given in wt. % of sulfur in Tables 4 and 5. In measurements based on different calibration curves, slight changes in the resulting values occur. The results of all calibrations used are almost the same, but in the calibration of the steel, repeated measurements lead to a smaller amount of scattering, which is also due to a larger number of calibration points with concentrations closer to the actual sulfur concentration in the limestone samples.

IOP Conf. Series: Materials Science and Engineering **379** (2018) 012033 doi:10.1088/1757-899X/379/1/012033

**Table 4.** Sulfur amount in Forest Quarry in Líšeň sample.

| Calibration type | Average value | Value difference |
|---|---|---|
| Dolomite stone | 0.041 | ±0.007 |
| Steel | 0.042 | ±0.004 |
| Slag | 0.044 | ±0.012 |

**Table 5.** Sulfur amount in Quarry stone sample.

| Calibration type | Average value | Value difference |
|---|---|---|
| Dolomite stone | 0.062 | ±0.008 |
| Steel | 0.061 | ±0.004 |
| Slag | 0.065 | ±0.014 |

## 5. Conclusion

The aim of this work was to analyse limestone using the available methods and to evaluate the influence of the applied method on the results of the analyses. The best results were achieved by melt-decomposition into the solution and subsequent ICP-OES analysis. With acid decomposition and microwave pressure decomposition, a higher amount of undiluted fraction containing Mg and Si mainly resulted in a distortion of the result. Samples were also measured using XRF. The best results were obtained by measuring in the form of pearls in the vacuum by means of calibration, however the measurement in the powder form by means of calibration can be also used for standard quality control. The sulfur content was measured by combustion sulfur analysis in solids. The smallest value scatter was achieved by calibration on steel. This calibration contained the largest number of calibration points at near sulfur levels as in limestone. Other calibrations are also useful for determining the sulfur content of limestone. XRD was used to determine the phase composition. This method is suitable for measuring crystalline phases in contents above 1%.

## Acknowledgments

This work was financially supported by the project Materials Research Centre at FCH BUT – Sustainability and Development. REG LO1211 with financial support from National Program for Sustainability I (Ministry of Education Youth and Sports).

## References

[1] Těhník V and Nečas R 2010 Uplatnění vápencové suroviny z hlediska kvality v různých průmyslových odvětvích *IX. Odborná konference MALTOVINY 2010* (Brno: VUT)

[2] ČSN 721216:1975 Chemical Analysis of Limestone

IOP Conf. Series: Materials Science and Engineering **379** (2018) 012034 doi:10.1088/1757-899X/379/1/012034

# Determination accuracy of analysis refractory materials by X-ray fluorescence

**M Janča[1], P Šiler[1], T Opravil[1] and J Kotrla[1]**

[1] Brno university of technology, Faculty of Chemistry, Materials Research Centre, Purkyňova 464/118, Brno Cz-612 00, Czech Republic

Email: xcjancam@fch.vut.cz

**Abstract**. This work was focused on the finding of the right way how to analyze refractory materials by X-ray fluorescence. This method can provide accurate results which can be easily repeated, what is the reason why this method is commonly used in industry. Using of the fluorescent spectrometric analysis comes with one issue, which is the presence of many distorting interferences. From this particular reason it is necessary to identify these interferences and then find the ideal solution for their elimination. The choosing of the right method of sample preparation is basic point for making of well analyze results. The differences are shown on the refractory materials samples. All the values were compared according to standards compositions. In this case we compare different conditions of measurement, under air and vacuum. And different preparation form in powder or fusion into pearl.

## 1. Introduction

### *1.1. X-ray spectrometry*

In this method, the inner electron is pulled out of its atomic orbit via the absorption of shortwave radiation (photon) with a sufficient amount of energy [1]. The photon energy must be greater than the energy that binds the electron to the core of the atom. If the inner electron is thrown out of the atom, the electron from the higher energy level moves to its place. At this transition, the photon is emitted by the specific energy characteristic of the atom [2]. This fluorescence radiation is called a characteristic X-ray radiation, the energy of the radiated photon being equal to the energy difference of the orbital represented by the transition of the electrons [3].

*1.1.1. Energy disperse x-ray spectrometry (EDXRF).* EDXRF is one of two main types of X-ray fluorescence techniques used for elemental sample analysis. With EDXRF spectrometers, all elements in the sample are excited simultaneously and energy dispersive detector combined with multiple channel analyzer together, collects the fluorescence radiation emitted from the sample. Then analyzer divides it on the basis of the different energy of the characteristic radiation of the various elements present in the sample. Resolution of EDXRF spectrometer depends on the radiation detector (typically in the range 150–600 eV). The benefits of this system are its simplicity, fast operation, the absence of moving fine parts and high efficiency [6]. It requires x-ray source, which is found in most laboratory instrument with a power of 50–60 kV tube. Next crucial component is X-ray detector, which is designed to convert the intensity of energy levels detected into electrical pulses [7]. In addition to the source and the X-ray detector, its filters, targets and collimators are also important. These types of filters are found in almost all EDXRF devices. Their function is to absorb or, on the contrary, excite and transfer the energy of some X-ray sources more than other. These filters ensure that if we are interested in measuring a certain type of atoms on an EDXRF device, by using the filter, we will prefer to measure these interest elements and to suppress the measured values of similar properties as the interest element, thus avoiding the partial merging of these values [8]. Energy disperse spectrometry device is schematically shown at Figure 1.

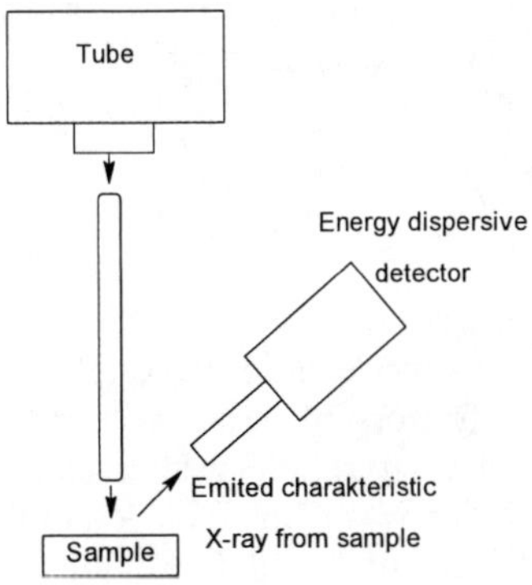


**Figure 1.** EDXRF device.

## 2. Experimental part

All samples were measured using the XENEMETRICS EX-6600 SSD energy dispersing device. It is energy dispersive X-ray spectrometer which has secondary polarizing filters, the primary targets for measurement and specification and secondary collimator. This instrument can also measure without standardized methods. It is able to determine the qualitative and quantitative composition of elements in the sample from fluoride to uranium. However, to improve the accuracy of measurement, it is necessary to calibrate the device for specific material [4]. For the measurement, was sample modified into two forms, powdered and fused into a pearl. For comparison, the samples were measured under various conditions in the presence of air or vacuum. The evaluation of measurements was carried out using the Fundamental Parameters software and through calibration using purchased standards of refractory materials The Okayama Ceramics Research Foundation [5].

### *2.1. Sample preparation*

For determination of refractory materials has been chosen two different forms of sample. It was the form as powder and fusion into borate pearls.

*2.1.1. Powder measurement.* Measured material was first milled on a mill and then filled into capsules for measurement on an XRF spectrometer. These capsules are on the side adjacent to the radiation source and the detector covered with the radiation permeable film.

*2.1.2. Measurement in the form of fused pearls.* The advantage of measuring in the form of pearls against the measurement of powder samples is to suppress the matrix effect and thus to obtain more accurate results. It is important to choose right mix of flux that will be used. For this reason, so-called acidity indexes are used to tell us what proportion of the flux to use.

### *2.2. Device parameters for individual measurement*

For analysis, it was important, among other things, to find suitable measurement parameters for each sample form (powder / pearl) and the environment in which the measurements were to take place (air / vacuum). The main set parameters are the time of measurement and setting of the X-ray lamp. For

IOP Conf. Series: Materials Science and Engineering **379** (2018) 012034 doi:10.1088/1757-899X/379/1/012034

the X-ray lamp, the amount of supply voltage and current is set. The voltage value has been set to 20 kV. The largest energy required for excitation is iron (6.4 kV). Normally, it is recommended to multiply the maximum value of element excitation by 1.6 times during measurements, but the 20 kV setting is recommended based on experimental experience [9]. The measurement settings are shown in Table 1.

**Table 1.** Table values of electric current and voltage.

| Sample | Voltage [kV] | Eletric current [μA] |
|---|---|---|
| Powder in air | 20 | 1900 |
| Powder in vacuum | 20 | 1200 |
| Pearl in air | 20 | 1100 |
| Pearl in vacuum | 20 | 1100 |

### *2.3. Calculation of acidity index of individual standards*

For calibration, the standards of The Okayama Ceramics Research Foundation were used [10]. This set of standards contains ten samples of refractory materials with different percentage composition of oxides. The theoretical acidity index for each sample were calculated as shown in the Table 2, on the basis of which the average acidity index was then displayed in the table on which the melt blend was assembled. The resulting composition of a suitable melting mixture has to be verified and optimized based on experiments.

Example of acidity index calculation for standard JRRM 301:

Silica SiO2:

$$\text{Acidity index} = \frac{\text{Number of oxygen atoms}}{\text{Number of other atoms}} = \frac{2}{1} = 2 \tag{1}$$

Calculation of the contribution to the acidity index of the sample by silicon dioxide

$$SiO_2 \text{ in JRRM 301} = \frac{x\,[\%]}{100\,[\%]} \cdot Ai_{(SiO_2)} = \frac{43.9}{100} \cdot 2 = 0.88 \tag{2}$$

Calculation of theoretical sample acidity index

$$\text{JRRM 301} = \sum Ai_{(xO_y)} = 1.69 \tag{3}$$

**Table 2.** Table of calculated acidity index for sample JRRM 301.

| Oxide | Percentage of sample [%] | Oxide acidity index | Theoretical acidity index of the oxide in the sample |
|---|---|---|---|
| $SiO_2$ | 43.9 | 2 | 0.88 |
| $Al_2O_3$ | 46.8 | 1.5 | 0.70 |
| $Fe_2O_3$ | 3.53 | 1.5 | 0.05 |
| $TiO_2$ | 1.03 | 2 | 0.02 |
| MnO | 0.01 | 1 | 0.00 |
| CaO | 0.79 | 1 | 0.01 |
| MgO | 0.69 | 1 | 0.01 |
| $Na_2O$ | 0.17 | 0.5 | 0.00 |
| $K_2O$ | 2.01 | 0.5 | 0.01 |
| $B_2O_3$ [a] | 0.87 | 1.5 | 0.01 |
| | | **Theoretical sample acidity index:** | 1.69 |

[a] Uncertified value

Theoretical acid numbers were calculated for all standards and are shown in Table 3. The average value of the acidity index of the refractory material was approximately 1.5. This value suggests that this material will have to be fused using more basic lithium metaborate. However, the metaborate itself undergoes crystallization, which is why the mixture was prepared by mixing lithium metaborate with lithium tetraborate in a 3:1 ratio. The total weight was 5 g of flux and 0.5 g of the powder sample. The thus prepared weights were placed in a platinum crucible in a fusion machine where the sample fused and the melt was then poured onto the platinum dish in which the pearl was cooled.

**Table 3.** Table with calculated acidity index of standards.

| Sample | JRRM 301 | JRRM 302 | JRRM 303 | JRRM 304 | JRRM 305 | JRRM 306 | JRRM 307 | JRRM 308 | JRRM 309 | JRRM 310 | **Average acidity index** |
|---|---|---|---|---|---|---|---|---|---|---|---|
| Acid number | 1.69 | 1.67 | 1.66 | 1.64 | 1.57 | 1.56 | 1.52 | 1.55 | 1.51 | 1.49 | **1.59** |

## 3. Results and discussion

### 3.1. *Measurement of standards by the fundamental parameter method*

Samples were measured and the percent of the individual oxides in the sample was calculated using the Fundamental parameters software. These values vary considerably from the certified standards. By using this method, inaccurate results were obtained, as can be seen in the Figure 2.

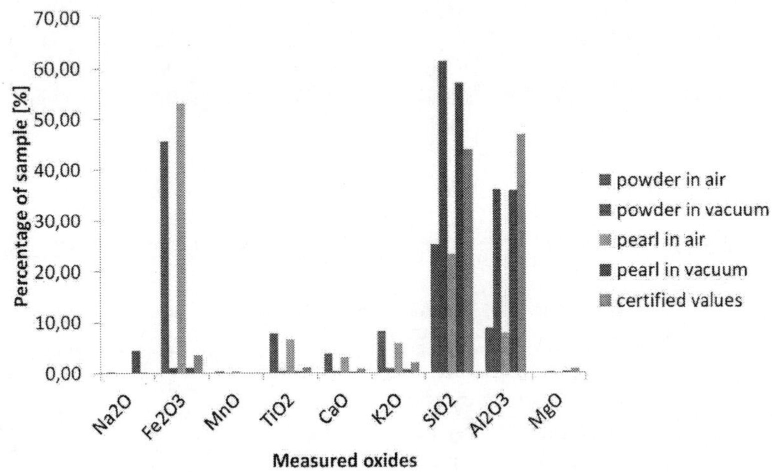


**Figure 2.** Measured values calculated by fundamental parameters method of JRRM 301 sample.

### *3.2. Calibration by using standards*

Calibration was made on the basis of the measured spectra of a set of standardized samples. In the measured spectrum, the certified values of oxides in the standards were always entered manually. These values were used to construct the calibration curve by linear regression.

**Table 4.** Sample JRRM 301, powder in air.

| Oxides in standard | Percentage of oxides [%] | Calculated percentage of oxide by FP [%] | Relative deviation by FP [%] | Absolute deviation by FP [%] | Calculated deviation after calibration [%] | Relative deviation after calibration [%] | Absolute deviation after calibration [%] |
|---|---|---|---|---|---|---|---|
| $Na_2O$ | 0.17 | 0.13 | 22.19 | 0.04 | 0.56 | 228.99 | 0.39 |
| $Fe_2O_3$ | 3.53 | 45.64 | 1192.94 | 42.11 | 2.41 | 31.68 | 1.12 |
| MnO | 0.01 | 0.28 | 2702.70 | 0.27 | 0.00 | 69.82 | 0.01 |
| $TiO_2$ | 1.03 | 7.78 | 655.57 | 6.75 | 0.68 | 33.97 | 0.35 |
| CaO | 0.79 | 3.91 | 395.07 | 3.12 | 0.55 | 30.37 | 0.24 |
| $K_2O$ | 2.01 | 8.23 | 309.61 | 6.22 | 1.54 | 23.58 | 0.47 |
| $SiO_2$ | 43.90 | 25.14 | 42.73 | 18.76 | 43.78 | 0.26 | 0.12 |
| $Al_2O_3$ | 46.80 | 8.80 | 81.19 | 38.00 | 50.11 | 7.08 | 3.31 |
| MgO | 0.69 | 0.08 | 88.95 | 0.61 | 0.36 | 47.30 | 0.33 |

**Table 5.** Sample JRRM 301, powder in vacuum.

| Oxides in standard | Percentage of oxides [%] | Calculated percentage of oxide by FP [%] | Relative deviation by FP [%] | Absolute deviation by FP [%] | Calculated deviation after calibration [%] | Relative deviation after calibration [%] | Absolute deviation after calibration [%] |
|---|---|---|---|---|---|---|---|
| $Na_2O$ | 0.17 | 0.13 | 26.45 | 0.04 | 0.35 | 108.77 | 0.18 |
| $Fe_2O_3$ | 3.53 | 0.97 | 72.56 | 2.56 | 2.31 | 34.55 | 1.22 |
| MnO | 0.01 | 0.01 | 41.01 | 0.00 | 0.00 | 93.45 | 0.01 |
| $TiO_2$ | 1.03 | 0.33 | 68.27 | 0.70 | 0.69 | 33.09 | 0.34 |
| CaO | 0.79 | 0.25 | 68.90 | 0.54 | 0.56 | 29.42 | 0.23 |
| $K_2O$ | 2.01 | 0.82 | 59.33 | 1.19 | 1.41 | 29.62 | 0.60 |
| $SiO_2$ | 43.90 | 61.33 | 39.71 | 17.43 | 44.29 | 0.89 | 0.39 |
| $Al_2O_3$ | 46.80 | 35.97 | 23.15 | 10.83 | 49.87 | 6.57 | 3.07 |
| MgO | 0.69 | 0.21 | 69.40 | 0.48 | 0.51 | 26.00 | 0.18 |

**Table 6.** Sample JRRM 301, pearl in air.

| Oxides in standard | Percentage of oxides [%] | Calculated percentage of oxide by FP [%] | Relative deviation by FP [%] | Absolute deviation by FP [%] | Calculated deviation after calibration [%] | Relative deviation after calibration [%] | Absolute deviation after calibration [%] |
|---|---|---|---|---|---|---|---|
| $Na_2O$ | 0.17 | 0.10 | 40.60 | 0.07 | 0.42 | 147.05 | 0.25 |
| $Fe_2O_3$ | 3.53 | 53.14 | 1405.27 | 49.61 | 2.66 | 24.68 | 0.87 |
| MnO | 0.01 | 0.28 | 2694.71 | 0.27 | 0.01 | 24.46 | 0.00 |
| $TiO_2$ | 1.03 | 6.57 | 538.20 | 5.54 | 0.73 | 29.19 | 0.30 |
| CaO | 0.79 | 3.07 | 289.15 | 2.28 | 0.55 | 30.71 | 0.24 |
| $K_2O$ | 2.01 | 5.75 | 185.92 | 3.74 | 1.61 | 19.82 | 0.40 |
| $SiO_2$ | 43.90 | 23.28 | 46.98 | 20.62 | 46.83 | 6.69 | 2.93 |
| $Al_2O_3$ | 46.80 | 7.81 | 83.31 | 38.99 | 46.65 | 0.31 | 0.15 |
| MgO | 0.69 | 0.00 | 99.95 | 0.69 | 0.53 | 22.93 | 0.16 |

**Table 7.** Sample JRRM 301, pearl in vacuum.

| Oxides in standard | Percentage of oxides [%] | Calculated percentage of oxide by FP [%] | Relative deviation by FP [%] | Absolute deviation by FP [%] | Calculated deviation after calibration [%] | Relative deviation after calibration [%] | Absolute deviation after calibration [%] |
|---|---|---|---|---|---|---|---|
| $Na_2O$ | 0.17 | 4.43 | 2503.34 | 4.26 | 0.42 | 149.16 | 0.25 |
| $Fe_2O_3$ | 3.53 | 1.11 | 68.58 | 2.42 | 2.93 | 17.08 | 0.60 |
| MnO | 0.01 | 0.01 | 42.95 | 0.00 | 0.01 | 44.68 | 0.00 |
| $TiO_2$ | 1.03 | 0.29 | 72.10 | 0.74 | 0.82 | 20.14 | 0.21 |
| CaO | 0.79 | 0.20 | 75.18 | 0.59 | 0.59 | 24.78 | 0.20 |
| $K_2O$ | 2.01 | 0.59 | 70.43 | 1.42 | 1.79 | 10.90 | 0.22 |
| $SiO_2$ | 43.90 | 57.15 | 30.18 | 13.25 | 45.25 | 3.07 | 1.35 |
| $Al_2O_3$ | 46.80 | 35.92 | 23.26 | 10.88 | 47.79 | 2.12 | 0.99 |
| MgO | 0.69 | 0.32 | 53.93 | 0.37 | 0.39 | 44.15 | 0.30 |

## 4. Conclusion

For powder calibration, the measurement was more accurate than the fundamental parameter method, and in the vacuum the measurement was more accurate than in the air. Highest absolute deviations were up to 6% in air and 3% in vacuum measurements. Parameters for setting the electric current for powder was about 700 mA higher in air than when it was measured in a vacuum. This difference is mainly due to the interaction of the primary radiation of the lamp with the molecules contained in the air.

For calibration in the form of a pearl, the results were further refined compared to those measured in powder form. This refinement is mainly due to the suppression of the matrix effect by dissolving the sample in the melt. As with other methods, it was true that vacuum measurements were more accurate than in the air. Highest absolute deviations were up to 3% in air and 1,5% in vacuum measurements. Parameters for the pearl sample are the same as in the air and in the vacuum. However, this setting is lower by 100 μA than in the form of powder in the vacuum. This is because during measuring in powder form the radiation passes through the cover sheet and also there are different sizes of measured particles.

**Acknowledgements**
This work was supported by the project Materials Research Centre at FCH BUT - Sustainability and Development, REG LO1211, with financial support from National Programme for Sustainability I (Ministry of Education, Youth and Sports).

**References**

[1] Černohorský T 1997 *Atomová spektroskopie* (Pardubice: Univerzita Pardubice)
[2] Okada T, Manabu K, Akio F, Hiroshi T and Shunji K 2000 *Adv. Space Res.* V **25** pp 345–48
[3] Kaniu M I and Angeyo K H 2014 *Geoderma* **241-42** pp 32–40
[4] Koplík J 2018 Rentgenová fluorescenční spektrometrie *Chempoint* (*Online* http://www.chempoint.cz/rentgenova-fluorescencni-spektrometrie)
[5] Akbulut S 2014 *Journal of Analytical Atomic Spectrometry* **29** pp 853–60
[6] Jenkins R, Gould R W and Gedecke D 1981 *Quantitative X-ray Spectrometry* (New York: Marcel Dekker)
[7] XRF an introduction of technology. *Maxsell TM* 2009 (*Online* www.goldtester.in/introduction-of-XRF-technology.html)
[8] Bertin E P 1978 *Itroduction to X-Ray Spectrometic Analysis* (New York:Plenum Press)
[9] Lahav G, Shilstein S, Shchemelinin S, Ikher S, Halperin D, Chechik R and Breskin A 2015

*Physica Medica.* **31** pp 233–41
[10] Okayama Ceramics Research Foundation 2015 (*Online* http://occ.optic.or.jp/pages/display/english/index.html)

IOP Conf. Series: Materials Science and Engineering **379** (2018) 012035 doi:10.1088/1757-899X/379/1/012035

# Façade biological colonisation assessment

**I Kukletová[1] and P Buchta[1]**

[1] Výzkumný ústav stavebních hmot, a.s. Hněvkovského 65, 617 00 Brno

Email: kukletova@vustah.cz

**Abstract**. With increasing number of thermally insulated buildings, the number of façade surfaces massively colonised by microorganisms is also rapidly growing. Such colonisation may occur even within a few years after the thermal insulation installation and evolve into a serious issue in a relatively short time. Perceived by the public as an aesthetic problem only, biofilms in fact represent rather significant technological and health risks. The aim of this case study is to show assessment and evaluation of the causes of biotic attack, including distinguishing between direct growth on the façade and mere pollution by surrounding spores. In addition, the impact of the façade biological colonisation on the building inner air microbial pollution was studied.

## 1. Introduction

Currently, the number of façades affected by the growth of microbiological biofilm is increasing. This phenomenon is directly related to the trend of building insulation. Biofilms on façades are perceived by the public as an aesthetic problem only, but in fact, they represent serious technological and health risks.

Microorganisms are transported onto the façade surfaces through the surrounding air. Under suitable conditions, they multiply and produce visible coloured coatings, so-called biofilms. The composition of a biofilm depends on climatic conditions and numerous other factors, the most important being humidity. Micro-particles of organic matter, supplied together with air dust, serve as nutrients. The possibility and extent of façade colonization by microorganisms depend primarily on the surface material, implementation of problematic details, location of the building, sunlight, presence and type of surrounding greenery, pH of the adjacent soil, and last but not least the purpose and way of using the building. A prerequisite for the growth and multiplication of microorganisms on facades in the outdoor environment is moisture. Particularly on insulated houses, the heat is prevented from escaping the building, the top layers of insulation remain cold and moisture occurs due to condensation. In European conditions, rain water also contributes to overall humidity on the northwest and northern sides of buildings, as there's no warming and drying sunlight on the surfaces and even the disinfecting effects of UV radiation are suppressed [1]. Research shows that the first organisms colonizing the surfaces are bacteria, followed by algae. Soon after algae, microscopic fibrous fungi, later mosses and lichens appear and prepare the soil for the emergence of higher plants. On building stone, the mostly occurring green algae genera are *Gloeocystis, Tetracystis, Cystomonas, Murielle, Chlorella, Stichococcus, Scenedesmus, Klebshormium, Apatococcus, Trebauxia*) [2]. Among the microscopic fungi, the degradation of the façades is mainly due to the representatives of the genus *Alternaria*. This genus is considered to be the main cause of brown and black stains on marble, sandstone, limestone and on facades of building structures. Another important genus is *Cladosporium*, which is very common on wet and cold building stone, including facades, and also soil moulds, such as *Acremonium, Mucor*, or *Penicillium* , naturally spread by the wind [2].

On the surface of facades, biofilm acts mechanically, but mainly chemically due to production of organic acids that react with the surface and damage it. Together with dust particles, biofilm retains water. During freezing in winter, the water converts into ice, plaster particles are released and thus the surface of the façade is degraded.

From moulds growing on the surfaces, a major amount of spores is released, spread through the air and enter the indoor environment of apartments and houses during ventilation. These microscopic reproductive particles settle in the respiratory system and cause allergic and consequently asthmatic seizures to many people. Some types of fungi may also occasionally cause other human diseases, such as skin diseases or inflammation of cornea or other internal tissues of the eye. The incidence of health problems always depends on the immunological condition of affected people and the amount of spores in the environment [3, 4].

In the atmosphere, up to several tens of thousands spores per cubic meter can occur, depending on the season and the vicinity of the spore source, such as the infested façade. In housing areas with more sources, the risk of bio colonization on wet spots rises. Higher concentrations of spores in the air also complicate the diagnosis of upper plaster surface attack, microbiological examination of the surface is distorted by air contamination and indication of actual mould infestation is difficult. For this reason, a combined sampling methodology called "swab-bulk-swab" was developed at the Klokner Institute, which takes into account the above-mentioned contamination. [5]

The aim of the paper is to show a way of assessing and evaluation of the biotic infestation causes, including the distinction between direct attack of the facade and mere contamination from the surrounding environment. The effect of the contact thermal insulation system colonization on indoor air microbial pollution is also monitored.

## 2. Sampling and determination methods

The studied object was a school building with different degree of façade infestation over the outer surface. The building is V-shaped with the arc part directed to north-west and surrounded by a park. The school is situated in urban area. Figure 1 shows the building floor plan, aerial photograph is shown in Figure 2.

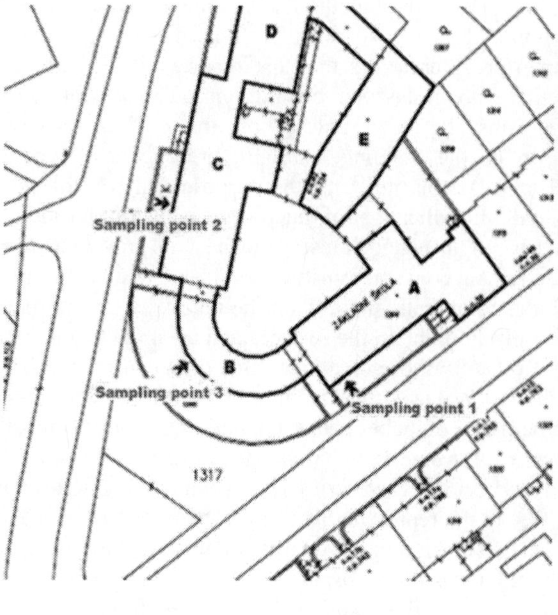


**Figure 1.** The building floor plan.

**Figure 2.** Aerial photograph.

IOP Conf. Series: Materials Science and Engineering **379** (2018) 012035 doi:10.1088/1757-899X/379/1/012035

To determine the biological colonization of the facade, 3 sampling points were selected, see Figure 1. Samples were taken from these sites by the "swab-bulk-swab" Method [5].

At first, a swab was taken from the surface of the sampling site with a sterile tampon soaked in sterile demineralized water. Subsequently, using a sterile stainless steel spatula, a sample of about 2 g of the plaster was taken. Finally, one more swab was performed on the abraded spot with a sterile tampon soaked in sterile demineralized water. Both swabs were immediately inoculated onto Sabouraud agar supplemented with chloramphenicol and cultured at 20–25°C in a thermostat for 5 days. From the plaster sample, 0.2 g was shaken in 1.8 ml of physiological saline solution and afterwards, 0.1 ml was inoculated on the surface of Sabouraud agar with chloramphenicol and cultured at 20–25°C in a thermostat for 5 days. After cultivation, the CFU value (colony-forming unit) per 1 g of plaster was determined. The bulk plaster sample was also cultured in the Bold Basal Medium [6] nutrient solution to detect green algae contamination. Swab cultures were assessed visually and individual colonies and genera were identified. Air sampling was also carried out to determine microbial air pollution at the sampling points 1 and 2, both inside the building in the room and outside the building in the immediate vicinity of the façade. The sampling was performed by an aeroscope, following the recommended standard procedures [7].

## 3. Results and evaluation

First, growth of colonies from both swabs (before and after the plaster was abraded) was visually evaluated. Cultivation dishes of swabs before abrasion are in Figures 3–5, those after abrasion in Figures 6–8.

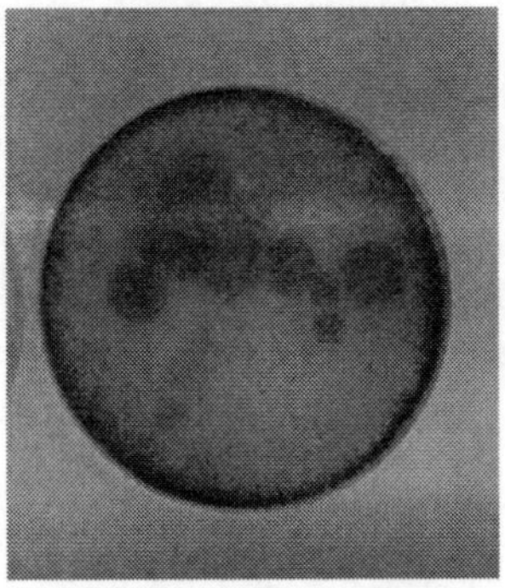

**Figure 3.** Sampling spot 1.

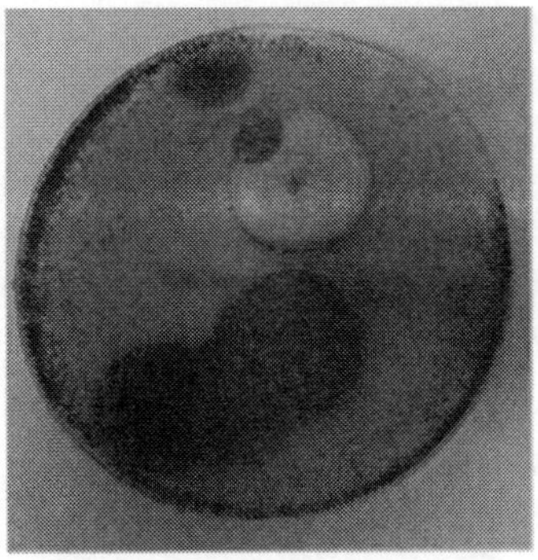

**Figure 4.** Sampling spot 2.

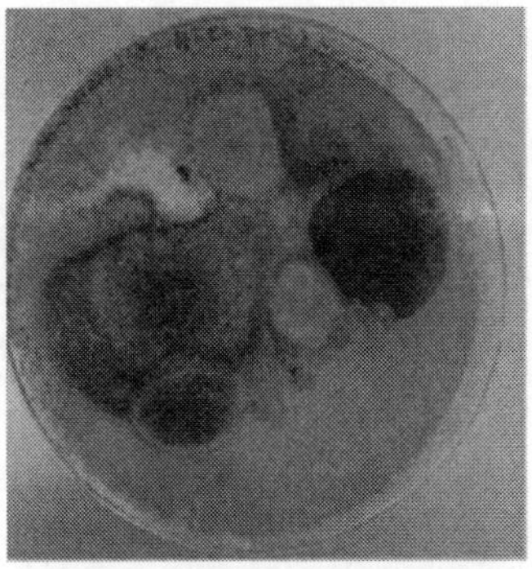

**Figure 5.** Sampling spot 3.

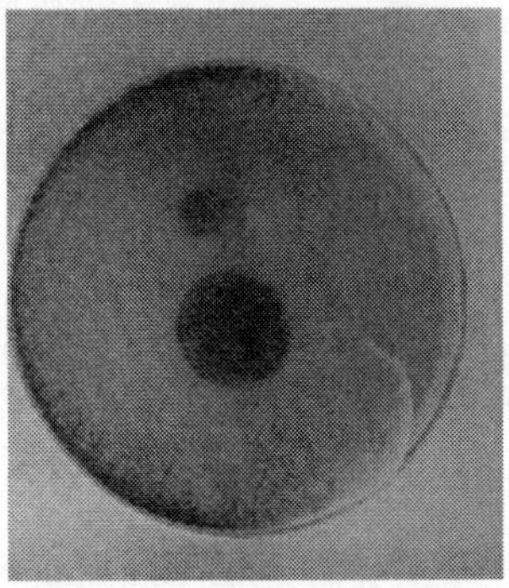

**Figure 6.** Sampling spot 1.

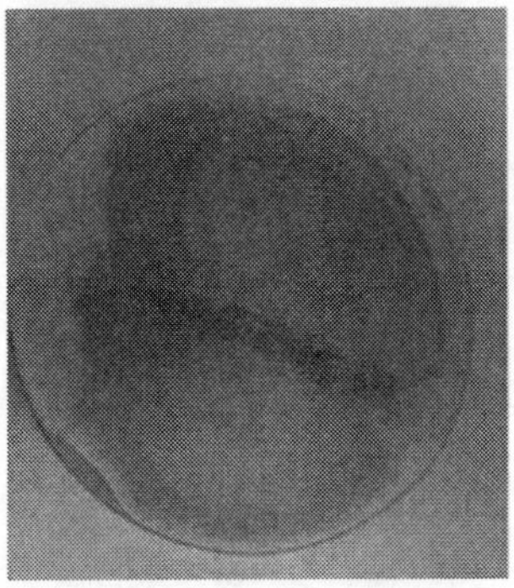

**Figure 7.** Sampling spot 2.

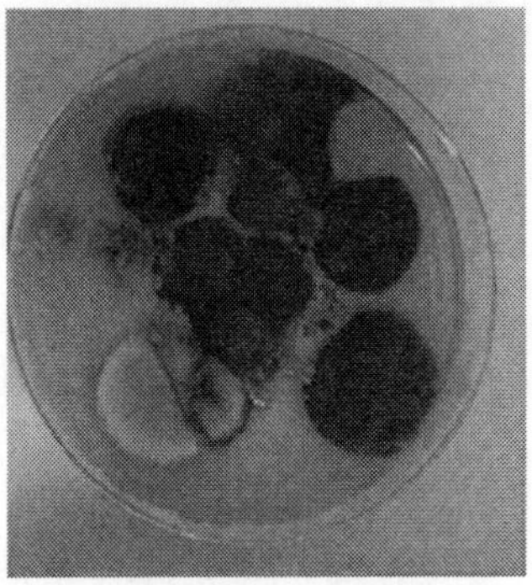

**Figure 8.** Sampling spot 3.

For both the first and the second sampling points, the difference between the two swabs cultivation is evident. The contamination of the second sampling point after abrasion was minimal, indicating a fast growing species. The difference between findings at individual sampling points confirms high contamination only on the surface of the façade, under the surface the infestation is lower.

At the third sampling point, a relatively strong colonization through the entire layer of plaster occurs, biofilm was evident even during sampling. Identification of microbiological genera is presented in Table 1.

**Table 1.** Identification of microorganisms found in swabs.

| **Sampling spot** | **1** | **2** | **3** |
|---|---|---|---|
| Genera identified before plaster abrasion | *Rhizopus sp., Cladosporium sp., Ulocladium sp., Aureobasidium sp.* | *Rhizopus sp., Cladosporium sp., Acremonium sp., Stachybotrys sp., Ulocladium sp., Aureobasidium sp.* | *Cladosporium sp., Stachybotrys sp., Chrysonilia sp., Trichoderma sp.* |
| Genera identified after plaster abrasion | *Rhizopus sp., Cladosporium sp., Epiccocum sp., Chrysonilia sp.* | *Chrysonilia sp.* | *Cladosporium sp., Ulocladium sp., Acremonium sp., Stachybotrys sp* |

Table 2 presents the average number of moulds and yeasts in the plaster bulk, their identification and also the degree of algae contamination. The microbial image of the plaster bulk corresponds with the results obtained by swabs.

**Table 2.** Plaster bulk genera identification and algae contamination.

| **Sampling spot** | **1** | **2** | **3** |
|---|---|---|---|
| Average mould and yeast CFU/g of plaster | $1.1 \cdot 10^3$ | $7 \cdot 10^2$ | $1 \cdot 10^5$ |
| Genera identification | *Ulocladium sp.* | *Cladosporium sp., Ulocladium sp., Stachybotrys sp., Phoma sp.* | *Cladosporium sp., Ulocladium sp., Acremonium sp., Auerobasidium sp., Penicillium sp., Phoma sp., Stachybotrys sp., Chaetomium sp.* |
| Algae contamination | infested | not infested | strongly infested |

The results show that the façade was least infested at sampling point 2, where the contamination under the plaster was minimal and the green algae contamination unproved. However according to literature, a healthy plaster contains up to 1 x $10^2$ CFU/g [8] only, so the sampling spot must be considered infested by mould growth. On the other hand, at the sampling point 3, the façade was massively colonized, which corresponds with its location. Sampling point 3 is found in the northwest arc of the object, adjacent to the park, where vegetation of bushes closely surrounds the building.

As a supplement to the analyses, microbial contamination of air was determined outside and inside the building at sampling points 1 and 2. The outdoor sampling results corresponded with the condition of the plaster as at sampling point 1 and 2 the airborne contamination was higher than 600 CFU/$m^3$ and 237 CFU/$m^3$, respectively. The indoor air contamination in the classes adjacent to the sampling points exhibited only small variability between each other which was most certainly caused by different use of space and cleaning. Slightly higher occurrence of colonies was observed in sampling spot 2, 230 CFU/$m^3$, while 180 CFU/$m^3$ were found in sampling spot 1. Monitoring of the remediation and repair impact on contamination of the indoor environment will be the subject of further research.

## 4. Conclusions

The analyses prove infestation of the façade through the plaster at all sampling points, i.e. on the surface, in the plaster bulk and also under the top layer. Among the sampling spots, spot 2 was the least colonized, while spot 3 was massively infested and therefore created a new source of further air contamination. These findings correspond with the location of the sampling spots, as spot 3 is located on the north-west side of the building and surrounded with adjacent park vegetation. Based on the analyses, replacement of the contact insulation system layer was recommended together with elimination of trees and bushes from the closest vicinity of the building.

The results confirm contamination of the plaster surface by airborne spores and the factual microbial infestation of the plaster layer is represented by samples from plaster bulk and the second swab.

**Acknowledgements**

This paper was elaborated with the institutional support for long-term development of research organizations by the Ministry of Industry and Trade of the Czech Republic.

## References

[1] Büchli R and Raschle P 2004 *Řasy a houby na fasádách* (Ostrava: MISE, s.r.o.) p 108 ISBN 978-80-254-8786-0

[2] Wasserbauer R 2001 *Biologické znehodnocení staveb* (Praha: ABF) p 257 ISBN 80-86165-30-2

[3] Klánová K 2013 *Plísně v domě a bytě: odstraňování a prevence* (Praha: Grada) p 104 ISBN 978-80-247-4790-3

[4] Flannigan B, Samson R A and Miller J D 2001 *Microorganisms in home and indoor work environments: diversity, health impacts, investigation and control* ( Boca Raton [Fla.]: CRC Press) p 529 ISBN 0415268001

[5] Valeš 2016 *Odběr vzorků pro laboratorní diagnostiku biodeteriogenů fasád kombinovanou metodou „stěr – obrus – stěr"* Report from the project No. TA 04010837 (Praha: ČVUT – Kloknerův ústav) p 29

[6] ČSN EN 15458 (672032) Paints and varnishes - Laboratory method for testing the efficacy of film preservatives in coating against algae

[7] Klánová K 2002 *Standardní operační postupy pro vyšetřování mikroorganismů v ovzduší a pro hodnocení mikrobiologického znečištění ovzduší ve vnitřním prostředí* (Praha: SZÚ Praha) p 21

[8] Klánová K 2005 Plísně na zateplené fasádě *Tepelná ochrana budov* **1** p 15–17

IOP Conf. Series: Materials Science and Engineering **379** (2018) 012036 doi:10.1088/1757-899X/379/1/012036

# Analysis of temperature degraded concrete at high temperature by applying of non-linear characteristics

**L Pazdera[1], L Topolář[1], M Hoduláková[1], R Dvořák[1] and K Mikulášek[1]**

[1] Brno University of Technology, Faculty of Civil Engineering, Czech Republic

Email: pazdera.l@fce.vutbr.cz

**Abstract**. Concrete structures are commonly exposed to thermal loads as a result of the structure, ambient conditions, the heat of hydration, or exposure to fire. Therefore, there has been a growing interest in the research of advanced monitoring and analysis of concrete structures subjected to thermal load. Non-linear characteristics have been used to identify thermal damage evolution in concrete structures. The present paper investigates the effects of a high temperature on selected physical properties of concrete. Concrete properties were monitored and analyzed in several thermal steps up to 1200°C. Concrete specimens were heated in a programmable laboratory furnace at a heating rate of 5°C/min and loaded at six temperatures, 200°C, 400°C, 600°C, 800°C, 1000°C, and 1200°C, with each maintained for 60 minutes.

## 1. Introduction

Interest in the behaviour of concrete under high temperatures has recently been fuelled by fires of industrial, government- or private-owned buildings, tunnels, and other building structures. An interesting principle of non-destructive testing of heat-degraded concrete at high temperatures is presented [1]. The method used involves measuring the wave propagation through material structures and its effect on higher harmonics. Therefore, the frequency spectrum of each structure was computed using fast Fourier transform. Structural Health Monitoring aims to improve the knowledge of the safety and maintainability of civil structures and infrastructures [2, 3].

There are changes expected if concrete is heated up to [4]:

| | |
|---|---|
| 200°C | slow capillary water loss and reduction in cohesive forces as water expands, ettringite dehydration, C-S-H gel dehydration, gypsum decomposition ($CaSO_4$ $2H_2O$), physically bound water loss, |
| 400°C | break up of some siliceous aggregates (flint), |
| 600°C | portlandite decomposition $Ca(OH)_2 \rightarrow CaO + H_2O$, quartz phase change $\beta-\alpha$ in aggregates and sands, |
| 800°C | second phase of the C-S-H decomposition, formation of $\beta$-$C_2S$, |
| 1000°C | dolomite decomposition, calcite decomposition $CaCO_3 \rightarrow CaO + CO_2$, carbon dioxide release, ceramic binding initiation which replaces hydraulic bonds, |
| 1200°C | basalt melting, |
| over 1200°C | total decomposition of concrete, melting. |

Poor material homogeneity and, in some cases, shape complexity of some units used in the building industry, are heavily restricting the applicability of "classical" ultrasonic methods [5, 6]. The dynamic material response of samples with damaged material integrity was demonstrated by the formation of different harmonics in the response frequency spectra [7]. The pulse echo method monitors changes in the material structure based on a change in the response to a pulse generated. Using a mechanical impact to induce microstrain is advantageous for concrete testing because it allows for testing of larger concrete specimens offering potential field transportability [8].

## 2. Experimental set-up

This part describes the experiment and the evaluation of the generated ultrasound signals when testing their passage through the material of concrete specimens after a heat load.

The tests were performed on beams with a cross section of 100 mm by 100 mm and length of 400 mm as shown in Figure 1. Three types of beams were made according to Table 1 with the kind of the coarse aggregate having been thus changed. Starting at a normal temperature of 22°C, the specimens were heated at a heating rate of 5°C/min up to a pre-set temperature maintained for 60 minutes and then placed in a room with a normal temperature of 22°C.

**Table 1.** Mixture monitored.

| | A<br>(kg/m³) | B<br>(kg/m³) | C<br>(kg/m³) |
|---|---|---|---|
| cement CEM I 42.5R | 345 | 345 | 345 |
| sand 0/4 | 813 | 813 | 813 |
| coarse aggregate 4/8 | | | 1010 |
| coarse aggregate 8/16 | 1010 | | |
| coarse aggregate 11/22 | | 1010 | |
| water | 176 | 176 | 176 |
| super plasticizer | 3 | 3 | 3 |

The pre-set temperatures were 200°C, 400°C, 600°C, 800°C, 1000°C, and 1200°C. The beams were tested at a normal temperature.

**Figure 1.** Experimental set up.

The specimens were excited by a sensor placed in the centre of a square side. Impulses were generated by the ultrasound equipment PUNDIT. The acoustic emission system XEDO was applied to detect the waves. Three piezoelectric sensors were placed on different sides of the specimen. The first sensor was situated near the exciter on the perpendicular side, the second one on the opposite side of the specimen being 50 mm away from the opposite side on the side perpendicular to the first one and the third sensor was on the same side as the first one but 50 mm away from the opposite edge.

Twenty specimens, made after thermal degradation by each temperature, were monitored by these tests. The errors of the measured and computed values do not exceed five percentage points.

## 3. Results

The results are presented as an analysis of the first eigenfrequency computed from the frequency spectra of the specimen response characteristic as shown in the following tables and figures.

**Table 2.** Cube compression strengths of three different mixtures.

| Temperature (°C) | A (MPa) | B (MPa) | C (MPa) |
|---|---|---|---|
| 20 | 65 | 75 | 50 |
| 200 | 63 | 63 | 43 |
| 400 | 54 | 57 | 48 |
| 600 | 46 | 41 | 28 |
| 800 | 31 | 31 | 24 |
| 1000 | 14 | 8 | 6 |
| 1200 | 16 | 10 | 28 |

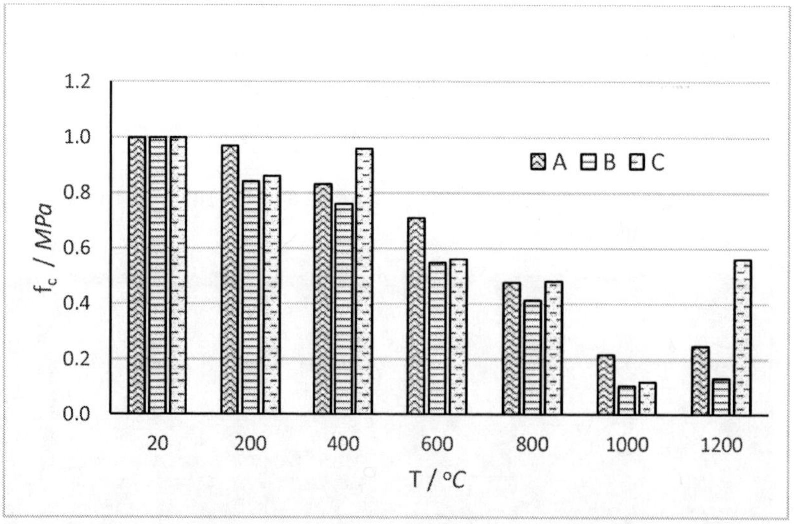


**Figure 2.** The cube compression strength ($f_c$) on the degraded temperature (T).

From the values of the compression strength on cubes (see Table 2 and Figure 2), it can be concluded that, for all temperatures, mixture B shows the biggest loss of strength relative to a reference temperature. Interesting is the behaviour of mixture C where there is a decrease as expected if heated to 200°C, however, with an increase if heated to 400°C, which could be accounted for by the fractions of the aggregate used being near. Then, an increase in the relative compression strength should also be pointed out at a temperature of 1200°C; for particular mixtures, this increase is caused by a new crystalline form, Wollastonite, being built at a temperature of about 1100°C. This increase is the greatest in mixture C and, again, this could be accounted for by the vicinity of the fractions of the aggregate used.

For analysing the behaviour of the samples by the pulse-echo method, the first harmonic frequencies were calculated of the signal monitored using fast Fourier transform. The values of the eigenfrequencies found for each sample at various degrees of degradation are shown in Table 3. The changes of these values relative to the basic value of the sample set at a temperature of 20°C can be seen in Figure 3.

**Table 3.** Frequencies of three different mixtures.

| Temperature (°C) | A (kHz) | B (kHz) | C (kHz) |
|---|---|---|---|
| 20 | 43.9 | 90.3 | 45.4 |
| 200 | 45.4 | 93.2 | 44.4 |
| 400 | 41.9 | 85.0 | 42.8 |
| 600 | 17.6 | 28.3 | 16.1 |
| 800 | 16.6 | 22.9 | 14.2 |
| 1000 | 7.3 | 14.6 | 14.6 |
| 1200 | 14.6 | 24.7 | 25.4 |

-

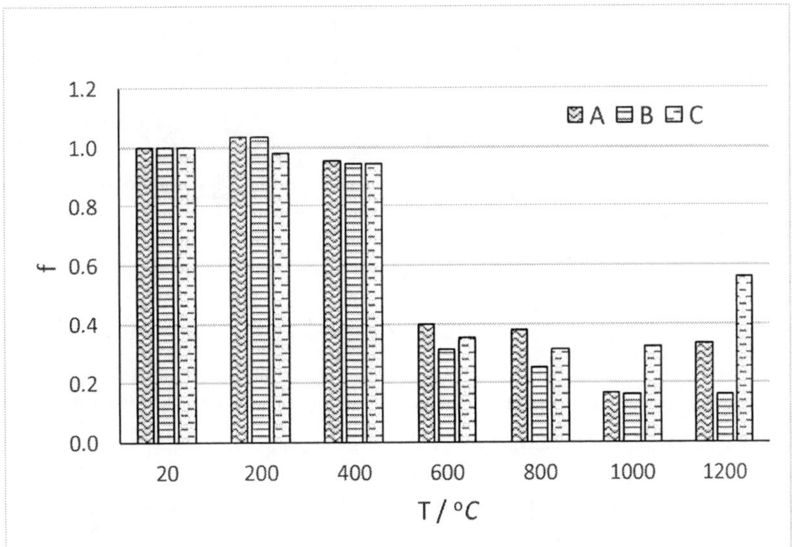


**Figure 3.** The relative eigenfrequency (f) on the degraded temperature (T).

The frequency value increasing after temperature increase to 200°C probably causes an improvement of the structure bonds. This occurred in mixtures A and where there is more space between the aggregate particles for self-repair of the cement due to the ambient humidity during the test. In mixture C with more finer aggregate particles, no such effect has been observed. As the temperature increases further, a decrease is apparent in the frequency values for all mixtures. Only for the last temperature of 1200°C, again, a reaction can be observed to the appearance of Wollastonite of the mixture, which is again most marked in mixture C.

A similar sample behaviour pattern can also be observed if the decrease of the higher harmonics is linearized for particular samples. Here, the expected exponential decrease of the harmonic component amplitudes is linearized by the spectrum being transformed into decibels. The values can be found in Table 4 with their relationships to the basic-temperature sample in Figure 4.

**Table 4.** Slopes of three different mixtures.

| Temperature (°C) | A (dB/kHz) | B (dB/kHz) | C (dB/kHz) |
|---|---|---|---|
| 20 | -0.494 | -0.414 | -0.487 |
| 200 | -0.586 | -0.583 | -0.415 |
| 400 | -0.661 | -0.448 | -0.390 |
| 600 | -1.122 | -0.644 | -0.568 |
| 800 | -1.728 | -0.471 | -0.774 |
| 1000 | -2.383 | -1.842 | -1.647 |
| 1200 | -1.391 | -1.390 | -0.763 |

Note that Table 4 shows the values of the attenuation $k$ for the equation

$$S_n = c \cdot \exp(k \cdot f_n) \quad (1)$$

where $S_n$ is the amplitude of the frequency spectrum, $f_n$ is the frequency, n is the number of the harmonic and c and k are constants. It is clear that $f_n = n \cdot f_1$ where $f_1$ is the first harmonic [9]. A change in the first harmonic along with the slope of the attenuation of the harmonics account for the changes in the observed samples.

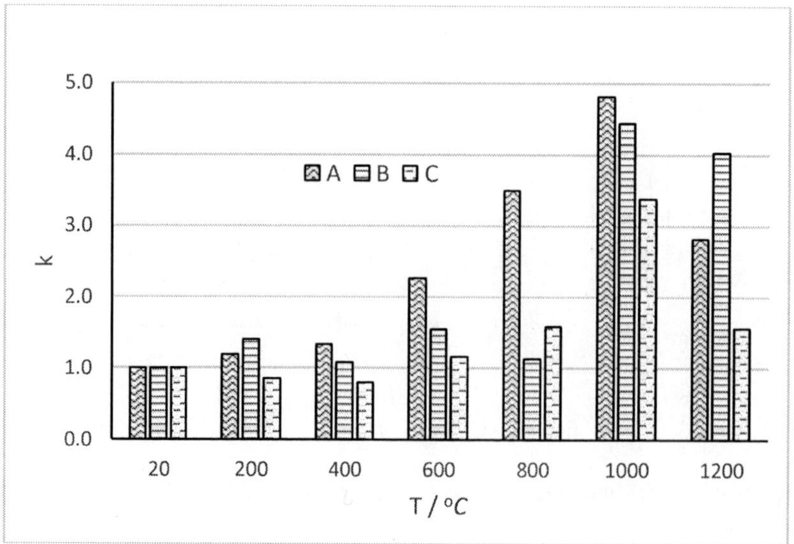


**Figure 4.** The relative slope of linear approximation (k) on the degraded temperature (T).

## 4. Conclusion

The pulse-echo method seems to be well-suited for determining the degree of damage to the heat-degraded concretes. As shown, the results of a non-destructive pulse-echo method can be well correlated with the destructive pressure test. This comparison is mainly evident in Figures 4 (the non-destructive testing of the pulse-echo method) and 3 (the destructive pressure test) when the characters are similar.

In the near future, we expect the methodology of these techniques to be developed and applied to various materials testing procedures. This method may also help create or improve a numeric model of wave propagation in concretes of different structures. From the evaluation of these samples, it can be assumed that, with the increasing size of the coarse aggregates, the thermal resistance of the concrete mixture decreases.

It has been shown that analyzing high-frequency dynamic characteristics generated by a pulse signal may indirectly influence the nature of an observed structure. The changes in the compression strength of heat-degraded samples are correlated with the changes in the dynamic properties of a structure.

**Acknowledgement**
This paper has been written under the project GAČR No.16-02261S supported by Czech Science Foundation and the project No. LO1408 "AdMaS UP - Advanced Materials, Structures and Technologies", supported by Ministry of Education, Youth and Sports under the „National Sustainability Programme I".

**References**
[1] Felicetti R, Monte F Lo and Pimienta 2017 A new test method to study the influence of pore pressure on fracture behaviour of concrete during heating *Cement and Concrete Research* **94** pp 13–23
[2] Ditommaso R and Ponzo FC 2015 Automatic evaluation of the fundamental frequency variations and related damping factor of reinforced concrete framed structures using the Short Time Impulse Response Function (STIRF) *Engineering Structures* **82** pp 104–112
[3] Rashidyan S, Ng TT and Maji A 2017 Estimating the Depth of Concrete Pier Wall Bridge Foundations Using Nondestructive Sonic Echo *Journal of Nondestructive Evaluation* **36**(3) art. No~56
[4] Piasta J 1989 Heat deformations of cement paste phases and the microstructure of cement past *Materials and Structures* **17**(6) pp 415–420
[5] Neild SA, Williams MS and McFadden PD 2003 Nonlinear vibration characteristics of damaged concrete beams *Journal of Structural Engineering-Asce* **129**(2) pp 260–268
[6] Miyamoto A, King Mw and Fujii M 1991 Analysis of Failure Modes for Reinforced-Concrete Slabs under Impulsive Loads *ACI Structural Journal* **88**(5) pp 538–545
[7] Korenska M and Manychova M 2010 New possibilities of non-destructive testing of ceramic specimen integrity *Ceramics-Silikaty* **54**(1) pp 72–77
[8] Jin J, Moreno MG, Riviere J and Shokouhi P 2017 Impact-Based Nonlinear Acoustic Testing for Characterizing Distributed Damage in Concrete *Journal of Nondestructive Evaluation* **36**(3) arc. No~51
[9] Pazdera L, Smutny J and Topolar L 2011 Applying High Order Statistics Analysis at Non Destructive Evaluation of Concrete Tiles *Experimentalni Analyza Napeti - Experimental Stress Analysis* (14 EAN) pp 309–314

# Experimental evaluation of the effect of carbon fibres on acoustic emission parameters obtained during compressive strength tests of alkali-activated slag composites

**L Topolář[1], P Rovnaník[1] and P Schmid[1]**

[1] Faculty of Civil Engineering, Brno University of Technology, Veveří 331/95, Brno 602 00, Czech Republic

E-mail: Libor.Topolar@vutbr.cz

**Abstract**. The generation of acoustic emission signals is directly associated with formation of cracks in materials during loading. This paper deals with possibilities of acoustic emission method application as the tool for the identification of structural damage in alkali-activated composite materials during compressive strength test. In experimental part, the three piezoelectric sensors were occupied for the continuous record of emission signals of stressed material feedback on applied mechanical load in real time. Detection of specific acoustic emission signals in the course of deformation of the test samples indicates that irreversible structural changes occur in the composite. Four different mixtures of alkali-activated slag mortars were prepared, the first one was reference without the addition of carbon fibres. The others contain the carbon fibres in amount 0.5, 1.0 and 2.0 % from the weight of the slag.

## 1. Introduction

The application of waste products, some of which show hydraulic properties, has grown as an another possible to Portland cement in civil engineering applications [1, 2]. Alkaline activation has become a beneficial procedure to obtain high strength composites without ordinary Portland cement [3]. The advantage is a reusing a waste product and also lower environmental and economic costs related to clinker production, can be reduced. For these reasons, new and more environmentally sustainable composites are being developed [4, 5].

The major advantage composites of alkaline-activated slag are their resistance against aggressive agents for their low porosity [6]. Their main disadvantage is shrinkage drying. The addition of short fibers to brittle composites represents a commonly used technique to improve their mechanical properties, especially those related to their tensile and shrinkage behaviors [7]. Fibers can be made on different materials such as steel, PE, PVA, PP, AR glass fibers or even carbon [8].

Carbon fibers offer among other things, an advantage such as that they can be used for electrical properties improvements of composites [9]. This improvement lies include the measurement of strain or damage sensing, heating control, EMI shielding and so on. And some application has been successfully applied in civil engineering structures, always using with cement composites [10, 11]. To achieve these functionalities, a certain level of electrical conductivity is necessary. Therefore, as concrete is a bad electrical conductor, conductive admixtures are needed. Several researchers have focused on these conductive admixtures in order to achieve a better electrical behavior without deterioration the composite's mechanical properties [12].

This paper deals with possibilities of acoustic emission method application as the tool for the identification of structural damage in alkali-activated composite materials during compressive strength test. Acoustic emission (AE) is a term used for describing elastic waves generated by abrupt localized changes in a stress field within a solid [13]. These waves propagate away from a source and cause transient surface displacements that can be measured with piezoelectric transducers see Figure 1.

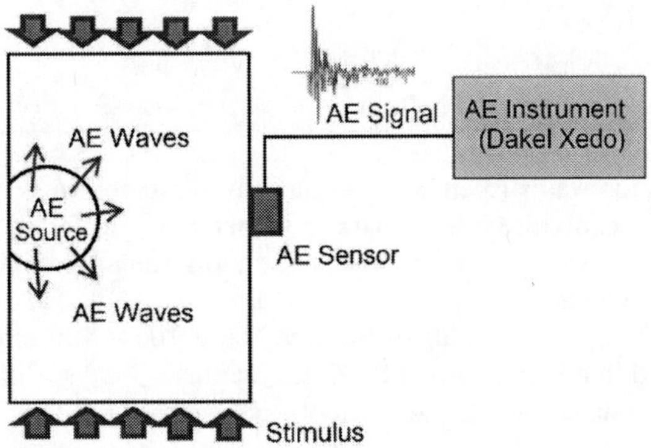


**Figure 1.** Principle of Acoustic Emission measurement.

Acoustic emission signals which were received during the compressive strength test were analyzed. The attention was focused on three basic parameters: the number of events, and the amplitude, duration of the AE signals (Figure 2). Different types of cracks generate different AE signals, and these differences can be related to the properties of the material [14, 15].

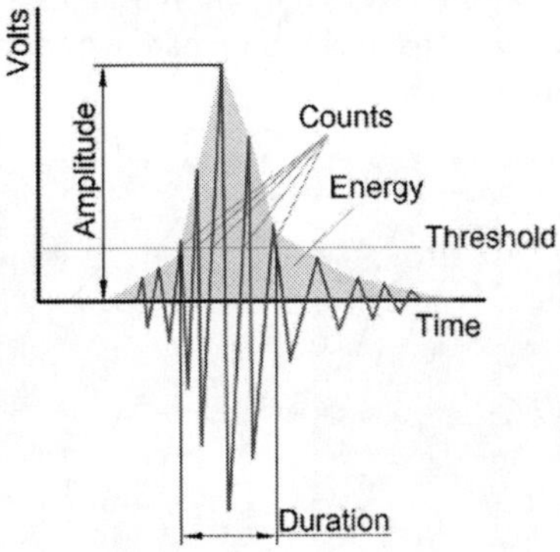


**Figure 2.** Parameters of Acoustic Emission signal.

## 2. Experimental Setup

### *2.1. Materials*

Principal samples for the testing of electrical properties were prepared by alkaline activation of ground granulated blast furnace slag. The slag was supplied by Kotouč (CZ) and its fineness was 383 $m^2/kg$ by Blaine. The average grain size of the slag obtained by laser granulometry was $d_{50}$ = 15.5 μm and $d_{90}$ = 38.3 μm. Solid sodium water glass SUSIL MP 2.0 (Vodní sklo, CZ) having a $SiO_2/Na_2O$ ratio of 2, was used as alkaline activator. Since all properties were tested on mortars, standard quartz sand with maximum grain size of 2.5 mm was used as aggregate. Carbon fibres Tenax A HT C124 having the average length of 3 mm were admixed to basic AAS mortar in order to enhance the electrical properties. The amount of carbon fibres added was 0.5, 1, and 2 % with respect to the mass of the slag. The composition of mixtures is presented in Table 1.

**Table 1.** Composition of AAS mixtures.

| Component | Slag (g) | SUSIL (g) | CF (g) | Sand (g) | Water (ml) |
|---|---|---|---|---|---|
| REF | 450 | 90 | - | 1350 | 185 |
| CF 0.5 | | | 2.25 | | 200 |
| CF 1.0 | | | 4.5 | | 218 |
| CF 2.0 | | | 9 | | 230 |

Sodium silicate activator was suspended and partially dissolved in water. Then, slag and quartz aggregate was added and the mixture was stirred in a planetary mixer for about 5 min to prepare a fresh mortar. Finally, carbon fibres were added into the slurry and further mixed for about 5 minutes in order to separate fibres from the clusters.

The mixes were cast into cubic moulds of the size 100 × 100 × 100 mm. After 24 h the hardened specimens were immersed in a water bath at 20°C for another 27 days. Before mechanical testing, all specimens were dried for 5 days at 80°C to get rid of excess moisture.

*2.2. Methods*

When loading into the destruction, each sample was placed in a testing FORM+TEST hydraulic press with a range of 0 – 200 kN (Figure 3). In addition, three acoustic type-IDK 09 emission sensors with 35dB preamplifiers were fixed to each sample using beeswax. Acoustic emission signals were taken by measuring equipment DAKEL XEDO. Universal measurement and diagnostic system DAKEL-XEDO was developed by ZD Rpety-Dakel company. XEDO is a modular system. One communication card and up to 15 various input cards can be located in a metal box. Communication between cards within a box is realized by hi-speed bus. An allows sampling of the signal from one sensor (speed up to 8 MSamples/sec), enumerates standard acoustic emission parameters, process emission events parameters for possible emission source localization.

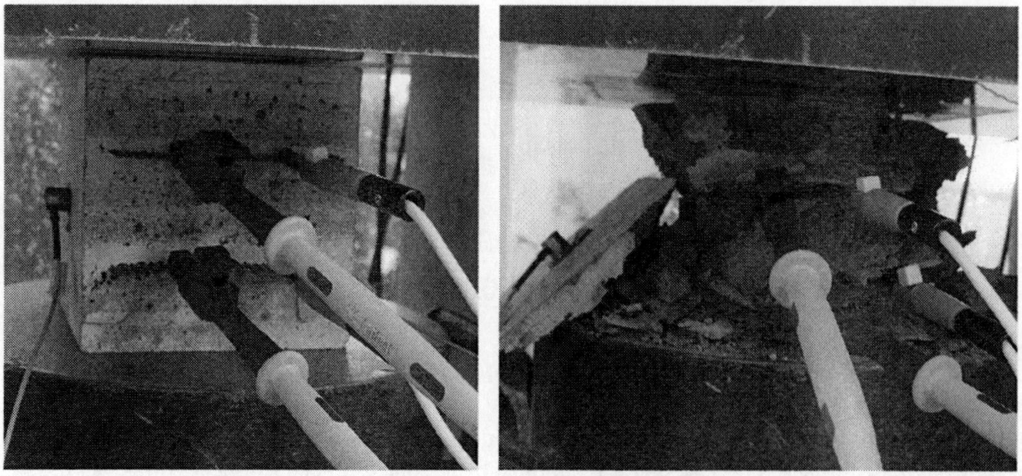

**Figure 3.** Sample during and after the test.

## 3. Results and Discussions

The Figure 4 shows maximal force achieved for each mixture. The increasing amount of carbon fibres caused a decrease in the maximal force.

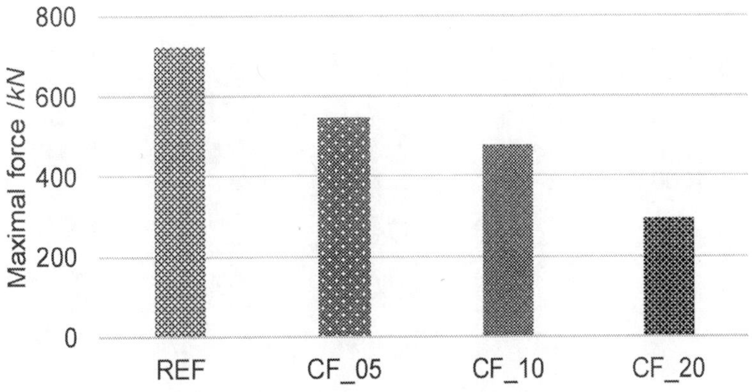


**Figure 4.** Maximal force for each mixture.

The AE parameter values measured during sample loading until destruction (Figure 5–7) already show the different response of the samples to loading.

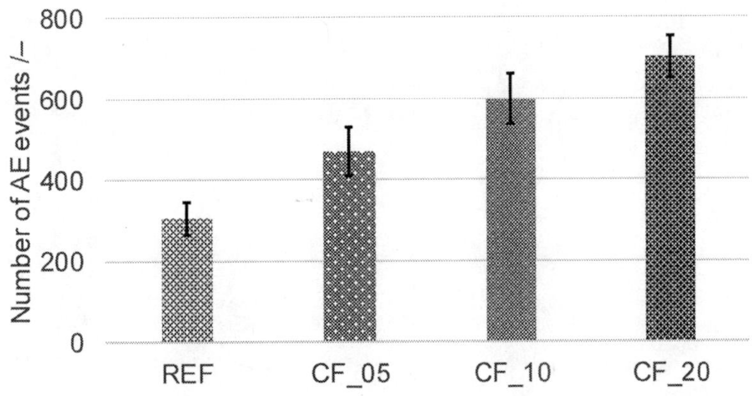


**Figure 5.** Number of AE events for each mixture.

If the AE sensor captures a signal over a definite level, an AE event is detected and recorded. The number of AE events accords to the material's ability to resist destruction. The occurrence of a large number of cracks generates a rather large number of events prior to destruction. The lowest number of events was shown by the reference sample (Figure 5). As the amount of carbon fibers increases, the number of AE events becomes higher. It can be assumed that the greater the number of acoustic events detected during the process of destruction, the smaller is the resistance against complete destroy.

The amplitude is the greatest measured voltage in a caught AE signal (Figure 2). This is a very significant parameter in AE inspection because it determines the detectability of the AE signal. Signals with small amplitudes (below the operator-defined minimum threshold) are not detected and not even recorded. The higher amplitudes indicate more significant structural damage to the sample. The amplitude values varied between 1852 and 2216 mV. With the addition of carbon fibers, the amplitude slightly decreased, which is probably caused the easier crushing of the matrix by the samples with addition of carbon fibers. The highest amplitude was recorded by the reference samples (Figure 6).

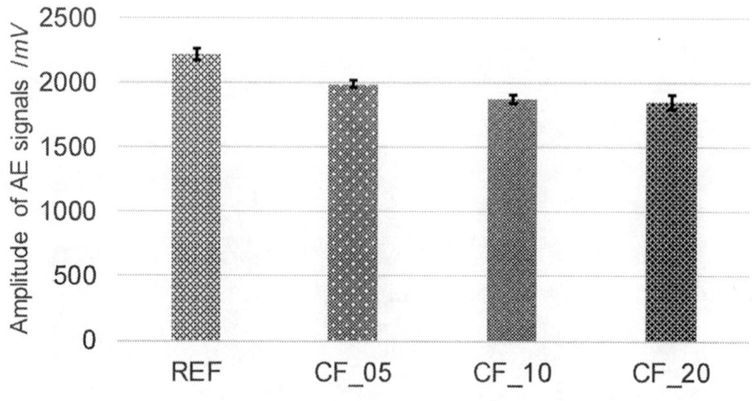


**Figure 6.** Amplitude of AE signals for each mixture.

The duration of AE signal is the time difference between the crossing of the first and last threshold (Figure 2). As the amount of carbon fibers increases, the value of the duration of AE signal becomes slightly higher too (Figure 7), but this increase is not significant.

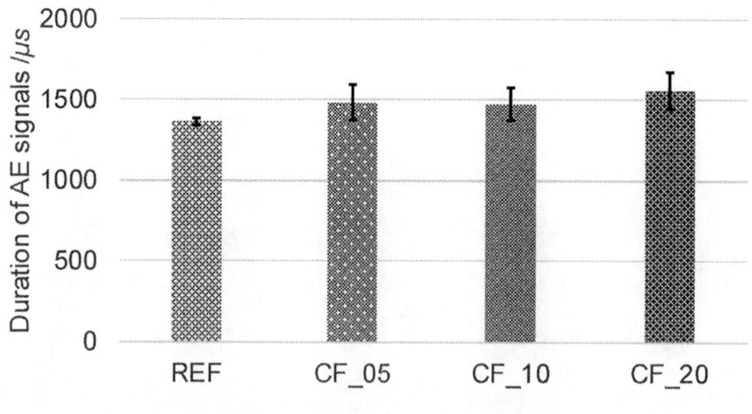


**Figure 7.** Duration of AE signals for each mixture.

## 4. Conclusions

Studying the properties of alkali-activated composites under loading is equally important for civil engineering as the study of steel in mechanical engineering. To have a good understanding of the changes in microstructure under loading is instrumental in understanding the behaviour of alkali-activated-based building materials. Using acoustic non-destructive methods in the civil engineering is not easy as many building materials are heterogeneous. This paper presents the experiments focused on a detailed analysis of selected acoustic emission parameters captured during the compressive strength test of alkali-activated composites. The acoustic emission method is often used for the failure of a structure at an early stage of damage long before the structure breaks down. The acoustic emission method proves to be an efficient instrument for detecting the appearance of micro-cracks when fine-grained alkali-activated composites are loaded.

The conclusion may be summarized as follows:

- a large number of the acoustic emission events are caused by the increased number of micro-cracks in the samples; the samples without and/or with lower amount carbon fibers showed smaller acoustic emission activity during the test, and thus probably fewer micro-cracks generated,
- the higher amplitudes indicate more significant structural damage to the sample especially at the end of the test,
- the addition of carbon fibers has not significantly influence on duration of AE signals.

**Acknowledgement**
This outcome has been achieved with the financial support of the Czech Science Foundation under project No 16-00567S and the Ministry of Education, Youth and Sports of the Czech Republic under the "National Sustainability Programme I" (project No. LO1408 AdMaS UP).

**References**

[1] Ravikumar D and Neithalath N 2012 Effects of activator characteristics on the reaction product formation in slag binders activated using alkali silicate powder and NaOH *Cem. Concr. Compos.* **34** 7
[2] Bilim C and Duran C 2012 Alkali activation of mortars containing different replacement levels of ground granulated blast furnace slag *Constr. Build. Mater.* **28** 1
[3] Fernández-Jiménez A and Palomo A 2009 Properties and uses of alkali cements *Rev. Ing. Constr.* **24** 3
[4] Rashad A M 2013 A comprehensive overview about the influence of different additives on the properties of alkali-activated slag – A guide for Civil Engineer *Constr. Build. Mater.* **47**
[5] Natali A, Manzi S and Bignozzi M C 2011 Novel fiber-reinforced composite materials based on sustainable geopolymer matrix *Proc. Eng.* **21**
[6] Puertas F, de Gutiérrez R, Fernández-Jiménez A, Delvasto S and Maldonado J 2002 Alkaline cement mortars. Chemical resistance to sulfate and seawater attack *Mater. Constr.* **52** p 267
[7] Melo Neto A A, Cincotto M A and Repette W 2008 Drying and autogenous shrinkage of pastes and mortars with activated slag cement *Cem. Concr. Res.* **38** 4
[8] Alcaide J S, Alcocel E G, Puertas F, Lapuente R and Garcés P 2007 Carbon fibrereinforced, alkali-activated slag mortars *Mater. Constr.* **57** p 288
[9] Baeza F J, Chung D D L, Zornoza L, Andión L G and Garcés P 2010 Triple percolation in concrete reinforced with carbon fiber *ACI Mater. J.* **107** 4
[10] Baeza F J, Galao O, Zornoza E and Garcés P 2013 Multifunctional cement composites strain and damage sensors applied on reinforced concrete (RC) structural elements *Materials* **6** 3
[11] Veselý V, Keršner Z and Merta I 2017 Quasi-brittle Behaviour of Composites as a Key to Generalized Understanding of Material Structure *Procedia Engineering* **190**
[12] Baeza F J, Galao O, Zornoza E and Garcés P 2013 Effect of aspect ratio on strain sensing capacity of carbon fiber reinforced cement composites *Materials & Design* **51**
[13] Ono K 2005 J. Current understanding of mechanisms of acoustic emission *Strain Anal. Eng. Des.* **40** 1
[14] Iwanami M, Kamada T and Nagataki S 1997 Application of AE technique for crack monitoring in RC beams *Cem. Concr.* **51**
[15] Li Z and Xi Y 1995 Application of Acoustic Emission Technique to Detection of Concrete Cracking and Rebar Corrosion *In: NDT-CE: Int. Symposium Non-Destructive Testing in Civil Engineering* (Berlin) pp 613–620

IOP Conf. Series: Materials Science and Engineering **379** (2018) 012038 doi:10.1088/1757-899X/379/1/012038

# Measurement and evaluation proposal of early age shrinkage of cement composites using shrinkage-cone

**B Kucharczyková[1], D Kocáb[1], I Rozsypalová[1], O Karel[1], P Misák[1] and T Vymazal[1]**

[1] Brno University of Technology, Faculty of Civil engineering, Veveří 331/95, Brno, Czech Republic

Email: barbara.kucharczykova@vutbr.cz

**Abstract**. The paper deals with an experimental analysis focused on the determination and evaluation of the early age shrinkage strain using Shrinkage-Cone. The measurements were performed on the cement pastes and mortars with a different *w/c* ratio. The authors propose a new approach to the measurement evaluation based on the determination of *time zero* at the time of estimated initial setting determined as the time when the internal temperature of specimens started to rise. In this case, also the effect of the thermal expansion is taken into account in the evaluation.

## 1. Introduction

The early as well as long-term shrinkage process in cement based composites is an actual problem especially of civil engineers and producers of cement based materials (concrete, mortars, floor compounds etc.). There are many standardized and non-standardized testing techniques which are used for estimation/determination of real value of shrinkage process during the whole time of material ageing, unfortunately the results are often inconsistent, especially because of different equipment, dimensions of test specimens and initial time of the start or evaluation of measurement [1, 2, 3]. Along the development of a high performance concretes and fine grained materials with a low water-cement ratio (w/c), the early age volume changes, caused by the chemical and autogenous shrinkage are studied in detail [4, 5]. The brief-summary of most frequently used testing procedures intended for the experimental measurement of the chemical and autogenous shrinkage can be found e.g. in [6]. Currently, three testing methods for early-age shrinkage measurement standardized by the ASTM are available - Standard test method for chemical shrinkage of hydraulic cement paste [7], Standard test method for autogenous strain of cement paste and mortar [8] and Standard test method for change in height at early ages of cylindrical specimens of cementitious mixtures [9]. The first method is usable for measurements on the pastes only, the second can be used also for the mortars and the third method is proposed to measure on the pastes, grouts, mortars and concretes.

Currently, dealers of the testing instruments are increasingly offering the Shrinkage-Cone. It is a test equipment intended especially for measurement of early age volume changes due to chemical and physical processes, such as the chemical and autogenous shrinkage and expansion [10]. The Shrinkage-Cone method is suggested to be an alternative method to above mentioned standard methods.

The main advantage of the Shrinkage-Cone method in comparison to the other above mentioned methods is the geometric shape of the mould used for measurement which ensures that the relative length and volume changes can be calculated simultaneously based on the measured changes in the height of the cone-shaped specimen [10]. Other advantages are for example the sufficient measurable volume, filling procedure and handling with the equipment.

The main aim of performed experiments was to verify the usability of the test method for fine-grained materials of different consistency and to propose another approach to the evaluation of the measurement using the Shrinkage-Cone equipment. Unfortunately, only small quantity of experimental data is presently available which could be compared to each other. There were not found any data for the measurement performed on the cement paste and also there were not found any results for measurements performed on the cement pastes with different w/c tied together with the measurement on the cement mortars.

## 2. Experimental analysis

### *2.1. Materials*

The measurements were performed for cement pastes and equivalent cement mortars with a different w/c ratio. For manufacturing of both composites CEM I 52.5 R Portland cement was used. Three different w/c ratios for pastes with values of 0.33, 0.40, 0.50 and two different for mortars with values of 0.40 and 0.50 were chosen for the experiment (it is not possible to prepare mortar with w/c = 0.33 without the addition of plasticizer). The standardized quartzite sand with a grain size distribution of 0–2 mm (CEN 196-1 [11]) was used for mortars manufacturing. The aggregate-cement ratio in mortar mixtures was always 3:1. The mixing procedure of the prepared fresh mixtures was based on the standard EN 196-1 [11]. Three measurements were performed for each paste and mortar to obtain the average curve of the strain development.

### *2.2. Shrinkage Cone*

The Shrinkage-Cone is measurement equipment, which allows to measure shrinkage and expansion of the building materials, which are prepared by mixing of solid and fluid [10] components to prepare a fresh mixture that harden due to the chemical processes. The touch-less measurement system allows to start the measurement immediately after the fresh material is poured into the mould. It is important to specify that the geometric shape of the cone-shaped mould ensures that the change of the recorded distance (change in the height of the cone-shaped specimen) corresponds to the length change of the test specimens because the ratio of the radius and height is constant for any volume (the relative length change can be then calculated). Based on this fact, the volume change of the measured material can be calculated as the third power of the ratio of the respective heights. For further information about the Shrinkage-Cone and details about the measurement procedure refer to the technical sheet of equipment available in [10].

For the purpose of the experiment, the shrinkage cone with a nominal height of 125 mm and volume of 682 $cm^3$ was used. To minimize the friction, the cone-shaped Teflon foil (PTFE) was placed into the cone before its filling. During the pouring, the appropriate method of compaction (respecting the workability) was applied to remove the entrapped air and to ensure the required rigidity and smooth top surface of the fresh mixture. To avoid the water evaporation, the top surface was coated by the oil (according to the manufacturer recommendation). Finally, the reflector made of PP coated with the aluminum sheet was placed in the middle of the top surface and the temperature-sensitive sensor was inserted into the fresh mixture. The manufacturing process is illustrated in figures 1–6 below. The measurement data was automatically stored in data loggers with a period of 10 second and 1 minute in the case of distance and temperature measurement, respectively.

The strain-curves presented herein represent relative length changes calculated from the ratio of $\Delta h/h$ in [‰], where $\Delta h$ is the measured change in height (a change in the height of the specimen) and $h$ is the nominal height of the cone-shaped mould. The presented strain curves represent the average value obtained from three measurements performed for each cement paste and mortar.

IOP Conf. Series: Materials Science and Engineering **379** (2018) 012038 doi:10.1088/1757-899X/379/1/012038

**Figure 1.** Cone-mould with PTFE foil.

**Figure 2.** Filling of the mould.

**Figure 3.** Application of the paraffin-oil on the top surface of fresh mixture.

**Figure 4.** Placing of the mould into the measurement frame with laser sensor.

**Figure 5.** Arrangement of measurement - apparatus for shrinkage and temperature measurement.

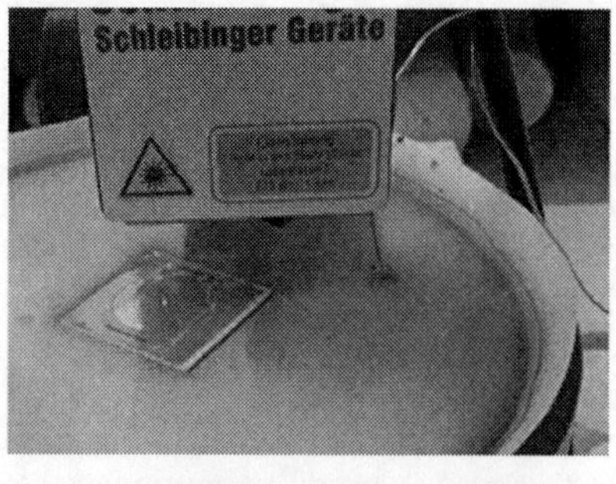


**Figure 6.** Detail of the reflector and internal temperature sensor position.

## 3. Results and discussion

The results of the pilot study are presented in figures below. Before the results are discussed, it is important to specify that the descending trend of the ε-curves indicates the vertical expansion and theirs ascending trend indicates vertical shrinkage. For the purpose of comparison of the results with the standardized methods or other measurement methods based on the measurement of the changes in length of the test specimens, only the changes in height of Shrinkage-cone are presented in the article. Based on this outputs, the volumetric change can be then calculated.

Figure 7 and Figure 8 show the results of measurements performed on cement pastes. Note, that the consistency of cement pastes with w/c = 0.5 was quite slurry. The strain displayed in Figure 7 was calculated as an overall deformation obtained during the whole time of measurement. As is shown in the figure, the cement paste with w/c = 0.33 started to expand immediately after the start of measurement in contrast to the two remaining pastes, which started to shrink slightly for approx. 0.5 hours and after that time they also started to expand. The rapid increase in expansion was slowed down in the estimated time of the initial setting time (the time at which the internal temperature started to rise), between 2.5 and 3.5 hours after start of measurement. When the internal temperature reached its maximum, all cement pastes which were investigated started to shrink, which occurred 6.5, 8.0 and 9.0 hours after the start of measurement for cement pastes with w/c = 0.33, 0.40 and 0.50 respectively.

It is necessary to realize that the behaviour of the material before the time of initial setting is strongly influenced by the stability and consistency of the fresh mixture. The early phenomena are strongly affected by the bleeding, settlement of the particles, air bubbles etc. The initial "shrinkage" can also indicates the settlement of reflector placed on the upper surface of the fresh mixture. For the purpose of measurement evaluation, Eppers and Mueller [12, 13] proposed the *time-zero* which corresponds to the initial setting time determined by an alternative procedure employing an automatic needle penetration apparatus. In their study, the *time-zero* of UHPC was setup between 5 to 6 hours [12]. When respecting their approach, the whole period of expansion (if exists) would be neglected.

A different approach is proposed in this study. The authors propose the *time-zero* at the time of the estimated initial setting determined as the time when the internal temperature of specimens started to rise. In this case, also the effect of thermal expansion can be taken into account in the evaluation.

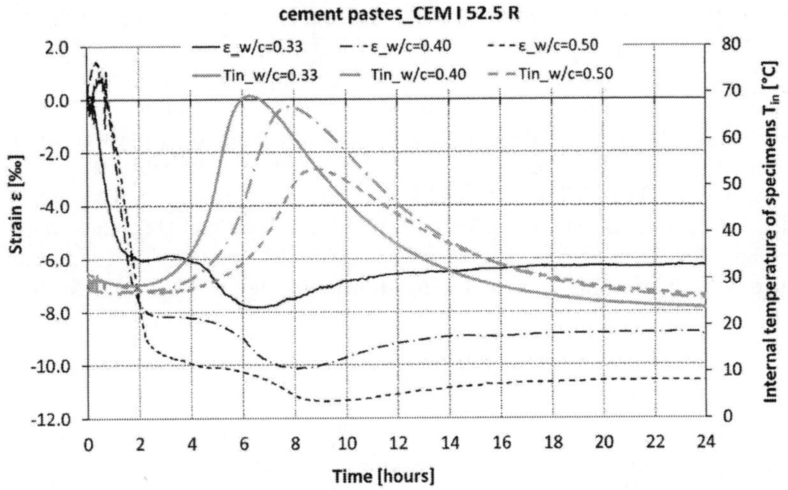


**Figure 7.** Calculated strain of cement pastes with time-zero at start of measurement.

The results with a new *time-zero* are presented in Figure 8. Using this approach, approx. 1/5 to 1/6 of the initial expansion is taken into account in the measurement evaluation. The maximum value of shrinkage that comes after the initial expansion does not exceed the value of initial expansion. This is quite an important fact from the microstructure-cracking formation point of view.

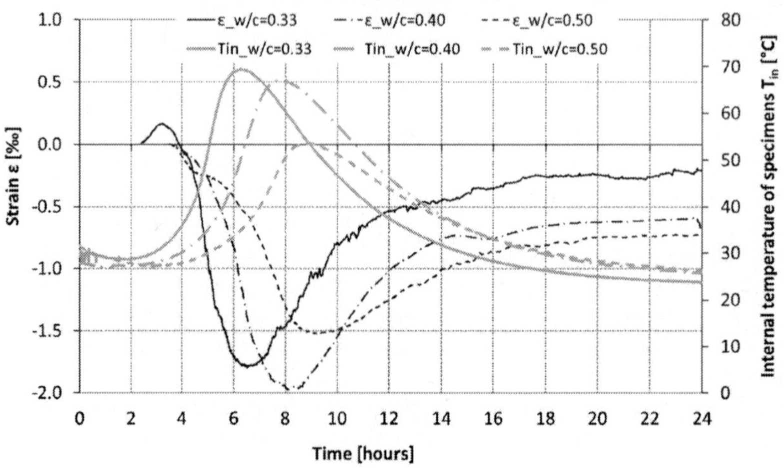


**Figure 8.** Calculated strain of cement pastes with time-zero at the estimated initial setting time.

The results obtained from testing of the cement mortars are displayed in Figure 9 and Figure 10. Two ways of evaluation was also performed. Figure 9 displays the evaluation of the strain with time-zero corresponding to the start of measurement. Quite different behaviour, compared to the cement pastes, was observed. As is shown in figure, both cement mortars started to shrink immediately with the maximum value reached approx. 1 hour after the start of measurement. A higher value of shrinkage was recorded for the mortar with a higher *w/c* ratio. After the initial shrinkage, both cement mortars started to expand slightly. When the time-zero is moved to the estimated time of mortars' initial setting the whole part of the initial shrinkage is neglected (refer to Figure 10) and the evaluated strain started with expansion which turned to shrinkage at the time when the internal temperature of both mortars reached the maximum value. In this case, the shrinkage did not exceed the expansion. It is worth investigating in detail the factors which caused the initial surge of shrinkage which occurred during the first hour of measurement.

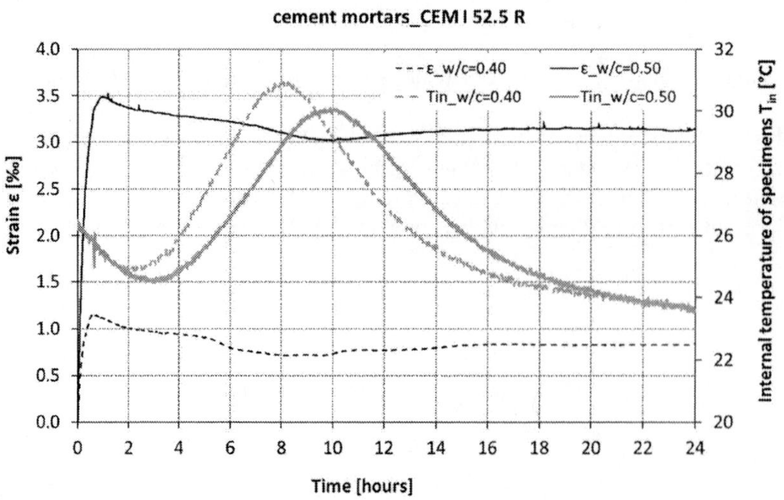


**Figure 9.** Calculated strain of cement mortars with time-zero at start of measurement.

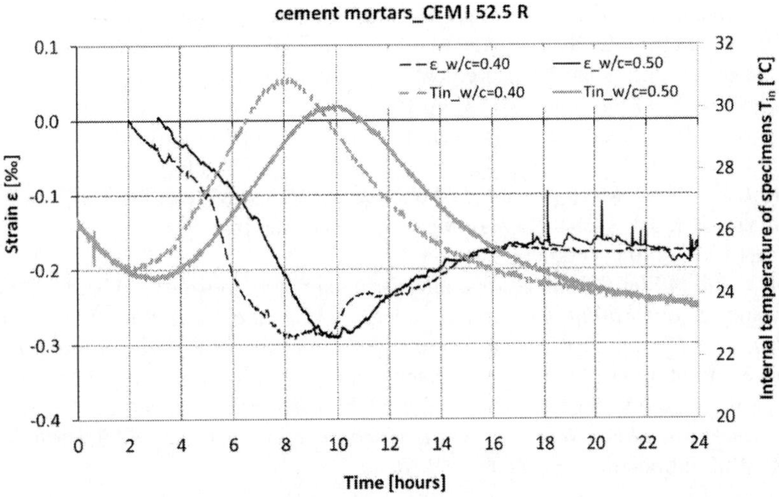


**Figure 10.** Calculated strain of cement mortars with time-zero at the estimated initial setting time.

## 4. Conclusion

Results presented in the paper are a part of the relatively extensive experimental and numerical analysis under a project aimed at the early-age characteristics of cementitious materials. In this project, the Shrinkage-Cone is considered to be a suitable piece of equipment for the measurement and evaluation of the chemical and autogenous deformation of the material during the early solidification process. This measurement device is also considered to be a supplemental measurement to the measurement of chemical shrinkage using the procedure in standard [7] and therefore the measurements were performed on the cement pastes.

The results of performed measurement showed low variability and sufficient sensitivity to measured characteristics. It was proved that this equipment is suitable for measurement of fresh mixtures of a high range of workability. In this context, the appropriate compaction method has to be applied to achieve the proper results with low variability. To achieve the low variability, it is important to ensure that the manufacturing and pouring process as well as the handling with the equipment is the same in all test cases. Nevertheless, there are still open procedural issues that can significantly affect the final measurement results. The measurements performed on the cement pastes with high fluidity can be affected with procedural errors caused by using an inappropriate method of covering the upper surface of the test specimen. The usage of the PP reflector for the measurement is also questionable.

**Acknowledgments**
This work was supported by the Czech Science Foundation [grant No. GA17-14302S].

**References**

[1] Neville A M 2011 *Properties of Concrete 5th* (England: Pearson)
[2] Reinhardt H W nad Grosse C U 2005 *Advanced testing of cement based materials during setting and hardening* (Bagneux: RILEM Publications)
[3] Kucharczyková B, Karel O, Daněk P, Kocáb D and Pőssl P 2017 Comparison of Measurements Methods Intended to Determination of the Shrinkage Development in Polymer Cement Mortars *Procedia Engineering* vol 195 pp 17–23
[4] Ed. by BENTUR A 2003 *Early age cracking in cementitious systems: report of RILEM Technical Committee 181-EAS 'Early Age Shrinkage Induced Stresses and Cracking in Cementitious Systems'* (Bagneux: RILEM Publ)
[5] Čítek D, Rydval M and Kolísko J 2016 Volumetric Changes of the UHPC Matrix and its

Determination *Applied Mechanics and Materials* vol 827 pp 215–218
[6] HU Z, SHI C, CAO Z, OU Z, WANG D, WU Z and HE L 2013 A review on testing methods for autogenous shrinkage measurement of cement-based materials *Journal of Sustainable Cement-Based Materials* vol 2(2) pp 161–171
[7] ASTM C1608-17 (West Conshohocken, PA: ASTM International)
[8] ASTM C1698-09 (2014) (West Conshohocken, PA: ASTM International)
[9] ASTM C827 / C827M-16 (West Conshohocken, PA: ASTM International)
[10] Shrinkage-Cone 2018-01-08 *Schleibinger Testing Systems*
[11] ČSN EN 196-1:2005 (Prague: ČNI)
[12] Eppers S and Müller C 2008 Autogenous shrinkage and time-zero of UHPC determined with the shrinkage cone *8th Int. Conf. on Creep, Shrinkage and Durability of Concrete and Concrete Structures (Proc.)* ed T Tada-aki et al (London: Taylor & Francis Group) pp 709–7014
[13] Eppers S and Müller C 2010 The shrinkage cone method for measuring the autogenous shrinkage - an alternative to the corrugated tube method *Int. RILEM Conf. on Use of Superabsorbent Polymers and Other New Additives in Concrete* ed O M Jensen, M T Hasholt and S Laustsen (RILEM Publications: SARL) pp 67–76

IOP Conf. Series: Materials Science and Engineering **379** (2018) 012039 doi:10.1088/1757-899X/379/1/012039

# New perspective material for building envelopes

**R Cechmanek[1], M Drdlova[1] and V Prachar[1]**

[1] Research Institute for Building Materials, Hnevkovskeho 30/65, 617 00 Brno, Czech Republic

E-mail: cechmanek@vustah.cz

**Abstract.** Thin-walled fibre-cement elements are widely used in residential and industrial buildings for their appearance and durability. Because they are made of light-weight materials, they are mostly used as cladding elements for façades of multi-storey buildings exposed to hard external conditions. Nevertheless, there are still demands from architects and building companies to decrease their weight for reducing manufacturing, transport and assembly costs. In one part of our research light-weighing possibilities of standard composition were evaluated by means of porous structure creation. In the second part the standard aggregate was substituted with available kinds of light-weight porous aggregates in a certain part or in a whole volume. As the limit criterion there was maximal particle size 1 mm for spraying of fibre-cement mixture by means of special air pressure equipment. From the range of possible fillers with different material properties granulated expanded glass worked the best. This paper brings not only a summary of laboratory experiments but also examples of successful realizations such as balcony panelboards, façade elements, bridge structure cladding elements or bulk flower pots. Even though these are not structural elements, it was necessary to evaluate also strength characteristics and resistance to environment for their durability in certain applications.

## 1. Introduction

Owing to the use of various kinds of air-entraining agents or light-weight fillers one can reduce the weight also of fibre-cement composites [1]. Mostly used light-weighing material is perlite [2].

The thin-walled elements with a matrix based on the high-value Portland cement with dispersed reinforcement in the form of alkali-resistant glass fibres are used in a range of applications such as large-sized cladding elements, architectural façade elements, the architectural elements of the urban equipment, balcony panelboards or light-weight walls. Thanks to its durability, low basic weight, colouring ability and shape diversity, glass fibre reinforced concrete may properly replace natural materials [3].

The effect of the use of various air-entraining agents or light-weight fillers on the fibre-cement composites was verified by means of the laboratory tests. Their practical applicability was examined in practice in regard of some elements made of the commonly manufactured glass-fibre-reinforced concrete (GFRC) e.g., channels for deposition of high voltage cables in the Prague Underground, manifold architectural element embedded in the urban space, especially various cladding elements.

The aim of our research was to decrease the total weight of the selected building materials and therefore to eliminate the necessity of using heavy machinery during their manufacturing process and further implementation. The light-weight should be achieved by means of lowering the bulk density and preserving the same dimensions of the used products.

Required bulk density corresponds to the decrease in weight amounting to ca. 25%, i.e. to the value of 1,600 kg/m$^3$ compared to the initial bulk density 2,050 kg/m$^3$ of the standard GFRC elements.

## 2. The possibilities of decreasing the weight of the cement composites

Within a preparation of these fibre-cement composites an exact amount of alkali-resistant glass fibres with length of 12 mm is added into fine-grained matrix consisting of cement, aggregate and admixtures in the last stage of a mixing process and in this way properly dispersed in the whole volume of the mixture. This technology requires the use of highly flowable mixtures free from bleeding and sedimentation of heavier mixture components [4].

By means of various kinds of light-weight aggregates [5] or air-entraining agents it is possible to decrease the weight of the thin-walled fibre-cement composite elements.

In the first part of our research the possibilities of decreasing the weight of the standard composition were evaluated by means of the creation of a porous structure. In the second part, the standard aggregate was substituted – in a certain part or in a whole volume – with the available kinds of light-weight porous fillers. From the range of possible fillers having different material properties, granulated expanded glass turned out to be the most suitable for the proper weight achievement.

Special air pressure equipment was used and maximum particle size 1 mm for spraying of fibre-cement mixture was applied as the limit criterion.

Several compositions were designed by means of the standard mixture modification and substitution of commonly used fine-grained silica sand. Decreasing of the bulk density of the final cement composite was proved, when the filler of distinctive lower density was compared to the standard silica sand.

## 3. Fibre-cement concrete mixture components

Fibre-cement composite consists of approximately 53% binder, 40% filler, 3% fine filler, 3% fibre reinforcement, 1% plasticizer (in weight percents). Several types of filler in different fractions were used to substitute 50 to 100% vol. of silica sand in the initial matrix according to their bulk density [6]. As an exception to the rule, there were so called microspheres that are very specific in their shape. Even a small addition to the mixture caused the increase of fresh concrete mass up to 2 times of its volume. As suitable amount of these micro-balls was approximately 3–4% of the filler weight.

For decreasing the weight of the fibre-cement the following composites were used:

- Hollow polymer microspheres
- Ceramic aggregate
- Expanded vermiculite
- Air-entraining agent

## 4. Technologies for GFRC production

The aim of all the laboratory experiments was to design a modified fibre-cement mixture suitable for the production in the premix technology, which is used in a production department of the Research Institute for Building Materials in Brno.

An exact amount of alkali-resistant glass fibres with length of 12 mm is added into fine-grained matrix consisting of cement, aggregate and admixtures in the last stage of a mixing process and in this way properly dispersed in the whole volume of the mixture. This technology requires the use of highly flowable mixtures free from bleeding and sedimentation of heavier mixture components.

The designed composition for premix is as follows:

- Portland cement CEM I 52.5 40–60%
- Fine-grained aggregate 30–50%
- Fine admixtures 0–10%
- Super-plasticizer 0–5%
- Alkali-resistant glass fibres 1–5%
- Water/cement ratio (–) 0.3–0.5

Due to the cooperation with the company DAKO Brno, Ltd. it was necessary to prepare the compositions for the spraying technology. It means that a continual glass fibre roving is chopped into individual fibres having 35 mm length and incorporated into a sprayed cement-sand wet mixture. The key parameter for passing through a nozzle of the spraying equipment is maximal particle size only 1 mm, therefore only suitable fillers with fractions lower than the limit could be used.

The standard composition consists of these materials:

- Portland cement CEM I 52.5 40–60%
- Fine-grained aggregate 40–60%
- Plasticizer 0–5%
- Alkali-resistant glass fibres 1–10%
- Water/cement ratio (–) 0.3–0.5

## 5. Evaluation of the technological properties of the fibre-cement composites

Basic physical-mechanical properties of the fibre-cement composites were evaluated on test standard specimens (250 × 50 × 10 mm) prepared with the premix technology after 28 days of curing in water. Firstly, bulk density, followed with absorptivity, flexural strength and impact resistance were evaluated for various compositions (see Table 1).

Bulk density and strength of the composites were determined in relation to each other. The better is the first one the poorer is the second one and vice versa. It is evident with flexural strength values, which are influenced by not only low strength of individual particles of the light-weight component, but also by higher level of water/cement ratio in the range of 0.32 to 0.56, which is needed for highly flowable and workable mixtures.

**Table 1.** Properties of light-weight compositions – laboratory.

| Composition | Bulk density (kg/m$^3$) | Flexural strength (MPa) | Absorptivity (%) | Impact resistance (kJ/m$^2$) |
|---|---|---|---|---|
| Standard | 2,050 | 11.0 | < 10.0 | 7.0 |
| Microspheres | 1,430 | 7.5 | 15.0 | 7.0 |
| Ceramic aggregate | 1,650 | 10.5 | 13.0 | 5.0 |
| Vermiculite 75 | 1,620 | 10.0 | 14.0 | 5.0 |
| Vermiculite 100 | 1,500 | 11.0 | 17.0 | 6.0 |
| Air-entraining agent | 1,760 | 9.0 | 13.0 | 7.5 |

## 6. Evaluation in production conditions

The next step was a verification of the parameters from laboratory in real production conditions. It was conducted in cooperation with the company DAKO Brno, Ltd. as three-stage gradual process of the production of the standard GFRC elements [7], [8].

### *6.1. First stage*

For the first stage two compositions of light-weighed fibre-cement composite were selected for the evaluation in the real production conditions. The first composition was designed with the addition of the hollow polymer microspheres and the second one with the air-entraining agent. Three thin-walled balcony panelboards were casted whose dimensions were 1000 × 900 × 20 mm. For impact resistance improvement two balcony panelboards were reinforced with PVA mesh put in the most exposed part; the last panelboard was reinforced with glass fabric. Impact resistance of the completed balcony panelboards, was examined in the Technical and Test Institute for Construction in Prague, Branch Brno, according to the legal standards applicable in the Czech Republic.

**Table 2.** Properties of light-weight compositions – first stage.

| Composition | Bulk density ($kg/m^3$) | Flexural strength (MPa) | Absorptivity (%) | Impact resistance ($kJ/m^2$) |
|---|---|---|---|---|
| Standard | 2,050 | 11.0 | < 10.0 | 7.0 |
| Microspheres | 1,460 | 6.0 | 15.0 | 4.5 |
| Air-entraining agent | 1,620 | 11.0 | 11.0 | 4.5 |

After the conducted tests it could be concluded that all the light-weight balcony panelboards assembled into a metal peripheral frame, corresponding to a real balcony construction, fulfilled the required lack of failure and/or any visible cracks on the panel board surfaces at the impact energy of 150 J.

*6.2. Second stage*

In the next stage the compositions with the addition of ceramic aggregate, expanded vermiculite and again with the air-entraining agent were evaluated in the production of the fibre-cement composites. For the evaluation of the both technologies, i.e. premix and spraying, several planar balcony panelboards reinforced with meshes, façade cladding elements and shaped bridge structure cladding elements were prepared for the comparison. The comparison was done for each variant of the weight-decreasing procedure/composition and regarded its advantages and disadvantages during the production and in the real-world application.

**Table 3.** Properties of light-weight compositions – second stage.

| Composition | Bulk density ($kg/m^3$) | Flexural strength (MPa) | Absorptivity (%) | Impact resistance ($kJ/m^2$) |
|---|---|---|---|---|
| Standard | 2,050 | 11.0 | < 10.0 | 7.0 |
| Ceramic aggregate | 1,290 | 6.5 | 15.0 | 5.0 |
| Vermiculite 100 | 1,550 | 10.5 | 17.5 | 7.5 |
| Air-entraining agent | 1,570 | 8.0 | 9.0 | 6.5 |
| Vermiculite 75 – spraying | 1,630 | 17.5 | 19.5 | 18.0 |
| Air-entraining agent – spraying | 1,850 | 18.5 | 13.5 | 19.0 |

Selected balcony panelboards were evaluated for the resistance to impact loading in the Technical and Test Institute for Construction in Prague, Branch Brno (TTIC) in the same way as the previous variants. The casted bridge structure cladding elements made of air-void fibre-cement composite were examined in the laboratory at the Brno University of Technology (BUT) for their resistance to wind loading. Façade panels made of both compositions by means of the spraying technology were subjected to durability examinations in a specially designed equipment simulating real weather conditions during a longer period of time (whole year), which included repeated exposition to water sprinkling, sun radiation and frost and defrost. The device was built in the testing laboratories of the Research Institute for Building Materials (RIBM). The results of the examination were positive.

*6.3. Third stage*

In the third stage of evaluation we carried out the optimization of the composition with vermiculite for better workability achievement, and, on the other hand, the optimization of the composition with the air-entraining agent – in that case to achieve lower bulk density.

Again, both technologies were used for the production of balcony panelboards, façade cladding elements, bridge cladding elements and a bulk flower pot as well. Despite the fact that the variant using the air-entraining agent was not successful because of the aerating effect was lost during the process of spraying, the weight of the elements with vermiculite was approximately 20% lower in comparison to the standard fibre-cement elements.

**Table 4.** Properties of light-weight compositions – third stage.

| Composition | Bulk density (kg/m$^3$) | Flexural strength (MPa) | Absorptivity (%) | Impact resistance (kJ/m$^2$) |
|---|---|---|---|---|
| Standard | 2,050 | 11.0 | < 10.0 | 7.0 |
| Vermiculite | 1,485 | 10.2 | 16.2 | 5.6 |
| Air-entraining agent | 1,722 | 8.4 | 11.9 | 5.6 |
| Vermiculite – spraying | 1,566 | 11.8 | 17.3 | 14.8 |
| Air-entraining agent – spraying | 1,886 | 12.7 | 12.1 | 12.9 |

At this stage of production several products were chosen for the further experiments. The casted balcony panelboard with vermiculite was examined in TTIC for the impact resistance. The very promising bridge cladding element with vermiculite was examined in the laboratory at BUT for wind resistance. The bulk flower pot served as a prototype for the production in the company DAKO Brno, Ltd. as well as for a small-scale production in RIBM.

Subsequently, the static evaluation of the planar façade panel and the shaped bridge cladding element made of light-weight composition with vermiculite was carried out. All the variants fulfilled the requirements according to the I. as well as the II. ultimate state.

## 7. Granulated expanded glass

Granulated expanded glass is a new perspective material for achieving distinctive lower weight for a range of applications, e.g. [9]. In the Research Institute for Building Materials it was, among others, used as a material suitable for absorbing blast energy [10], [11].

In the next stage of our research we assessed the possibilities of decreasing the weight of the materials by means of substitution of 50–100% vol. of silica sand with the granulated expanded glass [12], in different fractions (0.25–0.5 mm, 0.5–1 mm and 1–2 mm). Glass fibre reinforcement was used in amounts 2.0, 2.5 or 3.0% wt. of total dry mixture.

According to our expectations, the decrease in bulk density corresponded to the amount of the used filler. With 50% silica sand substitution we were able to reach only 1,800 kg/m$^3$ bulk density value, but 75% substitution caused substantial weight decrease to the level of 1,600 kg/m$^3$. The substitution of 100% of silica sand with the granulated expanded glass enabled us for the decrease in bulk density even to 1,500 kg/m$^3$.

Achieved flexural strength was in the interval 9–14 MPa. The higher amount of the expanded glass and the lower amount of glass fibres were applied, the lower values of the flexural strength were observed. The range of results varied between 4.5 and 6.5 kJ/m$^2$ for the impact resistance and between 10 and 15% for the absorptivity.

According to the results presented in table 5 we can conclude that the granulated expanded glass is the most suitable filler for the required bulk density of the GFRC.

**Table 5.** Properties of the cement composites with the weight-decreasing fillers.

| Filler (vol. %) | Bulk density ($kg/m^3$) | Flexural strength (MPa) | Absorptivity (%) | Impact resistance ($kJ/m^2$) |
|---|---|---|---|---|
| Expanded polymer microspheres 15–25 μm (3.5 [a]) | 1,434 | 7.57 | 15.13 | 7.32 |
| Expanded volcanic glass 0–1 mm (100) | 1,726 | 13.95 | 14.65 | 7.43 |
| Granulated expanded clay 0–2 mm (75) | 1,650 | 10.75 | 12.92 | 5.09 |
| Granulated expanded silica sand 0.6–2.5 mm (100) | 1,694 | 6.42 | 15.73 | 5.97 |
| Exfoliated vermiculite 0.25–0.71 mm (100) | 1,501 | 11.12 | 16.98 | 6.00 |
| Granulated expanded glass 0.25–0.5 mm (85) | 1,545 | 10.31 | 13.76 | 5.68 |

[a] This filler was added in the amount of 3.5% wt. from the total sand weight without its substitution.

### *7.1. Durability testing of the bridge cladding elements exposed to wind loading*

For this evaluation a mixture consisting of 3 fractions of granulated expanded glass (GEG) was used. The final fraction was counted as a weight ratio 33:33:33 of fractions 0.25–0.5, 0.5–1 and 1–2 mm.

Two compositions were designed with different sand substitutions and glass fibre additions. The first mixture marked as GEG I consisted of granulated expanded glass (75% substitution) and of 3.5% fibre content. The second mixture marked as GEG II consisted of granulated expanded glass (80% substitution) and the fibre content was decreased to 2.5% for better workability and fluidity of the fresh mixture [6].

For the sake of homogenization, a mixer with rotating central paddle was used in the same way as in the standard production of the glass-fibre reinforced concrete elements with silica sand.

The same mixtures with the expanded glass filler were adapted for a spraying technology and several cladding elements were made for further testing.

This technology has been successfully verified in the production of the company DAKO Brno, Ltd.

The promising bridge cladding elements were examined in the laboratory at BUT concerning their resistance to wind loading [13].

Tested elements were fixed in the same way as the standard cladding elements to simulate real conditions. One part of each element was fixed to steel profiles whereas the other part was not supported. In real construction the second part serves for covering of installation cables and pipelines. This part works as an overhanging end or rather a cantilever under uniform loading.

The effect of uniform loading caused by wind is simulated by the means of vacuum creation. Each element was fixed in a solid box and covered with transparent plastic foil. Air was sucked out of this space and induced atmospheric pressure had the same effect on the tested element as real uniform loading caused by wind. The setting of the durability test for each of the tested cladding elements is shown in Figure 1.

IOP Conf. Series: Materials Science and Engineering **379** (2018) 012039 doi:10.1088/1757-899X/379/1/012039

**Figure 1.** Setting of durability tests within wind loading.

Two cladding elements lightened with the expanded glass filler were compared to standard cladding elements with silica sand. After the conducted tests it could be concluded that the variant GEG II brought positive results, which is stated in the Table 6.

**Table 6.** Maximal loading and strain in cross section of individual elements.

| Element No. | Composition | Weight (kg) | Maximal loading $p_u$ (kN/m$^2$) | Maximal strain $\sigma_u$ (MPa) |
|---|---|---|---|---|
| T1 | Standard | 58.9 | 10.45 | 11.48 |
| T2 | GEG I | 41.3 | 7.55 | 10.20 |
| T3 | GEG II | 40.7 | 13.45 | 17.19 |

## 8. Conclusions

Both production technologies may allow the manufacturing of the assortment from the glass fibre reinforced concrete: balcony panelboards, cladding elements for façades or bridges and various architectural elements with lower weight.

By the means of using air-entraining agents, and various kinds of light-weight fillers – especially granulated expanded glass – it is possible to decrease bulk density of the fibre-cement composites up to 25%.

The fibre-cement composite with bulk density 1,600 kg/m$^3$ can be therefore obtained, as it was confirmed in the real production conditions. Subsequently, the reduction of transport costs was calculated at ca. 18%, reduction of assembly costs was calculated at ca. 25% and the reduction of material costs for a distinctive lighter superstructure was calculated at ca. 15%.

**Acknowledgments**
This article was supported by the Ministry of Industry and Trade of the Czech Republic as a part of the project FR-TI1/404 entitled "Light-weight fibre-cement composite elements".

**References**

[1] Ahsan A et al 2017 Comparison of Flexural Performance of Lightweight Fibre-reinforced Concrete and Normalweight Fibre-reinforced concrete *Periodica Polytechnica Civ. Eng.* **61** pp 498–504
[2] Okuyucu D, Uzal B, Turanli L and Tankut T 2011 Some characteristics of fibre reinforced semi-lightweight concrete containing unexpanded perlite both as aggregate and as a supplementary cementing material *Magazine of Concrete Research* **63** pp 1–10
[3] Vavruskova L, Cechmanek R, Chromkova I and Zavrel L 2016 Fibre cement composites that have significantly influenced global architecture *Monograph 537 Mechanics on Diagnosis and material issues in the 20th century architecture* ed J Mikula and K Korniejenko (Cracow: Cracow University of Technology Press) pp 57–82
[4] Cechmanek R, Prachar V, Lederer L and Loskot J 2015 Benefits of using light-weight elements in building industry *Monograph on Innovative, cost effective and eco-friendly fibre-based materials for construction industry* ed J Mikula and K Korniejenko (Cracow: Cracow University of Technology Press) pp 79–91
[5] Chandra S and Berntsson L 2002 *Lightweight Aggregate Concrete – Science, Technology and Applications* (Norwich: Noyes Publications, William Andrew Publishing)
[6] Cechmanek R, Prachar V and Loskot J 2014 Filler from Expanded Glass as a Light-Weighing Component of the Cement Composite with Fiber Reinforcement *Advanced Materials Research* **1000** pp 102–5
[7] Cechmanek R, Prichystalova E, Prachar V and Loskot J 2012 Light-weight fibre-cement composite – new variant of building element *Proc. Int. Conf. on Ecology and new building materials and products* (Brno: Vyzkumny ustav stavebnich hmot,a.s.) pp 144–7
[8] Cechmanek R, Prachar V and Loskot J 2012 Light-weighing possibilities of fibre-cement cladding elements *Proc. Int. Conf. on 19. Concrete days* (Hradec Kralove: Czech Concrete Society) pp 324–9
[9] Porfiri M and Gupta N 2009 Effect of volume fraction and wall thickness on the elastic properties of hollow particle filled composites *Comp. Part B – Eng.* **40** pp 116–73
[10] Gozzani G, Drdlova M, Prachar V and Popovic M 2016 Characteristics of glass based energy absorbers exposed to the blast load *Procedia Eng.* **151** pp 149–56
[11] Drdlova M, Cechmanek R, Holesinsky R and Popovic M 2016 The behaviour of the new glass based energy absorbers subjected to blast load *Proc. 22 Tech. Conf. on Experimental testing and modelling of brittle materials at high strain-rates* (Universite Grenoble Alpes) pp 142–7
[12] *Liaver – Technický list* (*Online* http://www.liaver.cz/repository/tds_Liaver_0,25-0,5_mm.pdf)
[13] Karmazinova M, Melcher J and Vesely J 2013 *Durability testing of bridge cladding elements exposed to wind loading* (Brno: Brno University of Technology)

IOP Conf. Series: Materials Science and Engineering 379 (2018) 012040 doi:10.1088/1757-899X/379/1/012040

# Effect of agents of organic origin on concrete degradation

**I Chromková[1] and R Čechmánek[1]**

[1] Research Institute for Building Materials, Hněvkovského 65, Brno, Czech Republic

Email: chromkova@vustah.cz

**Abstract.** Concrete and structural elements used within agricultural activities are affected by a range of factors, which have a significant effect on properties and durability of concrete. Among typical factors belong corrosive effect of silage juices, excrements of farm animals or also effect of microorganisms commonly present at soil. In the paper, there are stated results from testing of fine-grained concretes of different compositions, which were exposed to corrosive substances of biogenic origin. A great emphasis was especially put on observation of a silage effect on final physical-mechanical properties of concretes. Together with silage effects, effects of its components – acetic acid and lactic acid were also tested. These effects were compared with the corrosive effect of biogenic sulfuric acid. From current results, it was confirmed, that individual corrosive agents take part in a concrete degradation in a different extent. A concrete composition plays an important role in this process as well.

## 1. Introduction - theoretical

Concrete corrosion is characterized as a damage, deterioration and depreciation of concrete constructions by chemical and physical effects [1, 2, 3].

Rate of a concrete damage manifestation in a construction caused by chemical effects is dependent on a type of corrosive environment, which affects the given concrete. The effect on the concrete is caused by corrosive substances from the environment, transported to concrete components, which subsequently react with them. An important presumption for the chemical reaction is a presence of water, both in liquid and gaseous form. Building constructions could be eroded mainly by two types of a liquid environment, on the one hand by natural water and on the other hand by waste water [3, 4].

In most cases an acid corrosion is incurred by means of the present water environment. Because acids react mainly with calcium hydroxide and subsequently with calcium hydrosilicates and hydroaluminates to create calcium salts, a concrete corrosion rate is dependent on solubility of these salts and behaviour of a layer made of corrosion products. The more soluble the corrosion products are and the faster the wash out process is, the faster corrosion process is in the concrete mass as well. Binding components of a cement stone are changed by this process into a mass without binding properties.

Corrosion rate is determined on the one hand by a course of chemical reactions between the concrete and the corrosive environment and on the other hand by an exchange rate of the liquid medium close to the concrete surface. Gradual deterioration of the mass is therefore caused and it is accompanied with a creation of new reaction products. The layer of the reaction products is created gradually in dependence on a time of the corrosion action. The corrosion rate is considerable in the beginning but it decreases with growing of the reaction products. According to a range of experimental tests it can be stated, that the corrosion deterioration of the cement stone and the concrete is directly proportional to square root of the corrosion time [1, 3].

An important aspect for a progress and rate of the corrosion process is a concrete composition – a type of a binder system as well as a character of a used aggregate [3, 5, 6].

Many authors declare concrete results of their research on the corrosive environment effect on concrete constructions. Expression of the concrete deterioration degree is in these sources in the range of centimeters. For example, literature [7] states, that due to an effect of biogenic sulfuric acid in a sewage network the concrete deterioration is caused up to the depth of 30 mm after 5 years of exposure. It corresponds with research results [8], where a deterioration rate and a decrease of a surface layer is stated in a range of 3–6 mm/year.

## 2. Materials and methods

Aim of our experiment is to compare a deterioration degree of concrete specimens deposited for the exact time in corrosive solutions (media), which the most closely characterize corrosive agents arising within the agricultural activity or in the sewage network. The specimens made from the same concrete formula (and the same batch), deposited in water, were taken as a reference.

### *2.1. Testing methodology*

The base of our testing was patterned from the standard method [9] for verification of a corrosion resistance of conventional compact concrete, made with cements based on Portland clinker, in the liquid corrosive environments. The verification of a resistance degree to corrosive solutions is conducted on concrete specimens. In the beginning the concrete specimens are deposited in defined solutions. After taking them from the solutions (interval 6 months) selected evaluating criteria are observed. These criteria are subsequently compared with the same parameters found out on specimens made of the same concrete, which were deposited for the same time in the standardized way.

### *2.2. Evaluating criteria*

To basic evaluating parameters belong flexural strength, compressive strength, bulk density, water absorptivity and porosity. Evaluation of the deterioration degree of concrete surfaces was carried out also by means of polarizing and fluorescent microscopy. Testing of the deterioration degree of the concrete surfaces by means of a phenolphtalein indication solution was chosen as a supplementary method.

### *2.3. Verified mixtures of fine-grained concretes*

For the planned experiment concrete specimens (20 × 20 × 100 mm) from fine-grained mixtures were made. Mixture composition was derived from the formula conventionally used for production of concrete prefabricated elements for agricultural structures (silage boxes, mangers, reservoirs, dungheaps) in the particular production plant. For the specimen production the same raw materials were used: aggregate of fraction 0/4 mm, Portland cement CEM I 42.5R, fly ash and blast furnace slag in ratio 60:20:20% (standard formula – marked as A); combination of CEM I 42.5R and fly ash in ratio 60:40% (formula P); only Portland cement CEM I 42.5R (formula C). Ratio aggregate:binder was in all the formulas ca. 3:1.

Note: For the reason of verification of previous results the same formulas were prepared with using CEM II/B-S 32.5R in the same ratios with blast furnace slag and fly ash – the formulas were marked as AII, PII, CII.

Test specimens were 1 day after production deposited in the environment with 90% humidity for 27 days. After that the specimens were prepared for the exposition in the selected testing media. The specimens from the individual formulas were deposited into the media separately, i.e. for each formula and the testing medium a separated testing vessel (a fish tank) was prepared.

IOP Conf. Series: Materials Science and Engineering **379** (2018) 012040 doi:10.1088/1757-899X/379/1/012040

### *2.4. Testing media*

Within a solution of this research project we entered into a partnership with the Faculty of Natural Science, Department of Biochemistry in the Masaryk University in Brno. A selection of main initiators of corrosion actions in agricultural structures and a definition of tests solution representatives suitable for this experiment were carried out.

*2.4.1. Testing in the test solutions prepared in the laboratory.* This was the main direction of testing, for which the below described types of testing media were chosen:
*Acetic acid.* It is one of natural metabolites in living organisms. It can be usually found in plants, both as free acid and also as salts (acetates). It is contained in a significant amount in fermented fruits, as a subsequent fermentation product of a saccharide conversion.
*Lactic acid.* It is generated by a lactic fermentation of sugar, e.g. in milk, cheese and sauerkraut. This acid is together with the acetic acid a common component of the silage juices [10, 11].
*Biogenic sulfuric acid.* A representative of the corrosive medium, which is generated by actions of microorganisms occurring in soil and rocks [1]. A presence of the microorganisms (their type and amount) depends on a soil composition. In the case of biocorrosion by the biogenic sulfuric acid in the soil its progress and rate are dependent on a water amount, or rather on a soil humidity.
*Silage extract.* It is a leach from a silage prepared in a laboratory. Silage juices are composed of a cellular juice, which is released after mortification of plant cells, and of water, which is released by decomposition of organic substances. The juices released during an ensilage process are very corrosive, because they contain higher concentration of organic acids and other organic substances. pH of the silage juices is usually in values between 3.5–5 [10, 11].
*Distilled water.* It was used as a complementary corrosive medium.

Composition of the prepared testing solutions (stated in table 1) is based on findings and a research made by experts from the Masaryk University on real concentrations of individual corrosive agents present in the silage or in the excrements of farm animals.

Note: Methods and the used materials (and preliminary results of experiments as well) have been already reported in [12].

*2.4.2. Testing in a freshly prepared silage.* During the experiment another test was set up – this time the tested concrete specimens were deposited directly to a freshly prepared maize husks. Under exactly fixed conditions the silage was created in the course of 6 weeks. The silage preparation, an analysis of its quality as well as its subsequent observation during the test were carried out in a specialized laboratory.

## 3. Results and discussion

Research works during the experiment were divided into several directions.

### *3.1. Observation of a testing media condition*

*3.1.1. Testing solutions.* From the beginning of the experiments a measurement of media pH and a visual evaluation of their condition (turbidity, change of colour) in the same time were conducted. Shortly after the deposition of the concrete specimens into the test media a change of pH was noticed due to substantial concrete alkalinity (pH = ca. 13). Selected values of solution pH (from the time of deposition of the specimens) in the course of time are stated in Table 1. The use of biogenic sulfuric acid was based on a fact that sulfate (bio)corrosion, caused by this acid, is well known. The comparison of corrosion effects between organic (bio)products and a typical inorganic (bio)product brings a new approach of the study.

**Table 1.** pH of media with the deposited specimens made from the standard formula A.

| Medium | | pH | | | | |
|---|---|---|---|---|---|---|
| | | before deposition | 14 days of deposition | 90 days of deposition | 180 days of deposition | 360 days of deposition |
| MED **1** | distilled water | 5.50 | 12.05 | 11.95 | 11.46 | 12.10 |
| MED **2** | lactic acid | 1.31 | 4.75 | 5.66 | 7.40 | 7.14 |
| MED **3** | acetic acid | 3.40 | 10.99 | 11.62 | 11.75 | 10.89 |
| MED **4** | biogenic sulfuric acid | 1.20 | 1.38 | 6.86 | 6.82 | 6.09 |
| MED **5** | silage extract | 3.29 | 6.88 | 8.87 | 9.73 | 9.14 |

From the Table 1 it follows, that shortly after the beginning of the experiment a rapid growth of pH values occurred in the individual testing media. It is therefore obvious, that in the later time the acetic acid, the lactic acid and the biogenic sulfuric acid had no impact on the concrete specimens, but there was an effect of their salts.

*3.1.2. Silage.* In the specialized laboratory analyses of the ready-made silage were carried out (i.e. 6 weeks after its setting) and after that a silage quality was evaluated in the interval of 6 months (every time within taking of the concrete samples from the silage). Results (mean values from 3 parallel tests) are stated in Table 2.

**Table 2.** Analyses of the silage in the course of time.

| Parameter | Unit | Silage | | |
|---|---|---|---|---|
| | | 6 weeks | 6 months | 1 year |
| Dry matter of the original mass | % | 32.27 | 33.42 | 31.31 |
| pH | - | 3.80 | 3.94 | 4.10 |
| Ammonia | g/kg of the dry matter | 0.52 | 0.71 | 1.10 |
| Ethanol | % of the dry matter | 0.32 | 0.013 | 0.077 |
| Acetic acid | % of the dry matter | 3.94 | 0.87 | 1.87 |
| Lactic acid | % of the dry matter | 8.13 | 6.89 | 7.81 |

According to analyses results, the fresh maize silage was evaluated as a product of very high quality. From the analyses results of the silage in the age of 1 year with parameters of the fermentation process of the maize silage published in the literature [10] followed, that the silage keeps the same quality and all the time fulfils all the requirements (put on the silage from the point of view of farm animals′ nutrition).

Determined maximum content of the acetic acid reached a value of 2.1%, which corresponds with published values in [10]. For the common maize silage maximum values of the lactic acid content in a degree of 7% are stated, under right fermentation conditions even values over 8% are published [11].

Note: Ammonia content is not usually stated for the maize silages, the determined ammonia contents are low and show a growth according to the date of sampling as a consequence of a proceeding of proteolysis of nitrogen substances.

*3.2. Visual assessment*

Within each sampling of the concrete specimens from the test media their large photo documentation was carried out and their condition was visually evaluated.

*3.2.1. Samples from the testing solutions.* Each testing medium demonstrated a different effect on the tested concrete specimens. The specimens deposited for a period of 1 year in distilled water (MED 1) and the acetic acid (MED 3) showed no surface deterioration. A surface of the concrete specimens removed from the biogenic sulfuric acid (MED 4) showed signs of a severe deterioration – a well-marked porous structure of a cement binder, a deteriorated layer had a pasty character. In the lactic acid (MED 2) a change of colour occurred on the concrete specimens. Samples from the formulas C and P significantly changed to a rusty colour, the others had black coloured surfaces. Products of crystallization were noticeable on the specimens from the formulas C and P. It was about white fine little stars or flakes. Crystals appeared also on the specimens from the formula CII. In this case they have a regular crystallization structure and black-grey colour. On the tested specimens exposed in the silage extract (MED 5) a significant growth of hard reaction products of brown colour was noticeable (it was impossible to scrape them off). The most significant difference between the specimen state after their first and second sampling from the test media is in the extend of reaction products′ growth in the compositions C and CII. More detailed description of the concrete degradation observed in this experiment is given in [12].

*3.2.2. Samples from the silage.* The specimens deposited in the silage were after the end of exposition cleaned at first and after that a visual evaluation of their degradation degree were done. On the specimens there was a significant demonstration of a direct contact with the silage and its influence. A colouring in a different extend occurred on the surface (uneven spots), which is caused by effect of an uneven covering by the maize husks in the beginning of the experiment.

### *3.3. Testing of physical-mechanical properties*

The specimens made from fine-grained concretes taken from the test media were subjected to a set of tests in terms of physical-mechanical or physical-chemical properties.

*3.3.1. Samples from the testing solutions.* From the technological point of view the most important parameter for the assessment of the concrete resistance to the corrosion environment is a development of strength and other characteristic properties of concrete. Test results of the strength characteristics of the concrete specimens found out after 12 months of the deposition in the test solutions are depicted in Figure 1.

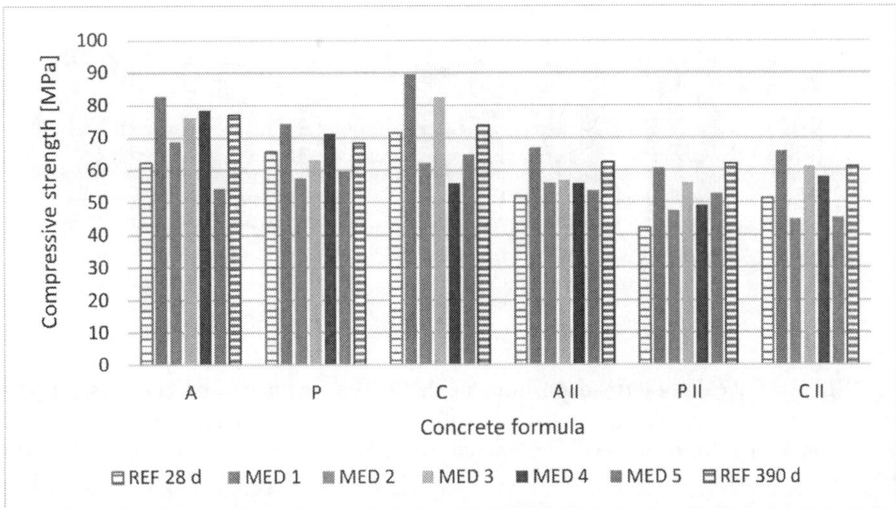


**Figure 1.** Effect of the testing media on values of compressive strength of the individual concrete formulas in comparison with strength of the reference specimens deposited in the standard conditions (formulas A, P, C prepared with CEM I 42.5R; formulas AII, PII, CII prepared with CEM II/B-S 32.5R) – deposition time 1 year.

*3.3.2. Samples from the silage.* The specimens taken from the silage were also evaluated preferentially from the point of view of the strength characteristics. Their development in the course of deposition in the silage is depicted in Figure 2 and 3.

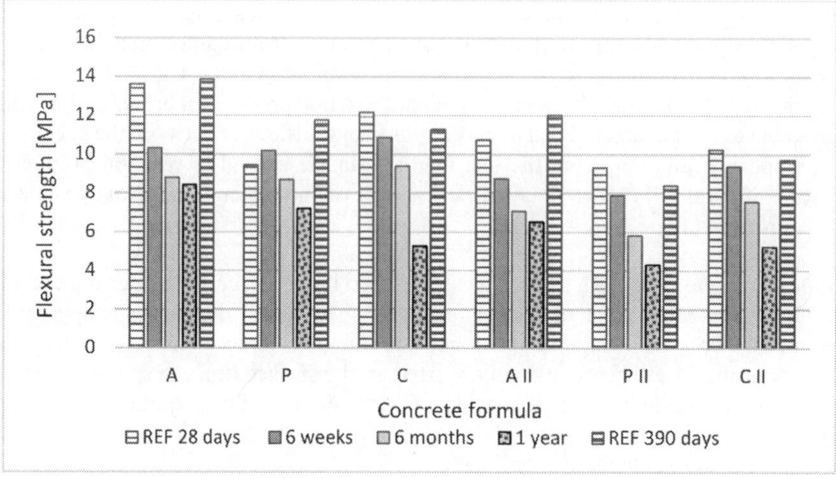


**Figure 2.** Effect of the long-term deposition in the silage on values of flexural strength of the individual concrete formulas in comparison with strength of the reference specimens deposited in the standard conditions (formulas A, P, C prepared with CEM I 42.5R; formulas AII, PII, CII prepared with CEM II/B-S 32.5R).

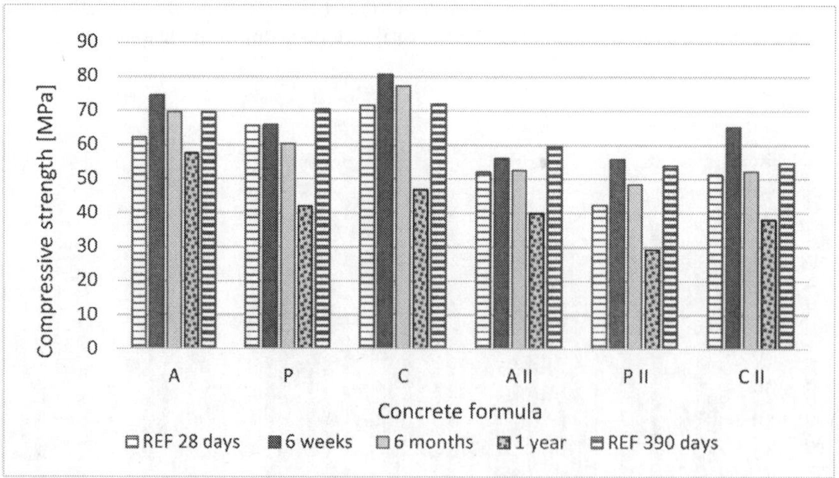


**Figure 3.** Effect of the long-term deposition in the silage on values of compressive strength of the individual concrete formulas in comparison with strength of the reference specimens deposited in the standard conditions (formulas A, P, C prepared with CEM I 42.5R; formulas AII, PII, CII prepared with CEM II/B-S 32.5R).

From the results depicted in Figures 2 and 3 it is obvious, that after 6 weeks from deposition into the maize husks due to its humidity the growth of the compressive strength within the concrete specimens of all formulas occurred (see the left hatched column and the column next to it). Due to the long-term deposition of the concrete in the silage a degradation of the surface layers of the concrete specimens occurred together with a significant infiltration of the silage juices to the concrete mass via its porous structure.

The specimens deposited for 1 year in the silage after the flexural strength test were subjected to another test with phenolphtalein, which enables to evaluate a rate and a depth of the degradation of the surface layer of the concrete specimens. The result is possible to observe in Figure 4. The photographs complete or confirm the results depicted in the graphs in Figure 2 and 3.

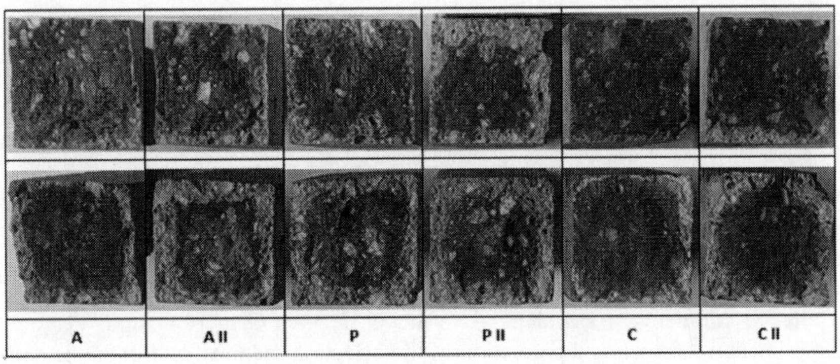


**Figure 4.** Comparison of the degradation of the surface layers of the specimens of the evaluated concrete formulas (reaction on a fraction area – the test with phenolphtalein) – the reference specimens deposited during their curing in the place with 90% humidity are above, the specimens deposited in the silage for 1 year are below.

## 4. Conclusions

Effect of the testing media prepared in the laboratory on the specimens deposited for 6 and 12 months is really significant. The surface deterioration is mostly visible on the specimens deposited in the lactic acid (MED 2), the biogenic sulfuric acid (MED 4) and in the silage extract (MED 5). On the specimens (especially from the formula C) deposited in the silage extract it was possible to observe the uneven growth of the solid reaction products of brown colour, which affected the degree of the strength characteristics comparing to the other formulas. The biggest influence on the degradation of the specimens was evaluated for the biogenic sulfuric acid, the inorganic (bio)product, which was used as a comparative agent with the organic (bio)products.

Comparing the lactic acid (MED 2) and the silage extract (MED 5), it can be concluded, that the effect of these testing media on the strength characteristics of the concrete specimens is very similar. An exception was the formula A, for which the medium 5 had the significantly negative effect.

From the testing of the concrete specimens deposited in the silage can be concluded, that the effect of the silage juices is substantial. The best resistance was demonstrated by the formula A. In the case of the formulas CII and AII prepared with CEM II their strength characteristics were comparable with the results for the formula P. We can therefore declare, that the formulas AII and CII are more resistant to the silage effect comparing with the formula P, which under standard conditions of curing demonstrated higher strength characteristics compared to the formulas AII and CII. The worst results after 1 year-long silage effect were detected for the formula PII.

Comparing the results of the strength characteristics of the concretes deposited in the silage extract prepared in the laboratory (or rather the lactic acid, which is a component of real silage juices) and the specimens deposited directly in the silage it is obvious, that the silage extract has not the same effect on the concrete as the silage. Only in the case of the formula A the effect of both media types was comparable. In the other cases the direct silage effect (1 year-exposition) on the concrete specimens had significantly more negative impact than the effect of the silage extract.

The real silage demonstrated higher effect on the degradation of the all concrete formulas comparing with the corrosive effect of the biogenic sulfuric acid, which in our experiment represented a completely different mechanism of corrosion and therefore served for the comparison of the different corrosive effects.

**Acknowledgements**
This paper was elaborated with the institutional support for the long-term development of research organizations by the Ministry of Industry and Trade of the Czech Republic.

**References**

[1] Ledererová J et al. 2009 *Biocorrosion effect on construction works* (in Czech) (Praha: Silikátový svaz)
[2] Chromková I, Leber P and Čechmánek R 2015 Protection of fibre-cement materials against biocorrosion In: *Innovative, cost effective and eco-friendly fibre-base materials for construction industry* (Kraków: Politechnika Krakowska)
[3] Dobrý o and Palek L 1988 Concrete corrosion in the construction practice (in Czech) (Praha: STNL)
[4] Prudil S 1980 Concrete durability in aggressive water (in Czech) (Brno: University of Agriculture in Brno)
[5] De Belie N, Monteny J, Beeldens A, Vincke E, Van Gemert D and Verstraete W 2004 Experimental research and prediction of the effect of chemical and biogenic sulfuric acid on different types of commercially produced concrete sewer pipes *Cement and Concrete Research. Elsevier Ltd. All rights reserved* **34** p 2223–2236
[6] Luptáková A, Eštoková A, Mačingová E, Kovalčíková M and Jenčárová J 2016 Biodeterioration of the Cement Composites *IOP Conference Series: Earth and Environmental Science: World Multidisciplinary Earth Sciences Symposium (WMESS 2016)*
[7] Wasserbauer R 2000 Biological depreciation of buildings (in Czech) (Praha: ARCH)
[8] Stein D 1999 Instandhaltung von Kanalisationen (Berlin: Ernst & Sohn Verlag)
[9] ČSN 73 1340:1988 Concrete constructions. Test of corrosion resistance of concrete. General requirements (in Slovak) (Praha: Publishing house of the Czech Office of Standards and Measurement)
[10] Třináctý J, et al. 2013 Evaluation of feed for dairy cows (in Czech) (Pohořelice: AgroDigest)
[11] Junges D et al. 2013 Additive containing homo and heterolactic bacteria on the fermentation quality of maize silage *Acta Scientiarum. Animal Sciences.* **35** 4 pp 371–377
Chromková I and Čechmánek R 2018 Influence of Biocorrosion on Concrete Properties, *Key Engineering Materials* **760** pp 83–90

IOP Conf. Series: Materials Science and Engineering **379** (2018) 012041 doi:10.1088/1757-899X/379/1/012041

# Testing of algae colonization growth risk on building materials

**I Kukletová[1] and I Chromková[1]**

[1] Research Institute for Building Materials, Hněvkovského 65, 617 00 Brno

Email: kukletova@vustah.cz

**Abstract.** Due to biological colonization, properties of building materials change. Green algae represent one group of the entire complex of biodeteriogens that interact with each other. For the purpose of testing algae growth risk on concrete with varied composition, a growth resistance testing methodology was developed based on ČSN EN 15458 (672032) *Paints and varnishes - Laboratory method for testing the efficacy of film preservatives in a coating against algae.* The modified method was employed to test ten samples of concrete with different composition, i.e. two types of cement and two types of classical fly ash were incorporated in the samples. The green algae growth was influenced by both the type of cement and the type of fly ash used. The results show that any addition of classic fly ash increases the risk of green algae colonization on the surface of concrete.

## 1. Introduction

Green algae and cyanobacteria require light, air humidity, and minerals to grow. They grow very well in places where water is retained (cornices, window sills, balconies, uneven zones of masonry, plinths), especially in the northern and north-eastern parts of buildings, including roofs. Algae also occur more frequently on spots where trees and bushes grow close to buildings and shade for a significant part of the day. Under favourable humidity conditions, extensive layers of different consistencies, thicknesses and colours appear [1]. Research shows that the first biofilm emerging on a surface consists of various types of bacteria followed by algae. Soon after algae, microscopic fibrous fungi colonize the biofilm, and later mosses and lichens are emerging to prepare the substrate for higher plants growth. On building stones, green algae are most common, namely the genera *Gloeocystis, Tetracystis, Cystomonas, Muriella, Chlorella, Stichococcus, Scenedesmus, Klebshormium, Apatococcus, Trebauxia*). Another large, frequently observed group consists of cyanobacteria (*Aphanothece, Lyngbya, Leptolyngbya, Nostoc, Synechococcus, Plectonema, Phormidium, Chroococcus, Myxosarcina, Pleurocapsa, Scytonema, Gloeocapsa, Apatococcus, Stichococcus, Phormidium*) [2]. Algae are able to grow in very extreme conditions, i.e. not only survive but even actively metabolize. Many species are known for their ability to survive for a long time at extreme temperatures and many also tolerate drying as well as high or low pH, strong gamma and UV radiation. On stone and building facades, algae can remain inactive for several years and return to biofilm under appropriate conditions. Their durability is due to presence of protective outer coverings which also serves as a reservoir of water that is held there by molecular forces. The cyanobacteria and algae shells also contribute to thickness and stability of the biofilm crusts they create on the building.

The surface of the building stone is mainly affected by algae metabolites. Algae synthesize polysaccharides, amino acids, vitamins, organic acids such as citric, glutamic, glycolic, oxalic, uronic and others, nitrogen bases, dyes and other substances. These products can be utilized by associated bacteria and moulds that produce more acids, especially formic acid, acetic acid, oxalacetic acid, malic acid, succinic acid, glyoxalic acid, gluconic acid, glucuronic acid and oxalic acid. This leads to acidolysis of the building stone and extensive formation of efflorescences containing increased concentrations of $Ca^{2+}$, $K^+$, $Na^+$, $Mn^{2+}$. The acid anions react with the cations of the stone to form water-soluble salts, which hydrate, increase the water content in the building stone pores and subsequently cause cation release from the building stone, so-called chelation. This leads to increased porosity and gradual decay of the building stone surface [2]. Algae also act mechanically, growing into small slits and expanding them. Some types of algae actively penetrate the building stone, create microgaps of various shapes and their secretions dissolve and leach carbonates present in the building stone. Fibrous green algae create slime-like coatings and crusts consisting of their fibres and plaster particles glued together. During the year, alternate drying and moistening of this biofilm occurs, as well as changes during freeze-thaw cycles, which together leads to removal/decorticating of the algae substrate. Thus, the top layer of the plaster or stone detaches from the deeper layers. The surface distortion process then proceeds in the following order: algae colonize humid cracks in stone sized from 0.1 to several millimetres in diameter, grow in dependence on the presence of water, carbonates around the cells precipitate in a reversible process $Ca(HCO_3)_2 \leftrightarrow CaCO_3 + H_2O + CO_2$, gradually cracks open by continuous carbonate precipitation and algae expansion with water inhibition, dust particles penetrate the cracks, algae die and the crack is colonized by heterotrophic bacteria and fungi, possibly insects, mites and spiders. The process leads to increased pressure on crack walls by newly produced organic components and to the removal of surface layer [3].

The aim of this research was to modify and optimize the method for evaluation of resistance of concrete samples to green algae. Subsequently, the modified method was used to test the effect of concrete composition, especially the addition of fly ash with a proven negative effect on the growth of water algae.

## 2. Green algae resistance testing methodology

The proposed methodology is based on ČSN EN 15458 (672032) Paints and varnishes - Laboratory method for testing the efficacy of film preservatives in coating against algae [4]. It is the only standard dealing with algae influence on building materials.

Testing bodies sized 75 × 75 × 10 mm were prepared. A standard formulation of fine-grained concrete was proposed for the test samples for the basic verification of the proposed methodology for testing the resistance to algae and moulds. Standard sands CEN I, CEN II and CEN III were chosen as fillers to eliminate impurities in concrete that could distort the results of the evaluation. Equal amounts of each sand were added into all mixtures. Two types of binders were used: CEM II S-M (S-LL) 32.5R cement and CEM I 42.5R cement. Therefore, two types of concrete mixtures (cement + sand mixture + water) were prepared and used as standards – standard 1 ST (cement CEM I) and standard 2 ST (cement CEM II). In the proposed P25 and P50 formulations, 25 and 50% of standard sand was replaced with fly ash selected from many samples obtained from different producers across the Czech Republic. Based on the results of previous analyses of the ashes, two types of classical fly ash were selected. These two ashes showed ecologically unfavourable properties, i.e. excessive levels of hazardous substances concentration in the leachate and unsatisfactory results of ecotoxicity testing. According to the authors' theory, these unfavourable properties should inhibit the algae growth on the concrete mixture surface. Ashes are indicated in the formulations as P1 and P2. The testing bodies were cured for 28 days and sterilized by UV light (30 min) prior to the test itself. Two bodies of each formulation were tested.

For the inoculation of testing bodies, an algal suspension of two algae species was used: *Nostoc commune* and *Stichococcus bacillaris*. The algae cultures were purchased from the collection of autotrophic organisms CCALA Třeboň (*492 Stichococcus bacillaris Naegeli, 118 Nostoc commune Vaucher*). The strains were grown and inoculation suspensions were prepared according to the standard (density of $10^6$ CFU/ml of the suspension). The Bold-Basal/Bristol Medium (BBM medium) was used both for growing strains and for testing. Composition of stock solutions for the BBM medium preparation is listed in Table 1.

**Table 1.** Stock solutions.

| Number of solution | Dissolved chemicals | Concentration |
|---|---|---|
| 1. | $NaNO_3$ | 25 g/1000 ml |
| 2. | $CaCl_2.2H_2O$ (or $CaCl_2.6H_2O$) | 5 g/1000 ml (or 3.72 g) |
| 3. | $MgSO_4.7H_2O$ | 7.5 g/1000 ml |
| 4. | $K_2HPO_4$ | 7.5 g/1000 ml |
| 5. | NaCl | 2.5 g/1000 ml |
| 6. | $KH_2PO_4$ | 17.5 g/1000 ml |
| 7. | EDTA+KOH | 5 g disodium EDTA + 3.1 g KOH /100 ml |
| 8. | $FeSO_4.7H_2O$ | 0.498 g + $H_2SO_4$.....0.1 ml /100 ml |
| 9. | $H_3BO_3$ | 1.142 g/100 ml |
| 10. | Microelements solution – 1000 ml | |
| | $ZnSO_4.7H_2O$ | 8.82 g |
| | $MnCl_2.4H_2O$ | 1.44 g |
| | $MoO_3$ (or $Na_2MoO_4.2H_2O$) | 0.71g (or 0,242 g) |
| | $CuSO_4.5H_2O$ | 1.57 g |
| | $Co(NO_3)_2.6H_2O$ | 0.49 g |

The BBM medium itself is prepared by mixing 10 ml of each of solutions 1–6 and 1 ml of solutions No. 7–10 with deionized water in a 1000 ml volumetric flask. For the test, a solid substrate is prepared from the nutrient BBM medium by addition of 1.5% of agar. This suspension is autoclaved and afterwards poured into Petri dishes to form circa 0.5 cm thick layer. The testing concrete bodies were placed on the surface of the layer, one into each dish. Both the body and the surface of the substrate were covered with a thin layer of mixed algal culture at a density of $10^6$ CFU/ml, as 10 ml of culture suspension was poured onto the test specimen and flowed down onto the substrate. Two cultivation controls were prepared. In the first case, the algal suspension inoculated only the agar, in the latter case a square of filter paper of the sample size was placed on the agar substrate.

Incubation was carried out at 23 ± 2°C, 1000 ± 200 lx, 16 h light/8 h dark cycle. The test lasted for 35 days. Since the 20th day of the test, the samples and the agar were sprayed with BBM nutrient medium to avoid drying out.

After 14, 21, 28 days and the end of the test, the algae growth on the test bodies was visually evaluated, both macroscopically and using a stereomicroscope. Gradual growth of culture in control dishes was evaluated and also, possible contamination with other types of microorganisms was observed. The final evaluation of the test results was carried out according to the standard by means of degrees 0–3 (0 no growth, 1 low growth, 2 medium growth, 3 heavy growth). For the comparison of the samples, the + and - signs were used for specification.

## 3. Results

In Figure 1, 2, 3, 4 a photodocumentation of sample 2-P2-25 during the test is listed. The algae growth was macroscopically observed after 14 days of duration, while the effect of cement mixture formulation on growth was observable after 21 days.

**Figure 1.** Development of algae growth during the test – sample 2-P2-25 14 days of cultivation.

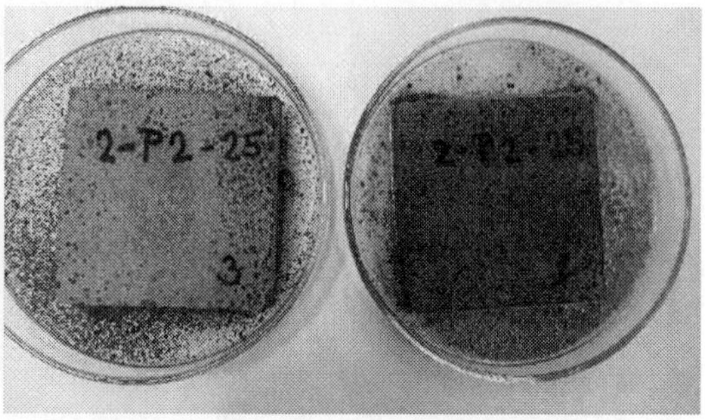


**Figure 2.** Development of algae growth during the test – sample 2-P2-25 21 days of cultivation.

**Figure 3.** Development of algae growth during the test – sample 2-P2-25 28 days of cultivation.

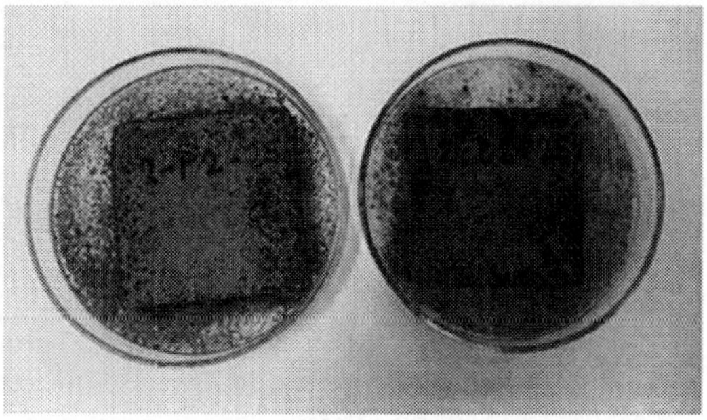

**Figure 4.** Development of algae growth during the test – sample 2-P2-25 35 days of cultivation.

In Figure 5, 6 condition of selected samples at the end of the test is listed. No undesirable contamination of the samples was detected.

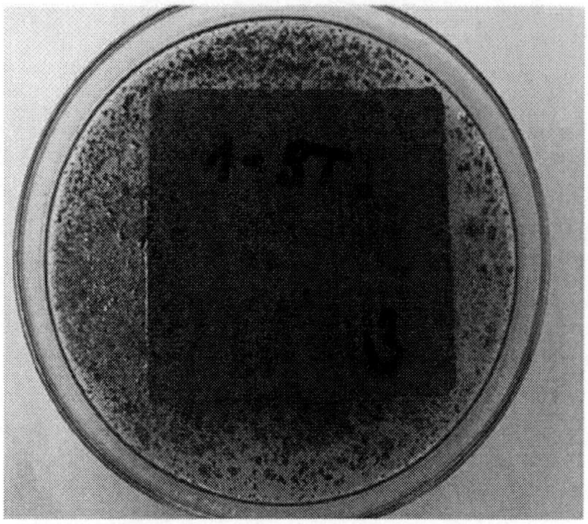

**Figure 5.** Algae growth on the last day of the test – sample 1 ST.

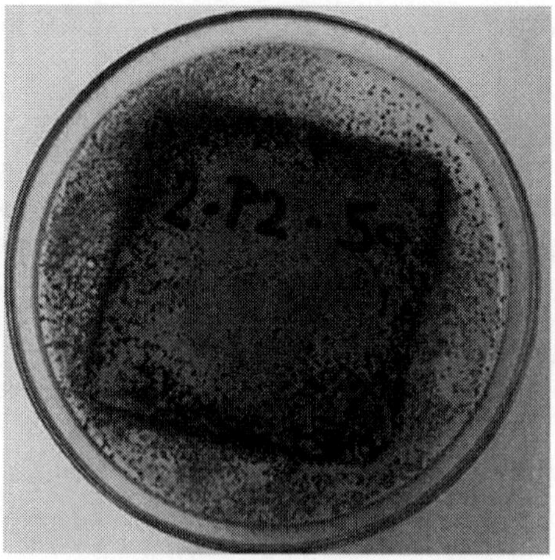

**Figure 6.** Algae growth on the last day of the the test – sample 2-P2-50.

The last day of the test, algae growth was evaluated using degrees 0–3 (0 no growth, 1 low growth, 2 medium growth, 3 heavy growth). For the comparison of the samples, the + and - signs were used for specification and the results were summarized in Table 2.

**Table 2.** Algae growth evaluation at the end of the test.

| Sample | Degree of growth |
|---|---|
| 1 ST | 0 |
| 2 ST | 2 |
| 1-P1-25 | 2- |
| 1-P1-50 | 2+ |
| 1-P2-25 | 1 |
| 1-P2-50 | 1+ |
| 2-P1-25 | 3 |
| 2-P1-50 | 3+ |
| 2-P2-25 | 2+ |
| 2-P2-50 | 2-3 |

The final results fully correspond with the concrete mixtures formulation. Both the type of cement and the type of fly ash influence the algae growth on the surface of the samples. Among the four studied concrete ingredients, the highest inhibition effect exhibited cement CEM I 42.5R and classical fly ash P1. Where these two components meet in the mixture, their combined inhibition effect is enhanced. The highest inhibition of growth in both sample series generally occurs in standard samples 1 ST and 2 ST as any fly ash addition into the mixture increases the possibility of algae growth on the surface.

## 4. Conclusions

Green algae represent one group of all the variety of biodeteriogens that participate on building bio corrosion. For the purposes of testing and evaluation the possibility of algae growth on different concrete surfaces, a new methodology was developed and used for testing 10 different concrete samples. The results show that both the type of cement and the type of fly ash incorporated into the concrete mixture influence the algae growth. The theory that fly ashes exhibiting ecotoxicity should suppress the growth when incorporated into concrete mixtures was disproven.

**Acknowledgements**
This paper was elaborated with the institutional support for long-term development of research organizations by the Ministry of Industry and Trade of the Czech Republic.

## References

[1] Büchli R and Raschle P 2004 *Řasy a houby na fasádách* (Ostrava: MISE, s.r.o.) p 108 ISBN 978-80-254-8786-0
[2] Wasserbauer R 2001 *Biologické znehodnocení staveb* (Praha: ABF) p 257 ISBN 80-86165-30-2
[3] Ledererová J 2009 *Biokorozní vlivy na stavební díla* (Praha: Silikátový svaz) p 273 ISBN 978-80-86821-50-4.
[4] ČSN EN 15458 (672032) Paints and varnishes - Laboratory method for testing the efficacy of film preservatives in coating against algae

# AUTHOR INDEX